Engineering Materials

For further volumes:
http://www.springer.com/series/4288

Masumi Saka
Editor

Metallic Micro and Nano Materials

Fabrication with Atomic Diffusion

Springer

Editor
Prof. Masumi Saka
Department of Nanomechanics,
Graduate School of Engineering
Tohoku University
Aoba 6-6-01, Aramaki
Aoba-ku, Sendai 980-8579
Japan
e-mail: saka@ism.mech.tohoku.ac.jp

ISSN 1612-1317 e-ISSN 1868-1212

ISBN 978-3-642-15410-2 e-ISBN 978-3-642-15411-9

DOI 10.1007/978-3-642-15411-9

Springer Heidelberg Dordrecht London New York

Library of Congress Control Number: 2010936451

Cover design: deblik, Berlin

Printed on acid-free paper

Springer is part of Springer Science+Business Media (www.springer.com)

Preface

Recently, various kinds of micro and nano metallic materials have been produced by using different chemical or physical techniques, and these materials have been found to possess excellent mechanical, electrical, optical and thermal properties compared with their bulk states. These micro and nano metallic materials are therefore expected to be key elements of future technologies and will be widely used to overcome worldwide issues in energy, food, health, etc. Needless to say, in many cases for particular applications, further research and development are needed to resolve the wide gap that exists between research status and their availability for practical applications.

This book covers many peripheral technologies that can be employed to effectively use existing micro and nano materials for future applications, as well as the relevant fabrication technologies. In this book we mainly examine physical methods for forming micro and nano metallic materials by controlling the diffusion of atoms, although many micro and nano metallic materials can be synthesized based on chemical reactions. Two different atomic diffusion phenomena that can be utilized for fabricating micro and nano metallic materials are introduced. The first of these is 'electromigration', which is driven by a high density flow of electrons, and the other is 'stress migration', which relies on a gradient of hydrostatic stress in the material. Techniques for evaluating the mechanical and electrical properties of these materials are also described with some practical examples. Moreover, 'modification' technologies for these materials such as cutting, welding and bending are described, along with some of their applications.

This book is consisted of six chapters. (I) Introduction: Scope of This Book. (II) Basis of Atomic Diffusion: Theoretical Background of Two Phenomena of Atomic Diffusion, i.e., Electromigration and Stress Migration. (III) Fabrication of Micro and Nano Metallic Materials: Typical Micro and Nano Metallic Materials Formed by Atomic Diffusion with Details of the Fabrication Procedures. (IV) Evaluation of Mechanical Properties: Techniques for Evaluating the Mechanical Properties of Micro and Nano Metallic Materials, such as the Concentrated-Mass Cantilever, some Kinds of Bending Tests, etc. (V) Evaluation of Electrical Properties: Techniques such as the Four-Point AFM Probe, Microwave AFM Probe, etc. for

Evaluating the Electrical Properties of Micro and Nano Metallic Materials. (VI) Modification of Nano/Micromaterials: Welding and Cutting Technologies with Joule Heat for Micro and Nano Metallic Materials, together with some Applications, and Bending of Metallic Nanowires to form Nanocoils. The authors hope that this book will play a role in making readers familiar with the key techniques in the field of micro and nano metallic materials.

The authors wish to express their gratitude to Professor Emeritus Hiroyuki Abé, Tohoku University, for his valuable advice and thankworthy encouragement to our research activities. Also we appreciate colleagues at Tohoku University; Dr. Shien Ri, Mr. Takao Shōji and Ms. Hiromi Sugawara, for their kind help in preparing the manuscript of this book. The techniques covered in this book are based on the research by the authors "Formation of Metallic Nanomaterials by Controlled Atomic Accumulation and Their Characterizations" supported by Grant-in-Aid for Scientific Research (S) 18106003.

Sendai, June, 2010 Masumi Saka

Hirosaki , June, 2010 Kazuhiko Sasagawa

Akita, June, 2010 Mikio Muraoka

Akita, June, 2010 Hironori Tohmyoh

Nagoya, June, 2010 Yang Ju

Contents

Introduction 1
1 Background 1
2 Typical Micro and Nano Materials 2
3 Importance of Fabricating Micro and Nano Metallic Materials 3
4 Fabrication of Micro and Nano Metallic Materials 4
4.1 Vapor-Liquid-Solid Growth 4
4.2 Template-Based Synthesis 5
4.3 Stress Migration 6
4.4 Electromigration 8
5 Importance of Evaluation and Modification 9
6 Summary 11
References 13

Basis of Atomic Diffusion 15
1 Electromigration 16
1.1 Historical Review of EM 16
1.2 Theory of EM (Huntington–Grone Equation) 16
1.3 Polycrystalline Structure of Metallic Thin Films 18
1.4 Bamboo Structure in Metallic Thin Films 18
1.5 Effect of Passivation on EM 19
1.6 Governing Parameter for EM Damage, AFD 19
1.7 Derivation Method of Characteristic Constants of EM 35
2 Stress Migration 42
2.1 Introduction 42
2.2 Historical Review of Typical SM-Induced Phenomena 43
2.3 Summary 46
References 48

Fabrication of Micro and Nano Metallic Materials 53
1 Introduction 54
2 Fabrication of Micro and Nano Metallic Materials by Utilizing EM 54
2.1 Effect of Temperature Gradient on Atomic Diffusion 54
2.2 Effective Atomic Accumulation Due to an Introduced Artificial Slit 58
2.3 Fabrication of Al Micro Spheres by Utilizing EM 59
2.4 Fabrication of Al Micro Belts by Utilizing EM 63
2.5 Effect of Purity of Source Material on Fabrication of Al Micro/Thin Materials. 65
2.6 Effect of Temperature on Controlling EM to Selectively Form Micro and Nano Materials 67
2.7 Numerical Simulation of Fabrication of Micro/Nano Materials 72
3 Fabrication of Micro and Nano Metallic Materials by Utilizing SM 81
3.1 Rapid and Mass Formation of Cu Nanowires by Utilizing SM 81
3.2 Fabrication of Ag Micro Materials by Utilizing SM 86
References 90

Evaluation of Mechanical Properties 93
1 Overview of Mechanical Tests 93
2 Contact Resonance of Concentrated-Mass Cantilevers 98
2.1 Short History of Atomic Force Microscopy and Dynamic Modes 98
2.2 Dynamics of Bending 102
2.3 Dynamics of Concentrated-Mass Cantilevers 107
2.4 Evaluation of Elastic Modulus 110
3 Large Bending Test of Nanowires 113
3.1 Introduction 113
3.2 Theory on Equilibrium Bent-Shape of Nanowires 114
3.3 Evaluation of Elastic Modulus and Strength 120
3.4 Diameter Dependency of Elastic Modulus 127
4 Small-Span Bending Test 130
4.1 Importance of Bending Test 130
4.2 Testing Scheme 130
4.3 Example of Small-Span Bending Test 132
References 138

Evaluation of Electrical Properties 143
1 Introduction 143
1.1 Conductivity Measurement of Metallic Nanowires 145

1.2 Geometry (Cross-Section Shape) of Nanowires 145
1.3 Measuring the Dimensions (Length and Cross-Section Area) of Nanowires 146
1.4 Measuring the Resistance of Nanowires 147
1.5 Determining the Grain Size of Nanowires 149
2 The FPP Method 149
2.1 The Advantage of the FPP Method 149
2.2 The Modified FPP Method 150
3 The Four-Point AFM Probe Method 151
3.1 Fabrication of the Four-Point AFM Probe 152
3.2 A Nanowire Sample and Its AFM Image 154
3.3 Measuring the Conductivity of the Nanowire 155
4 The Microwave AFM Method 157
4.1 Fabricating the Tip of the M-AFM Probe 159
4.2 Fabrication of the M-AFM Probe 159
4.3 Measuring Topography by M-AFM Probe 162
4.4 Microwave Image Obtained by the M-AFM 164
4.5 Measurement of Electrical Properties by M-AFM 166
References 169

Modification of Nano/Micromaterials 173
1 Welding and Cutting of Materials 174
1.1 Introduction 174
1.2 Heat Conduction in Thin Wires 177
1.3 Melting Phenomenon at Nanocontacts 180
1.4 Cutting of Ultrathin Pt Wires 184
1.5 Welding of Ultrathin Pt Wires 187
2 Application of Welding and Cutting 194
2.1 Manipulation of a Small-Scaled Object 194
2.2 Fabrication of a Small-Scaled Electromagnetic Ring 197
2.3 Fabrication of a Small-Scaled Thermoelectric Element 201
3 Permanent Bending of Brittle Nanowires for Formation of Nanocoils 205
3.1 Introduction 205
3.2 Residual Stress in Deposited Thin Films 206
3.3 Misfit-Strain-Induced Bending of Nanowires 210
References 217

Errata to: Metallic Micro and Nano Materials 221

Index 231

Introduction

Masumi Saka

1 Background

With the advancement of science and technology, and in particular the development of micro and nano technologies, the ability to restructure and alter the properties of materials has been realized by enabling the control of molecules and atoms. Micro and nano technologies are set to improve many aspects of our lives, such as information, medicine, materials, and so on. The evolution of micro and nano technologies has drawn much attention due to potential applications in the fields of manufacturing and processing, e.g., the electronic microcircuit industry. Wide-ranging research has also been carried out in this area, which is now considered to be one of the most promising areas for technical advances in the future.

Since Feynman presented the famous talk "There's Plenty of Room at the Bottom" in 1959, technologies that we can use to control atoms and molecules and to rearrange matter to suit our needs have experienced significant developments. Nanotechnology has ambitiously been called the next industrial revolution, a wholly different approach to the way human beings rearrange matter [22]. There are now many examples of nano-scale tools that are sensitive enough to measure the atomic-scale forces exerted by a single layer of atoms. These tools can also detect the bump on a surface made by a single atom, or even exert atomic-scale forces with such control that we can pick up individual atoms and move them exactly where we would like. Among these tools, the scanning tunneling microscope (STM) was invented in the early 1980s by Binnig and Rohrer after 3 years of collaboration at the IBM Research Laboratory in Zurich, Switzerland. It is the father of all scanning probe microscopes, and earned its creators the Nobel Prize in 1986 [22]. With these techniques, we can observe molecular and atomic

M. Saka (✉)
Department of Nanomechanics, Tohoku University, Aoba 6-6-01, Aramaki, Aoba-ku, Sendai 980-8579, Japan
e-mail: saka@ism.mech.tohoku.ac.jp

M. Saka (ed.), *Metallic Micro and Nano Materials*, Engineering Materials,
DOI: 10.1007/978-3-642-15411-9_1, © Springer-Verlag Berlin Heidelberg 2011

compositions in order to investigate and rearrange the structures of micro and nano materials. Since the advent of these tools, research into micro and nano technologies has developed rapidly.

Carbon nanotubes (CNTs) are typical of the materials that are attracting attention in the fields of micro and nano technologies. In the early years of the 1990s, Iijima [10] identified this new type of finite carbon structure using an arc-discharge evaporation method. A whole new era of intense research and experimental exploration began after the discovery of CNTs to understand their unique physical and electronic properties.[1]

In the past several decades, microelectronic industries have been spending tremendous effort in developing and commercializing the Moore's law, leading to not only many breakthroughs and revolution in information and communication technology, but also noticeable changes in the way of living of human being.[2] The properties of many materials on the micro and nano scales are totally different from those of their bulk, and this provides a wide platform to investigate and rebuild the structures of materials for physical and chemical applications in micro and nano electromechanical systems (MEMS/NEMS). In addition, micro and nano metallic materials are very important components in the microelectronic industry just now. Consequently, it is extremely desirable to have the capability to fabricate micro and nano metallic materials and to evaluate their properties.

2 Typical Micro and Nano Materials

Some typical micro and nano materials that are having a significant influence on the development of micro and nano technologies will be introduced in this section.

CNTs are one of the most important nano materials because they have very high aspect ratios and are potentially useful in many applications in nano technology, electronics, optics and other fields of material science due to their novel properties. Since the discovery of CNTs by Iijima [10], they have been the subject of intense interest. The CNTs can be divided into single-wall carbon nanotubes (SWCNTs) and multi-wall carbon nanotubes (MWCNTs). SWCNTs are formed by rolling up a single graphite 'sheet' into a tube with a cylindrical nano structure. MWCNTs are usually formed from several layers of graphene cylinders that are nested concentrically. The CNTs are the strongest and stiffest materials discovered to date in terms of tensile strength and elastic modulus, respectively. All of the CNTs are expected to have a high thermal conductivity, and all CNTs have unusual electrical

[1] Nanopackaging: nanotechnologies and electronics packaging, Morris (ed) (2008, p. 325), Carbon nanotubes: synthesis and characterization, Yadav, Kunduru and Prasad: With kind permission of Springer Science and Business Media.

[2] Nanopackaging: nanotechnologies and electronics packaging, Morris (ed) (2008, p. 533), Nanoelectronics landscape: application, technology, and economy, Zhang: With kind permission of Springer Science and Business Media.

properties due to the symmetry and unique electronic structure of graphene. Thus far, CNTs have been synthesized by arc-discharge technique [5], laser evaporation [4] or chemical vapor deposition [3].

There is a great deal of interest in the subject of tin (Sn) whiskers, due to the application of Sn solder in electronic manufacturing [29]. In the 1940s, Bell Labs initially observed the problem of Sn whisker formation during pure Sn electroplating and became very active in Sn whisker research. Their first related external publication was produced in 1951 by Compton et al. [2]. During the 1950s, the fundamentals of whisker formation were addressed, and most of the basic concepts [6, 8, 13, 28] are still in use today to describe whisker formation and growth. With the experience of over 50 years of history covering Sn whisker research, many highlights have been identified, e.g., the single crystal structure of the whiskers, spontaneous growth from electroplated Sn surfaces, lead (Pb) additions to Sn to mitigate against whisker growth. However, following the Restriction of Hazardous Substances (RoHS) directive that was issued by the European Union on July 1, 2006, the Sn-based solders have developed into a new generation of Pb-free solders to replace the traditional SnPb solders, and consequently Sn whisker growth on Pb-free solder has become a new reliability problem [40].

Zinc oxide (ZnO) is one of the most important functional oxide semiconductors due to its distinctive properties and its applications in transparent conducting oxide materials and gas sensors. Zinc oxide nanowires [9] have been grown by a vapor-phase transport process via the vapor–liquid–solid (VLS) mechanism, and zinc oxide nanobelts [18] have been synthesized by evaporating the desired commercial metal oxide powders at high temperature. ZnO micro and nano materials offer tremendous potential in providing electronic, photonic and spin-based functionality.

Al, Ag and Cu are the most popular materials for the metallization of integrated circuits. Al has been used as the interconnect conductor for 40 years. The industry has turned to Cu as the interconnect material because Cu is a better electrical conductor [34]. Ag also has great potential for application in the semiconductor industry. Al, Ag and Cu films can be easily deposited by vacuum evaporation, and their respective micro and nano materials have been synthesized by different approaches [11, 24, 25, 34].

3 Importance of Fabricating Micro and Nano Metallic Materials

Among the materials mentioned above, micro and nano metallic materials are particularly attractive for use in MEMS applications, which are considered to be a key technological development with outstanding potential in modern industry. Spearing [31] proposed the basic premise behind the concept of MEMS is that the efficiencies of high volume production and low unit cost achieved by the microelectronics industry over the past 50 years can be translated to devices in which mechanical and electrical components are integrated within a single silicon chip. It should be noted that MEMS are complex systems, which show impressive functionality and performance in applications covering various areas. MEMS also offer the opportunity to materials scientists and engineers to be able to characterize materials in ways that have not hitherto been possible [31].

Micro and nano metallic materials are also needed for small devices. In our research, various micro and nano materials, such as Al, Cu and Ag, have been investigated by fabrication, modification and evaluation. Nowadays, Al is still the main material that is used in integrated circuits because of its advantages in lithographic processes. Meanwhile, Cu has been considered as a better material to replace Al in some applications due to its high electromigration (EM) resistance. Ag is also a promising material in MEMS technology, since its electrical resistance is the lowest when compared to other metallic materials, so the effect of Joule heating can be significantly reduced. All of these materials are the most commonly used materials in our daily lives, and have big potential applications in MEMS and small-scale devices. These observations indicate the importance of work involving micro and nano metallic materials.

4 Fabrication of Micro and Nano Metallic Materials

The mechanisms for fabricating micro and nano metallic materials can be classified into two types. One type depends on chemical reactions, e.g., VLS growth and template-based synthesis. The other type uses physical phenomena, such as stress migration (SM), EM and plastic deformation. Let us briefly review these typical growth mechanisms, along with their applications.

4.1 Vapor–Liquid–Solid Growth

Wagner and Ellis [35] proposed a mechanism for crystal growth from vapor which they called the VLS mechanism in 1960s, and they studied the growth of Si whiskers using Au as an impurity. According to this mechanism, Si whiskers are formed at the interface between solid Si and a liquid Au–Si alloy. Wu and Yang [36] reported the growth of Ge nanowires using Au clusters as a solvent at high temperature, and they showed a real-time observation of this growth process in an in situ high-temperature transmission electron microscopy (TEM). Figure 1 shows a sequence of TEM images during the growth of a Ge nanowire in situ, which confirms the validity of the VLS crystal growth mechanism at the nanometer scale. These mechanisms have been used to synthesize various nanowires including elements [12, 35], oxides [15], etc.

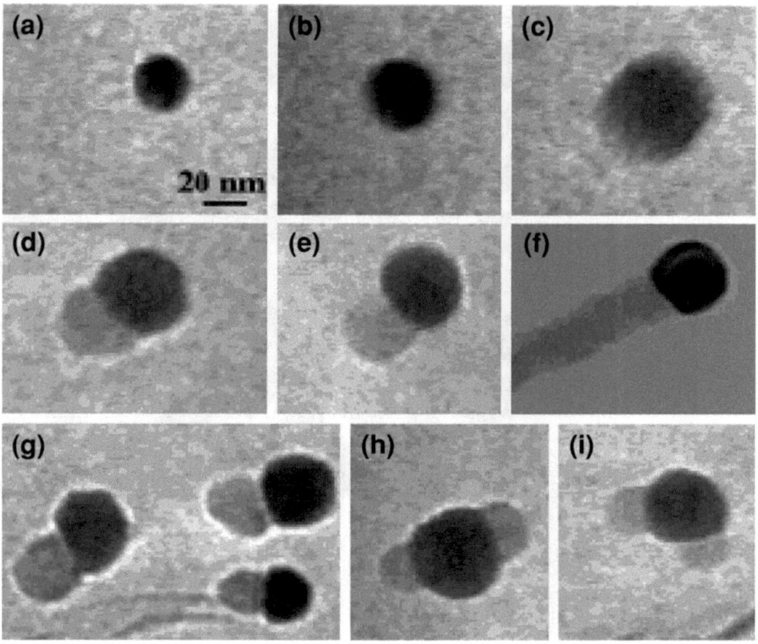

Fig. 1 In situ TEM images recorded during the process of nanowire growth. **a** Au nanoclusters in solid state at 500°C; **b** alloying initiated at 800°C, at this stage Au exists mostly in solid state; **c** liquid Au/Ge alloy; **d** the nucleation of Ge nanocrystal on the alloy surface; **e** Ge nanocrystal elongates with further Ge condensation and eventually forms a wire (**f**); **g** several other examples of Ge nanowire nucleation; **h**, **i** TEM images showing two nucleation events on single alloy droplet [36]

4.2 Template-Based Synthesis

Template-based synthesis is a convenient method for generating micro and nano structures, which was developed by Penner and Martin [19]. In this technique, various porous structures are used, and micro and nano materials can be synthesized within the pores. The anodic alumina nanoporous structures templates have received considerable attention in synthetic nanostructure materials due to their several unique structure properties such as controllable pore diameter, extremely narrow size distribution for pores diameter and their interval, ideally cylindrical shape of pores [41]. Figure 2 shows an example of the templates.

Mesophase structures self-assembles from surfactants provide another class of useful and versatile templates for generating 1D nanostructures in relatively large quantities; see Fig. 3. It is well known that surfactant molecules spontaneously organize into rod-shaped micelles (or inverse micelles) when their concentration reaches a critical value. These anisotropic structures can be immediately used as soft templates to promote the formation of nanorods when coupled with an

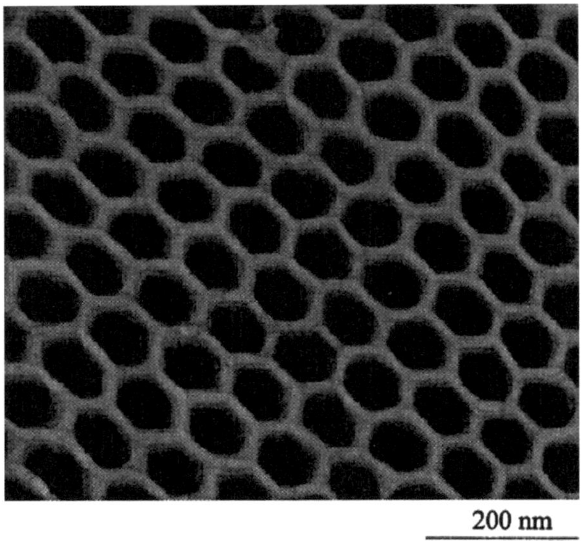

Fig. 2 AFM image of an anodic alumina membrane [41]

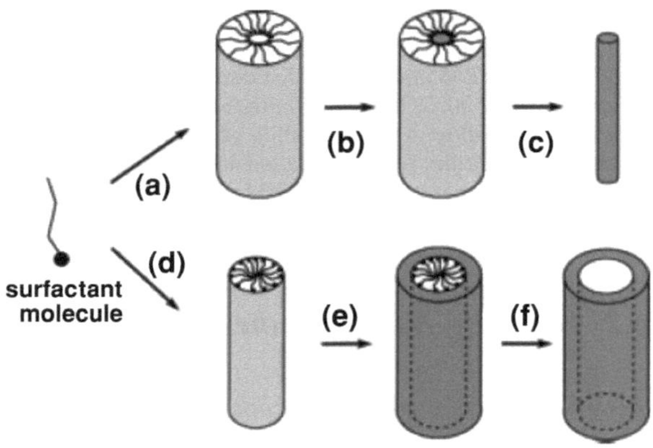

Fig. 3 Schematic illustrations showing the formation of nanowires by templating against mesostructures self-assembled from surfactant molecules. Xia et al. [38, p. 362]. Copyright Wiley-Verlag GmbH & Co. KGaA. Reproduced with permission

appropriate chemical or electrochemical reaction. The surfactant molecules need to be selectively removed to collect the nanorods as a relatively pure sample [38]. Micro and nano structures of Au, Ag, Pt, CuS, CuSe, CdS, ZnS and ZnSe have been grown based on the template-based synthesis technique [7, 21].

4.3 Stress Migration

SM induced metallic nanowhiskers have been studied extensively, such as Sn whiskers [14], Si nanowires [20], and Cu whiskers [25], etc. The most likely explanation for SM is based on the effects of compressive stress, which results from the different thermal expansion coefficients, on the formation of nanowhiskers. The mechanism may be easily understood by considering a simple phenomenon in a street where people migrate from a crowded area to an uncrowded area, as shown schematically in Fig. 4.

Lee and Lee [14] proposed a mechanism for the spontaneous growth of Sn whiskers. The driving force for the evolution of Sn whiskers is a biaxial compressive stress of about 8 MPa developed in Sn deposits by the formation of an intermetallic compound of Cu_6Sn_5, especially in grain boundaries of Sn films. The biaxial compressive stress gives rise to strains normal to the film plane, which are dependent on the Sn grain orientations.

The stress-driven formation of Si nanowires was presented by Prokes and Arnold [20]. The silicon wires grow directly from the silicon substrate, they do not need to be manipulated or aligned for subsequent applications. Wires in the

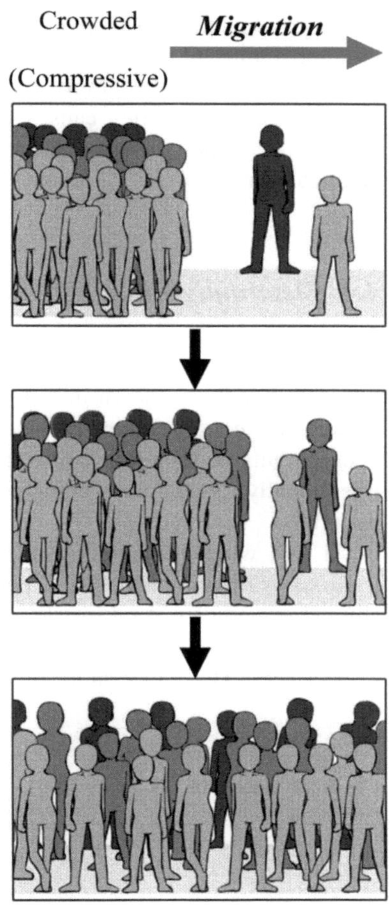

Fig. 4 The phenomenon of SM in a street where people migrate from a crowded and hence highly compressive area to an uncrowded area

20–50 nm diameter range with lengths over 80 μm can be easily grown by this technique. The critical parameters in the growth of these nanowires are the surface treatment and the carrier gas used [20].

Cheng et al. [1] reported a method of making nanowires of bismuth (Bi) with diameters ranging from 30 to 200 nm and lengths up to several millimeters. The nanowires are extruded spontaneously at the rate of a few micrometers per second at room temperature from the surfaces of freshly grown composite thin films consisting of Bi and chrome-nitride. The high compressive stress in these composite thin films is the driving force responsible for the nanowire formation. Shim et al. [30] reported that single crystalline Bi nanowires were found to grow on as-sputtered films after thermal annealing at 260–270°C. This was facilitated by relaxation of stress between the film and the thermally oxidized Si substrate that originated from a mismatch of the thermal expansion. The diameter-tunable Bi nanowires can be produced by controlling the mean grain size of the film, which is dependent upon the thickness of the film.

Saka et al. [25] reported the rapid and mass growth of Cu nanowhiskers on polycrystalline films. The diameters and lengths of the nanowhiskers are governed by temperature, film thickness, grain size and time. Cu atoms diffuse due to local stress gradients in the grains which originate in the material singularities and geometrical singularities caused at the grain boundaries by anisotropy of the grains. Weak spots in the oxide layers on the films act as initiation sites for the nanowhiskers.

4.4 Electromigration

EM is a physical phenomenon whereby metallic atoms are transported by the 'electron wind', as shown in Fig. 5. It is a serious problem in integrated circuits because of the appearance of voids and hillocks, as shown in Fig. 6, leading to open-circuits and short-circuits, respectively. On the other hand, it can be used intentionally to generate a compressive stress-release phenomenon to grow micro materials.

Al thin wires [23], micro-spheres [32] and micro-belts [17] have been synthesized by utilizing EM. The experimental sample was a passivated Al line with a slit and some holes at the anode end. The diameters and lengths of the formed Al thin wires are controlled by the sizes of the holes and current supply time, respectively. The temperature of the discharged atoms affects the shapes of the

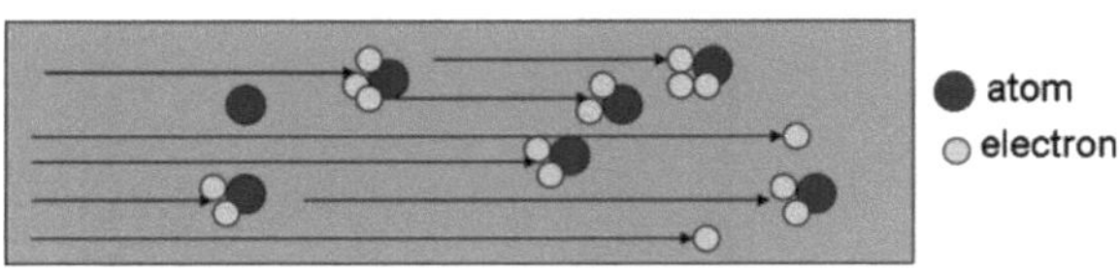

Fig. 5 Schematic of electromigration

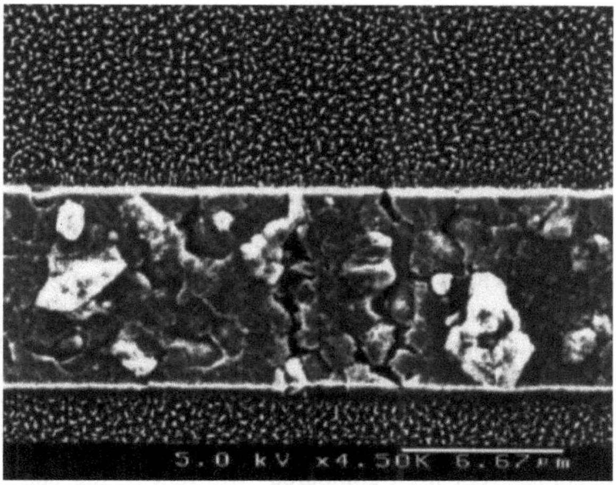

Fig. 6 Voids and hillocks due to EM in passivated polycrystalline Al lines. Reprinted with permission from Sasagawa et al., J. Appl. Phys. **91** (2002, p. 1888). Copyright © 2002, American Institute of Physics

micro-structural features formed from the metal. High, intermediate, and low temperatures of the atoms after discharge at the anode end were found to correspond to the formation of large micro-spheres, thin wires and small spheres, respectively [26].

The Al micro materials can be fabricated at predetermined positions, and the widths and thicknesses of Al micro-belts can be controlled by the sizes of the holes that are introduced. In addition, it has been found that the purity of the Al film has a significant effect on the fabrication of micro materials, resulting from the activation energy for grain boundary diffusion [16].

Al micro materials such as Al micro-spheres may provide some potential applications in the fields of light-weight structural materials, microstructured composites, electromagnetic-shielding materials, acoustical materials, and so on [32].

5 Importance of Evaluation and Modification

The unique properties of micro and nano materials, including mechanical, electrical, optical and thermal properties are expected to be utilized by employing them as structural, electrical, optical or thermal components, especially in MEMS or NEMS. When considering reliability for potential applications, it is important to evaluate the material properties. Differing from some traditional methods, some new techniques are introduced here for the evaluation of mechanical and electrical properties, including the concentrated-mass cantilever, the four-point atomic force microscope (AFM) probe, and so on.

Due to the difficulties in both synthesis and assembly, the as-assembled nanowires often do not have an ideal shape, structure, and configuration [39].

Therefore, these materials may need to be modified to obtain the desired structures by welding, cutting, integrating, and so on. A welding technique for 20 nm diameter Ge nanowires by thermal heating has been reported [37]. However, it has a limitation in terms of chemical and thermal stability as the lengths of the wires are reduced to nanometer scales. Xu et al. [39] demonstrated that a high-intensity electron beam can be applied to create holes, gaps, and other patterns of atomic and nanometer dimensions on a single nanowire, to weld individual nanowires to form metal–metal or metal–semiconductor junctions, and to remove the oxide shell from a crystalline nanowire. Figure 7 shows the welding of a Au nanowire to a Sn nanowire.

Let us introduce a method for welding conductive nanowires, whereby Pt nanowires with diameters of about 600–800 nm can be successfully welded end to end to form a continuous wire by Joule heating. The contact area between the two opposing nanowires is quite small, and the electrical resistivity in this vicinity becomes very high compared with that of the nanowire. Welded nanowires can be cut at a point on the wire by supplying a high-current to the wire using a cone-shaped probe [33]. The details will be discussed in chapter titled Modification of Nano/ Micromaterials.

The technique of modification is important for making use of micro and nano materials, by which these materials can be rearranged for some specific function. Some techniques concerning the remodeling of micro and nano materials will also be described in this book.

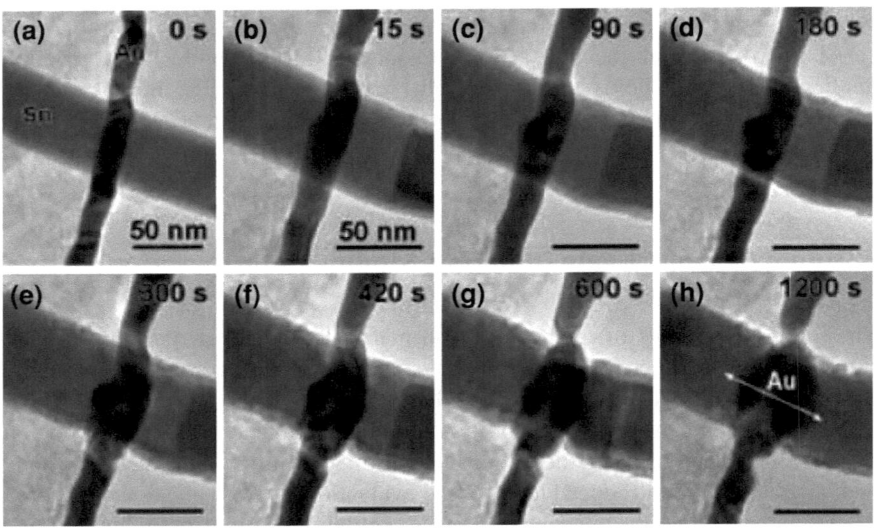

Fig. 7 TEM images showing the welding of a Au nanowire to a Sn nanowire with a high-intensity electron beam. Xu et al. [39, p. 1225]. Copyright Wiley-VCH Verlag GmbH & Co. KGaA. Reproduced with permission

6 Summary

Micro and nano materials, especially metallic materials, differ from conventional bulk materials due to not only the small size of their structures, but also their application in our daily life. Unusual electrical, optical, mechanical and thermal properties can be used in many fields to cause a revolution in industry. The field covered by this book is shown in Fig. 8, including the technique for fabrication and evaluation of micro and nano metallic materials, and their application in nanoscience and nanotechnology.

With the increasing demand for miniaturization in micro and nano devices, the investigation of fine structures including fabrication, evaluation and modification has become more and more necessary. Investigating how to treat micro and nano materials in a convenient and low-cost way is now an attractive field, and is also the main content of this book. We start with the basis of atomic diffusion, which can be employed as the foundation for the fabrication of micro and nano materials. In chapter titled Basis of Atomic Diffusion, two kinds of atomic diffusion, EM and SM, are demonstrated. A driving force and a protective layer are essential factors to fabricate micro and nano materials by both EM and SM, and therefore designed samples are employed under specified experimental conditions. The mechanism of EM is provided by using a formula for atomic flux divergence, where the effect of the structures of the samples on divergence is considered. With the help of the formula, readers are led to an easy understanding of the mechanism used to fabricate fine structures using EM. On the other hand, SM induced micro and nano materials that are formed using the mechanisms reported so far are also reviewed, and the trends of current work in this area are also discussed.

Chapter titled Fabrication of Micro and Nano Metallic Materials discusses the fabrication of micro and nano metallic materials by utilizing EM and SM, both of which can cause the stress release phenomenon. Growth mechanisms are described

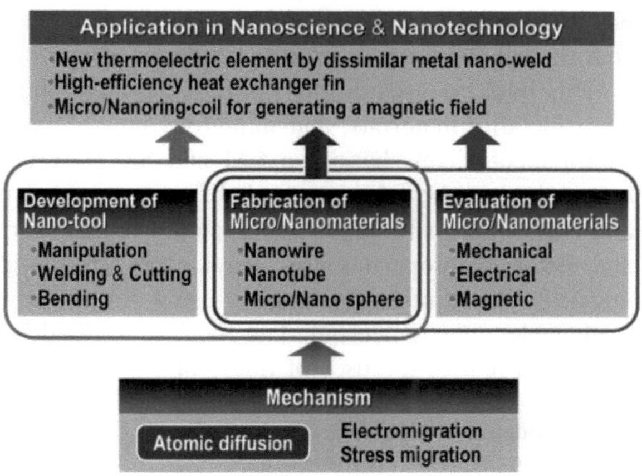

Fig. 8 Study of metallic micro/nanomaterials [27]

for Al thin wires, micro-spheres and micro-belts by using EM in the passivated films. The purity of the source materials and the temperatures at the anode end are shown to play a significant role in the growth process. Then, a numerical simulation of the formation of micro and nano materials by EM is presented. This simulation can help us to identify optimized experimental conditions for the fabrication of these materials. On the other hand, Cu nanowires and Ag micro materials are shown to be synthesized successfully by using SM. A natural oxide layer or an artificial layer on the surface is necessary for the discharge of metallic atoms to form these materials. This method, based on SM, is low cost without any chemical reaction, and it is shown that we can control the diameters and lengths of the formed nanowires.

Then, an evaluation of the mechanical properties of metallic thin wires is demonstrated in chapter titled Evaluation of Mechanical Properties. We show the use of a concentrated-mass cantilever and a bending method to evaluate the elastic modulus and bending strength of brittle nanowires, and we show an evaluation of the elastic–plastic properties of metallic materials by using an unsymmetrical, small-span bending test. The advantages of these methods are explained by comparing them with reported techniques in this field.

The electrical properties of micro and nano metallic materials are one of the most important aspects that govern their potential applications. In chapter titled Evaluation of Electrical Properties, a four-point AFM probe method is exhibited, which can effectively measure the resistivity of a metallic nano material. This method combines a conventional four-point probe with an AFM. A new-generation technique named the 'microwave AFM method' is shown, by which topography and the distribution of electrical properties can be measured simultaneously.

Finally, the modification of micro and nano materials is presented. The cutting and welding of metallic materials are essential techniques to meet the requirements of industrial applications. In chapter titled Modification of Nano/Micromaterials, a cutting and welding technique utilizing Joule heating is described. A constant current supply can cause sufficient Joule heating to melt the wire, and if we apply a force to separate the molten part, then the wire can be cut. On the other hand, two nanowires can also be joined together by this technique. This technique can be used to fabricate electromagnetic rings and thermoelectric elements, for example. In addition, a coating technique that is used to bend a nanowire into a helix is introduced.

This book focuses on the synthesis, evaluation and modification of micro and nano metallic materials, and numerous new techniques are demonstrated. Fabrication mechanisms based on EM and SM may be used to form other metallic materials, and the fabrication of different kinds of materials such as micro-tubes and core–shell materials is anticipated. Methods for the evaluation of mechanical and electrical properties also provide potential applications for micro and nano materials. In addition, techniques using Joule heating have promising futures in many fields of engineering. The topics described in this book will be helpful for readers to realize the uses of micro and nano technologies.

Acknowledgments The author wishes to express his thanks to Y. Lu and S. Ishikawa for their kind help in preparing the manuscript.

References

1. Cheng, Y.-T., Weiner, A.M., Wong, C.A., Balogh, M.P., Lukitsch, M.J.: Stress-induced growth of bismuth nanowires. Appl. Phys. Lett. **81**, 3248–3250 (2002)
2. Compton, K.G., Mendizza, A., Arnold, S.M.: Filamentary growths on metal surfaces—Whiskers. Corrosion **7**, 327–334 (1951)
3. Dai, H.: Carbon nanotubes: synthesis, integration, and properties. Acc. Chem. Res. **35**, 1035–1044 (2002)
4. Devetta, L., Canu, P., Bertucco, A., Steiner, K., Qin, L.C., Iijima, S.: Structure and formation of raft-like bundles of single-walled helical carbon nanotubes produced by laser evaporation. Chem. Phys. Lett. **269**, 65–71 (1997)
5. Ebbesen, T.W., Ajayan, P.M.: Large-scale synthesis of carbon nanotubes. Nature **358**, 220–222 (1992)
6. Eshelby, J.D.: A tentative theory of metallic whisker growth. Phys. Rev. **91**, 755–756 (1953)
7. Govindaraj, A., Deepak, F.L., Gunari, N.A., Rao, C.N.R.: Semiconductor nanorods: Cu, Zn, and Cd chalcogenides. Isr. J. Chem. **41**, 23–30 (2001)
8. Herring, C., Galt, J.K.: Elastic and plastic properties of very small metal specimens. Phys. Rev. **85**, 1060–1061 (1952)
9. Huang, M.H., Wu, Y., Feick, H., Tran, N., Weber, E., Yang, P.: Catalytic growth of zinc oxide nanowires by vapor transport. Adv. Mater. **13**, 113–116 (2001)
10. Iijima, S.: Helical microtubules of graphitic carbon. Nature **354**, 56–58 (1991)
11. Jana, N.R., Gearheart, L., Murphy, C.J.: Wet chemical synthesis of silver nanorods and nanowires of controllable aspect ratio. Chem. Commun. 617–618 (2001)
12. Kodambaka, S., Tersoff, J., Reuter, M.C., Ross, F.M.: Diameter-independent kinetics in the vapor-liquid-solid growth of Si nanowires. Phys. Rev. Lett. **96**, 096105 (1–4) (2006)
13. Koonce, S.E., Arnold, S.M.: Growth of metal whiskers. J. Appl. Phys. **24**, 365–366 (1953)
14. Lee, B.-Z., Lee, D.N.: Spontaneous growth mechanism of tin whiskers. Acta Mater. **46**, 3701–3714 (1998)
15. Li, S.Y., Lee, C.Y., Tseng, T.Y.: Copper-catalyzed ZnO nanowires on silicon (100) grown by vapor-liquid-solid process. J. Cryst. Growth **247**, 357–362 (2003)
16. Lu, Y., Saka, M.: Effect of purity on the fabrication of Al micro/thin-materials by utilizing electromigration. Mater. Lett. **63**, 2294–2296 (2009)
17. Lu, Y., Saka, M.: Fabrication of Al micro-belts by utilizing electromigration. Mater. Lett. **63**, 2227–2229 (2009)
18. Pan, Z.W., Dai, Z.R., Wang, Z.L.: Nanobelts of semiconducting oxides. Science **291**, 1947–1949 (2001)
19. Penner, R.M., Martin, C.R.: Controlling the morphology of electronically conductive polymers. J. Electrochem. Soc. **133**, 2206–2207 (1986)
20. Prokes, S.M., Arnold, S.: Stress-driven formation of Si nanowires. Appl. Phys. Lett. **86**, 193105 (1–3) (2005)
21. Rao, C.N.R., Deepak, F.L., Gundiah, G., Govindaraj, A.: Inorganic nanowires. Prog. Solid State Chem. **31**, 5–147 (2003)
22. Rogers, B., Pennathur, S., Adams, J.: Nanotechnology: Understanding Small Systems. Taylor & Francis, Boca Raton (2008)
23. Saka, M., Nakanishi, R.: Fabrication of Al thin wire by utilizing controlled accumulation of atoms due to electromigration. Mater. Lett. **60**, 2129–2131 (2006)
24. Saka, M., Ueda, R.: Formation of metallic nanowires by utilizing electromigration. J. Mater. Res. **20**, 2712–2718 (2005)
25. Saka, M., Yamaya, F., Tohmyoh, H.: Rapid and mass growth of stress-induced nanowhiskers on the surfaces of evaporated polycrystalline Cu films. Scr. Mater. **56**, 1031–1034 (2007)
26. Saka, M., Kato, K., Tohmyoh, H., Sun, Y.X.: Controlling electromigration to selectively form thin metal wires and metal microspheres. J. Mater. Res. **23**, 3122–3128 (2008)

27. Saka, M., Tohmyoh, H., Muraoka, M., Ju, Y., Sasagawa, K.: Formation of metallic micro/nanomaterials by utilizing migration phenomena and techniques for their applications. Mater. Sci. Forum **614**, 3–9 (2009)
28. Sears, G.W.: A growth mechanism for mercury whiskers. Acta Metall. **3**, 361–366 (1955)
29. Sheng, G.T.T., Hu, C.F., Choi, W.J., Tu, K.N., Bong, Y.Y., Nguyen, L.: Tin whiskers studied by focused ion beam imaging and transmission electron microscopy. J. Appl. Phys. **92**, 64–69 (2002)
30. Shim, W., Ham, J., Lee,K.-I., Jeung, W.Y., Johnson, M., Lee, W.: On-film formation of Bi nanowires with extraordinary electron mobility. Nano Lett. **9**, 18–22 (2009)
31. Spearing, S.M.: Materials issues in microelectromechanical systems (MEMS). Acta Mater. **48**, 179–196 (2000)
32. Sun, Y.X., Tohmyon, H., Saka, M.: Fabrication of Al micro-spheres by utilizing electromigration. J. Nanosci. Nanotechnol. **9**, 1972–1975 (2009)
33. Tohmyoh, H., Imaizumi, T., Hayashi, H., Saka, M.: Welding of Pt nanowires by Joule heating. Scr. Mater. **57**, 953–956 (2007)
34. Tu, K.N.: Recent advances on electromigration in very-large-scale-integration of interconnects. J. Appl. Phys. **94**, 5451–5473 (2003)
35. Wagner, R.S., Ellis, W.C.: Vapor-liquid-solid mechanism of single crystal growth. Appl. Phys. Lett. **4**, 89–90 (1964)
36. Wu, Y., Yang, P.: Direct observation of vapor-liquid-solid nanowire growth. J. Am. Chem. Soc. **123**, 3165–3166 (2001)
37. Wu, Y., Yang, P.: Melting and welding semiconductor nanowires in nanotubes. Adv. Mater. **13,** 520–523 (2001)
38. Xia, Y., Yang, P., Sun, Y., Wu, Y., Mayers, B., Gates, B., Yin, Y., Kim, F., Yan, H.: One-dimensional nanostructures: synthesis, characterization, and applications. Adv. Mater. **15**, 353–389 (2003)
39. Xu, S., Tian, M., Wang, G., Xu, J., Redwing, J.M., Chan, M.H.W.: Nanometer-scale modification and welding of silicon and metallic nanowires with a high-intensity electron beam. Small **1**, 1221–1229 (2005)
40. Zeng, K., Tu, K.N.: Six cases of reliability study of Pb-free solder joints in electronic packaging technology. Mater. Sci. Eng. R **38**, 55–105 (2002)
41. Zheng, M., Zhang, L., Zhang, X., Zhang, J., Li, G.: Fabrication and optical absorption of ordered indium oxide nanowire arrays embedded in anodic alumina membranes. Chem. Phys. Lett. **334**, 298–302 (2001)

Basis of Atomic Diffusion

Kazuhiko Sasagawa and Masumi Saka

Abstract Atomic diffusion, or more specifically, electromigration (EM) and stress migration (SM), are described in this chapter. The driving force of atomic diffusion is electron wind in EM and the gradient of hydrostatic stress in SM. In Sect. 1, the fundamental principles of EM are presented. For actual metal lines, which may have various microstructures and be covered with a passivation layer, EM behavior is explained. Then, a method for calculating the divergence of atomic flux due to EM is introduced, and the formulation process is described to help readers understand the application of the calculation method. The formula of the divergence *AFD* describes the behavior of EM damage well, which affects the reliability of silicon integrated circuits. Based on the *AFD* formula, a method for deriving the characteristic constants of EM in the line is introduced. In Sect. 2, the basic principles of SM are given through a brief review of typical SM-induced phenomena. The current research trends regarding nanomaterial production using SM and the mechanism of the production are discussed.

K. Sasagawa (✉)
Department of Intelligent Machines and System Engineering, Hirosaki University,
3 Bunkyo-cho, Hirosaki 036-8561, Japan
e-mail: sasagawa@cc.hirosaki-u.ac.jp

M. Saka
Department of Nanomechanics, Tohoku University, Aoba 6-6-01, Aramaki, Aoba-ku,
Sendai 980-8579, Japan
e-mail: saka@ism.mech.tohoku.ac.jp

M. Saka (ed.), *Metallic Micro and Nano Materials*, Engineering Materials,
DOI: 10.1007/978-3-642-15411-9_2, © Springer-Verlag Berlin Heidelberg 2011

1 Electromigration

1.1 Historical Review of EM

An electric field has long been known to induce the motion of ions in metals [23]. In 1961, Huntington and Grone reported the EM phenomenon of current-induced motion of scratches on a metal surface [37]. EM has been described as the transport of metal atoms, driven by momentum transfer from electron flow. Huntington and Grone proposed an equation (now known as the Huntington–Grone equation) that describes atomic flow. EM was first observed in bulk metals. In the late 1960s, EM was recognized as a failure mechanism of integrated circuits (ICs) [10]. Black [5] systematically studied EM in IC metal lines. The damage induced by EM is manifest as voids and hillocks, which are formed by the depletion and accumulation of metal atoms, respectively. The growth and linking of voids results in electrical discontinuity in the IC metal lines, which in turn leads to open-circuit failure. The lifetime of the metal line is primarily governed by EM damage, and therefore must be predicted quantitatively to ensure the reliability of ICs. Black formulated an empirical equation for predicting the mean time to failure, *MTF*. Black's equation is given by

$$MTF = Aj^{-n} \exp\left(\frac{Q}{kT}\right), \tag{1}$$

where A is a constant related to the line shape and line material, j is the input current density, n is the order of current density dependence, Q is the activation energy, k is the Boltzmann constant, and T is the absolute temperature. Today, this equation remains widely used to predict the lifetime of IC interconnects. Since the 1980s, EM has been extensively researched with respect to the reliability of ICs, because EM, along with SM, is recognized as the main failure mechanism. EM in IC metal lines has been studied numerically to predict the lifetime and failure site of the lines [1, 40, 45, 46, 58, 62]. Recently, Cu has begun to replace Al as a line material. In comparison with Al, Cu is expected to be more robust against EM failure; nevertheless, EM failure remains a major issue affecting the reliability of modern ICs.

1.2 Theory of EM (Huntington–Grone Equation)

Atomic diffusion in metal can be considered the diffusion of ionized particles [65]. The diffusion velocity $\boldsymbol{v}$ of the ionized particles caused by external force is given by

$$\boldsymbol{v} = \beta \mathbf{F}, \tag{2}$$

where β is mobility and **F** is the driving force. According to the Nernst–Einstein equation, β is expressed by

$$\beta = \frac{D}{kT}. \tag{3}$$

Here, D is the diffusion coefficient given by

$$D = D_0 \exp\left(-\frac{Q}{kT}\right), \tag{4}$$

where D_0 is prefactor.

The relationship between the flux of particles (i.e., the number of particles passing through a unit area per unit time) and the velocity of particles is given by

$$\mathbf{J} = N\boldsymbol{v}, \tag{5}$$

where **J** is the particle flux and N is the particle density. For atoms, **J** is the atomic flux and N is the atomic density. Using Eqs. (2), (3) and (5), the flux of the ionized particles caused by external force is given as

$$\mathbf{J} = \frac{ND}{kT}\mathbf{F}. \tag{6}$$

The driving force **F** consists of two types of forces, $\mathbf{F}_c$ and $\mathbf{F}_e$, in the case of EM. $\mathbf{F}_c$ is the force acting on an electric charge in electric field **E**, and $\mathbf{F}_e$ is the force transmitted by electron collisions. The forces $\mathbf{F}_c$ and $\mathbf{F}_e$ are simply expressed by

$$\mathbf{F}_c = q\mathbf{E} = C_1 e\mathbf{E} \tag{7}$$

and

$$\mathbf{F}_e = C_2(-e)\mathbf{E}, \tag{8}$$

where q is the ionic charge, e is the charge of an electron, and C_1 and C_2 are proportional constants. Denoting electrical resistivity by ρ and current density by **j**, electric field **E** is given by $\rho\mathbf{j}$, and the driving force **F** is given by

$$\mathbf{F} = \mathbf{F}_c + \mathbf{F}_e = (C_1 - C_2)e\mathbf{E} = Z^* e\rho\mathbf{j}, \tag{9}$$

where Z^* is the effective valence. Substituting Eq. (9) into Eq. (6), we obtain the Huntington–Grone equation:

$$\mathbf{J} = \frac{ND}{kT}Z^* e\rho\mathbf{j} = \frac{ND_0}{kT}\exp\left(-\frac{Q}{kT}\right)Z^* e\rho\mathbf{j}. \tag{10}$$

Fig. 1 Polycrystalline line

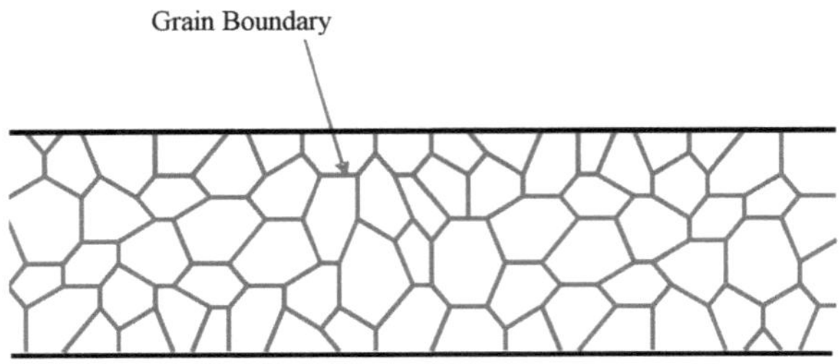

1.3 Polycrystalline Structure of Metallic Thin Films

EM in metal lines occurs along grain boundaries as well as within grains (lattice diffusion). When a metal line width is greater than several micrometers, there are several grains in the width direction. The thickness of a metallic thin film is generally smaller than the grain size. The microstructure of a polycrystalline line is shown in Fig. 1. The main path of atomic diffusion is considered to be along the grain boundaries, and lattice diffusion can be neglected in the case of EM in a polycrystalline line [6], because the diffusion coefficient in grain boundaries is much larger than that in the lattice [57].

1.4 Bamboo Structure in Metallic Thin Films

If the width of a metal line is less than about 1 μm, there is only one grain in the width direction. Such a metal line has a so-called bamboo structure. The microstructure of a bamboo line is shown in Fig. 2. In the bamboo line, it is assumed that the atomic flux in the grain boundary is negligible [73] and that lattice diffusion [20, 48, 49, 52, 68] including interface diffusion [12, 36, 52, 63] by EM is dominant. This assumption is based on the small number of grain boundaries, which are perpendicular to the longitudinal axis of the line [73, 75]. The velocity of atomic diffusion in a bamboo line is much slower than that in a polycrystalline line, because the diffusion coefficient for the lattice within the grain is much smaller than that for grain boundaries.

Fig. 2 Bamboo line

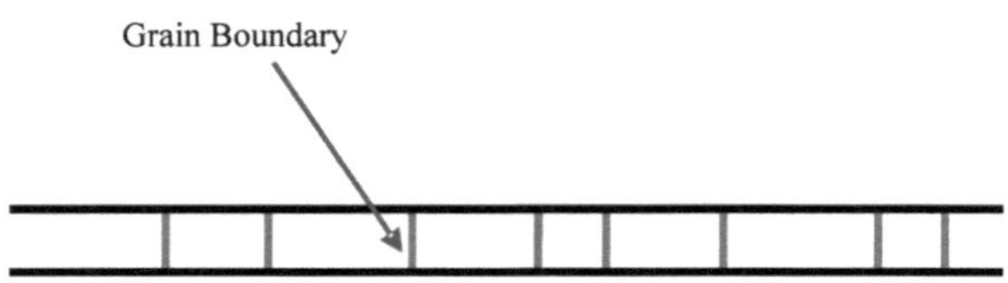

1.5 Effect of Passivation on EM

Metal lines used in packaged silicon ICs are covered with a passivation layer. In contrast to unpassivated lines, hillock formation is difficult to induce by EM in passivated lines. When hillock formation is suppressed, a gradient of mechanical stress builds up in the line, specifically, compressive stress is generated by atomic density increasing at the anode of the line and tensile stress is generated by atomic density decreasing at the cathode of the line. This gradient induces another form of atomic diffusion, called 'back flow', in the opposite direction of EM [7–9]. Consequently, EM in the passivated line is inhibited by the back flow. This also explains why the lifetime of covered metal lines is longer than that of uncovered lines [44, 60].

In a modification of the Huntington–Grone equation (10), the atomic flux in the passivated metal line is assumed to be represented [61] by

$$|\mathbf{J}| = \frac{ND_0}{kT}\exp\left\{-\frac{Q+\kappa\Omega(N-N_{\mathrm{T}})/N_0-\sigma_{\mathrm{T}}\Omega}{kT}\right\}\left(Z^*e\rho j^* - \frac{\kappa\Omega}{N_0}\frac{\partial N}{\partial l}\right), \tag{11}$$

where κ is the effective bulk modulus [42], N_{T} is the atomic density under tensile thermal stress σ_{T}, j^* is the component of current density in the direction of $\mathbf{J}$, N_0 is the atomic density at a reference condition, Ω is the atomic volume [$\cong 1/N_0$], and ρ [$=\rho_0\{1+\alpha(T-T_{\mathrm{s}})\}$] is the temperature-dependent resistivity; ρ_0 and α, respectively, are the electrical resistivity and the temperature coefficient at the substrate temperature T_{s}. $\partial N/\partial l$ is the atomic density gradient in the direction of $\mathbf{J}$. The effect on diffusivity of the stress generated in the metal line is given by the term $\kappa\Omega(N-N_{\mathrm{T}})/N_0-\sigma_{\mathrm{T}}\Omega$ in the exponential function [2, 50]. On the other hand, the effect of the back flow of atoms induced by an atomic density gradient is given by $(\kappa\Omega/N_0)\,\partial N/\partial l$ [7, 8, 42]. These two effects are taken into account in Eq. (11). The quantity N_0 is obtained under stress-free conditions at 300 K [74], and N_{T} can be approximated by N_0 [61].

There is a threshold current density of EM damage, j_{th}, below which no EM damage appears in the case of passivated and via-connected metal lines [9]. When $|j^*| \leq j_{\mathrm{th}}$, the driving force given in the last set of parentheses in Eq. (11) vanishes because the driving force of EM induced by j^* and that of back flow induced by $\partial N/\partial l$ are balanced.

1.6 Governing Parameter for EM Damage, AFD

1.6.1 Formulation of *AFD*

To gain insight into EM failure, a governing parameter for EM damage in metal lines has been identified [1]. The parameter governing EM damage was formulated on the basis of the divergence of the atomic flux induced by EM, and is denoted as *AFD*. An *AFD*-based method for predicting EM failure has been developed; this method allows the lifetime and possible failure site to be predicted accurately and universally.

The prediction of EM failure, that is, estimating the lifetime and failure location, has been attempted by using an empirical equation [6] and numerical simulations [40, 45, 46]. On the other hand, the evaluation of the threshold current density is also of great interest. The threshold value has conventionally been evaluated under the assumption that the product of the threshold and line length is constant [8, 47]. The predictions of lifetime and failure site, and the evaluation of j_{th} have been attempted separately in various works. With the introduction of the governing parameter for EM damage, *AFD*, the predictions and the evaluation can be carried out in a unified manner [1].

Formulations of *AFD* are considered for unpassivated polycrystalline lines, passivated polycrystalline lines, unpassivated bamboo lines and passivated bamboo lines, respectively. Metallic micro/nano structures can be fabricated by using EM in a polycrystalline line. Because the EM behavior in the passivated line includes the atomic density gradient in the specimen as described in Sect. 1.5, the formulation of *AFD* is more complex for passivated lines than unpassivated lines. The formulation of *AFD* for an unpassivated polycrystalline line is presented here to illustrate the fundamental concepts underlying *AFD*.

It can be assumed that metal atoms migrate along grain boundaries in a polycrystalline line. A model of the polycrystalline structure [57] is shown in Fig. 3, where d is the grain size. The thin metal lines have columnar grain structure. In this model, only one grain is assumed to be in the direction of line thickness. Let us consider the divergence of the atomic flux in the unit region enclosed by the rectangle in Fig. 3. The rectangle includes one triple point of grain boundaries with length $l\left[=\sqrt{3}d/6\right]$. Here, $\Delta\varphi$ is a constant related to the relative angle between grain boundaries, and θ is the angle between Grain boundary-I and the x-axis of the Cartesian coordinate system (x, y). The x and y components of the current density vector and the temperature at the triple point are denoted by j_x, j_y and T, respectively. Substituting the current density component along the grain boundary and the temperature at the end of each grain boundary into the Huntington–Grone equation (10), the atomic flux is obtained for the three points on the side of the rectangle. The sign of the flux is defined as positive for the direction outward from the unit region. After multiplying the effective width of the grain boundary δ and the unit thickness by every atomic flux at the ends of Grain boundary-I, -II and -III, the number of atoms migrating along the grain boundaries per unit time is summed. The sum is divided by the volume of the unit region, $A\left[=\sqrt{3}d^2/4\right]$.

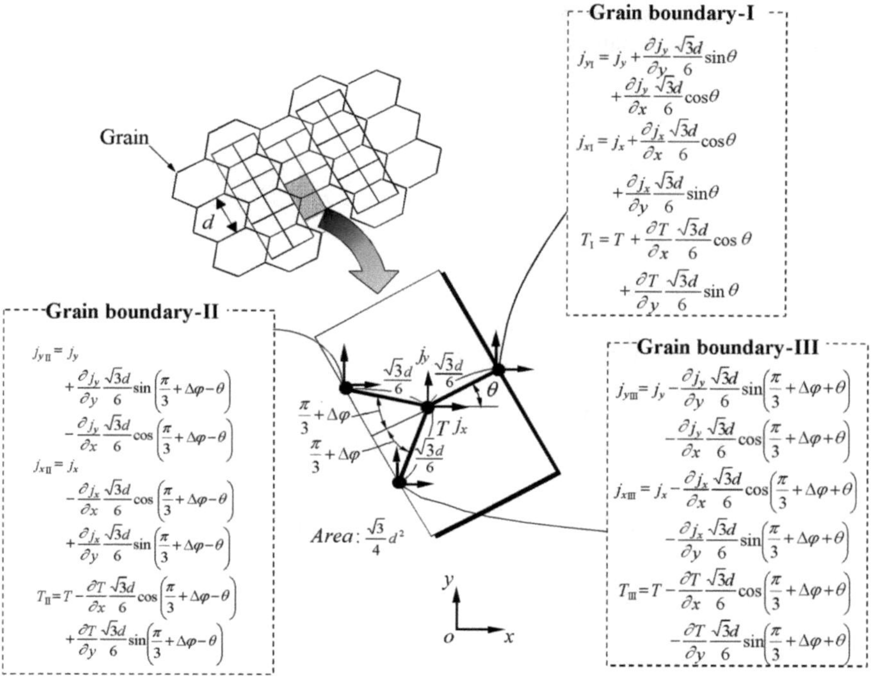

Fig. 3 Model of polycrystalline structure

Thus, the divergence of the atomic flux in the unpassivated polycrystalline line $AFD_{\mathrm{gb}\theta}$ is formulated as follows:

$$\begin{aligned} AFD_{\mathrm{gb}\theta} &= \frac{(J_{\mathrm{I}} + J_{\mathrm{II}} + J_{\mathrm{III}})\delta}{A} \\ &= \frac{C_{\mathrm{gb}}\rho\delta}{A}\left\{\frac{1}{T_{\mathrm{I}}}\exp\left(-\frac{Q_{\mathrm{gb}}}{kT_{\mathrm{I}}}\right)j_{\mathrm{I}}^{*} + \frac{1}{T_{\mathrm{II}}}\exp\left(-\frac{Q_{\mathrm{gb}}}{kT_{\mathrm{II}}}\right)j_{\mathrm{II}}^{*} + \frac{1}{T_{\mathrm{III}}}\exp\left(-\frac{Q_{\mathrm{gb}}}{kT_{\mathrm{III}}}\right)j_{\mathrm{III}}^{*}\right\}, \end{aligned} \tag{12}$$

where J_{I}, J_{II} and J_{III} are atomic fluxes, T_{I}, T_{II} and T_{III} are temperatures, and j_{I}^{*}, j_{II}^{*} and j_{III}^{*} are components of the current density along the grain boundary, defining outward as positive, at the end of each Grain boundary-I, -II and -III. The activation energy for grain boundary diffusion is denoted as Q_{gb}, and the constant C_{gb} is given by

$$C_{\mathrm{gb}} = \frac{ND_0eZ^{*}}{k}. \tag{13}$$

The atomic flux along each grain boundary at its end is calculated using each projection of current density components $(j_{x\mathrm{I}}, j_{y\mathrm{I}})$, $(j_{x\mathrm{II}}, j_{y\mathrm{II}})$ and $(j_{x\mathrm{III}}, j_{y\mathrm{III}})$, illustrated in Fig. 3, along the grain boundary for evaluating j_{I}^{*}, j_{II}^{*} and j_{III}^{*} in Eq. (12).

After that, we define $\pi/3 + \Delta\varphi - \theta$ as φ_1 and $\pi/3 + \Delta\varphi + \theta$ as φ_2. By using the Maclaurin series with natural number n, $AFD_{gb\theta}$ is transformed as follows:

$$
\begin{aligned}
AFD_{gb\theta} = \frac{C_{gb}\rho\delta}{A} \Bigg\langle & \frac{1}{T}\exp\left(-\frac{Q_{gb}}{kT}\right)\cos\theta \cdot \{1 - ()_{\mathrm{I}}\} \\
& \times \left[1 + \frac{Q_{gb}}{kT}()_{\mathrm{I}} + \frac{1}{2}\left\{\frac{Q_{gb}}{kT}()_{\mathrm{I}}\right\}^2 + \cdots + \frac{1}{n!}\left\{\frac{Q_{gb}}{kT}()_{\mathrm{I}}\right\}^n\right](j_x + X_{j_{x\mathrm{I}}}) \\
& + \frac{1}{T}\exp\left(-\frac{Q_{gb}}{kT}\right)\sin\theta \cdot \{1 - ()_{\mathrm{I}}\} \\
& \times \left[1 + \frac{Q_{gb}}{kT}()_{\mathrm{I}} + \frac{1}{2}\left\{\frac{Q_{gb}}{kT}()_{\mathrm{I}}\right\}^2 + \cdots + \frac{1}{n!}\left\{\frac{Q_{gb}}{kT}()_{\mathrm{I}}\right\}^n\right]\left(j_y + X_{j_{y\mathrm{I}}}\right) \\
& + \frac{1}{T}\exp\left(-\frac{Q_{gb}}{kT}\right)(-\cos\varphi_1) \cdot \{1 - ()_{\mathrm{II}}\} \\
& \times \left[1 + \frac{Q_{gb}}{kT}()_{\mathrm{II}} + \frac{1}{2}\left\{\frac{Q_{gb}}{kT}()_{\mathrm{II}}\right\}^2 + \cdots + \frac{1}{n!}\left\{\frac{Q_{gb}}{kT}()_{\mathrm{II}}\right\}^n\right](j_x + X_{j_{x\mathrm{II}}}) \\
& + \frac{1}{T}\exp\left(-\frac{Q_{gb}}{kT}\right)\sin\varphi_1 \cdot \{1 - ()_{\mathrm{II}}\} \\
& \times \left[1 + \frac{Q_{gb}}{kT}()_{\mathrm{II}} + \frac{1}{2}\left\{\frac{Q_{gb}}{kT}()_{\mathrm{II}}\right\}^2 + \cdots + \frac{1}{n!}\left\{\frac{Q_{gb}}{kT}()_{\mathrm{II}}\right\}^n\right]\left(j_y + X_{j_{y\mathrm{II}}}\right) \\
& + \frac{1}{T}\exp\left(-\frac{Q_{gb}}{kT}\right)(-\cos\varphi_2) \cdot \{1 - ()_{\mathrm{III}}\} \\
& \times \left[1 + \frac{Q_{gb}}{kT}()_{\mathrm{III}} + \frac{1}{2}\left\{\frac{Q_{gb}}{kT}()_{\mathrm{III}}\right\}^2 + \cdots + \frac{1}{n!}\left\{\frac{Q_{gb}}{kT}()_{\mathrm{III}}\right\}^n\right](j_x + X_{j_{x\mathrm{III}}}) \\
& + \frac{1}{T}\exp\left(-\frac{Q_{gb}}{kT}\right)(-\sin\varphi_2) \cdot \{1 - ()_{\mathrm{III}}\} \\
& \times \left[1 + \frac{Q_{gb}}{kT}()_{\mathrm{III}} + \frac{1}{2}\left\{\frac{Q_{gb}}{kT}()_{\mathrm{III}}\right\}^2 + \cdots + \frac{1}{n!}\left\{\frac{Q_{gb}}{kT}()_{\mathrm{III}}\right\}^n\right]\left(j_y + X_{j_{y\mathrm{III}}}\right) \Bigg\rangle,
\end{aligned}
\tag{14}
$$

where

$$()_{\mathrm{I}} = X_{T_{\mathrm{I}}} - X_{T_{\mathrm{I}}}^2 + \cdots\cdots + (-1)^{r+1}X_{T_{\mathrm{I}}}^r + \cdots\cdots, \tag{15}$$

$$X_{T_{\mathrm{I}}} = \frac{\frac{\partial T}{\partial x}l\cos\theta + \frac{\partial T}{\partial y}l\sin\theta}{T}, \tag{16}$$

$$X_{j_{x\mathrm{I}}} = \frac{\partial j_x}{\partial x}l\cos\theta + \frac{\partial j_x}{\partial y}l\sin\theta, \tag{17}$$

$$X_{\mathrm{j_{yI}}} = \frac{\partial j_y}{\partial x} l \cos\theta + \frac{\partial j_y}{\partial y} l \sin\theta, \tag{18}$$

$$()_{\mathrm{II}} = X_{\mathrm{T_{II}}} - X_{\mathrm{T_{II}}}^2 + \cdots\cdots + (-1)^{r+1} X_{\mathrm{T_{II}}}^r + \cdots\cdots, \tag{19}$$

$$X_{\mathrm{T_{II}}} = \frac{-\frac{\partial T}{\partial x} l \cos\varphi_1 + \frac{\partial T}{\partial y} l \sin\varphi_1}{T}, \tag{20}$$

$$X_{\mathrm{j_{xII}}} = -\frac{\partial j_x}{\partial x} l \cos\varphi_1 + \frac{\partial j_x}{\partial y} l \sin\varphi_1, \tag{21}$$

$$X_{\mathrm{j_{yII}}} = -\frac{\partial j_y}{\partial x} l \cos\varphi_1 + \frac{\partial j_y}{\partial y} l \sin\varphi_1, \tag{22}$$

$$()_{\mathrm{III}} = X_{\mathrm{T_{III}}} - X_{\mathrm{T_{III}}}^2 + \cdots\cdots + (-1)^{r+1} X_{\mathrm{T_{III}}}^r + \cdots\cdots, \tag{23}$$

$$X_{\mathrm{T_{III}}} = \frac{-\frac{\partial T}{\partial x} l \cos\varphi_2 - \frac{\partial T}{\partial y} l \sin\varphi_2}{T}, \tag{24}$$

$$X_{\mathrm{j_{xIII}}} = -\frac{\partial j_x}{\partial x} l \cos\varphi_2 - \frac{\partial j_x}{\partial y} l \sin\varphi_2, \tag{25}$$

and

$$X_{\mathrm{j_{yIII}}} = -\frac{\partial j_y}{\partial x} l \cos\varphi_2 - \frac{\partial j_y}{\partial y} l \sin\varphi_2. \tag{26}$$

As discussed later, the following relation holds:

$$|X_{\mathrm{T_I}}|, |X_{\mathrm{T_{II}}}|, |X_{\mathrm{T_{III}}}| \leq 1. \tag{27}$$

Each term in angle brackets$\langle\rangle$ in Eq. (14) is composed of the product of the curly brackets { } and square brackets [], both of which include a term of 1. We can calculate the values of the terms in each set of brackets and eliminate small terms.

1.6.2 Comparison of Terms in *AFD* Formula

There is no need to consider all terms in Eq. (14) in a general operation environment, and some small terms can be neglected. The value of each term is estimated under a general condition in the acceleration test of EM, that is, input current density $j_\infty = 0.1$–3.0 MA/cm^2, substrate temperature $T_s = 373$–676 K, grain size $d = 0.4$–0.8 μm and activation energy $Q_{gb} = 0.567$ eV [58]. The *AFD* values are calculated assuming the line shape shown in Fig. 4. The Cartesian coordinate system (x, y) shown in Fig. 4 is used. In the assessment of small terms, the metal line with a single bend between the current input/output, such as contact pads or vias is treated. In the angled metal line, the electric current distribution and

Fig. 4 Model of asymmetrically angled line

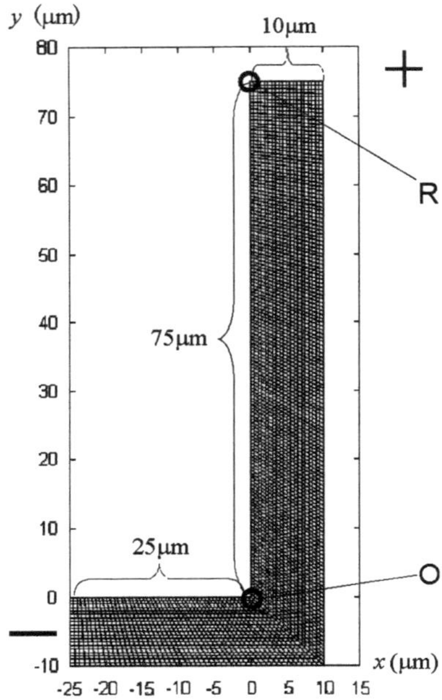

temperature distribution in the straight part differ from those near the corner, and the values of terms in the *AFD* expression differ between the straight part and the corner. The current density is known to be concentrated near the corner and the temperature gradient is very large near the corner of an asymmetrically angled line [56]. Therefore, the values of the terms in Eq. (14) are estimated in each straight part, R, and in the corner, O, in Fig. 4. Current distribution and the temperature distribution in each part are calculated by utilizing finite element (FE) analysis based on the governing equation given below.

Taking electrical potential as ϕ_e, the governing equation for electrical potential is given by

$$\nabla^2 \phi_e = 0, \tag{28}$$

where $\nabla^2 = \partial^2/\partial x^2 + \partial^2/\partial y^2$. Ohm's law is written as

$$\mathbf{j} = -\frac{1}{\rho_0} \mathrm{grad}\phi_e. \tag{29}$$

The equation of steady-state heat conduction is

$$\lambda \nabla^2 T + \rho_0 \mathbf{j} \cdot \mathbf{j} = 0, \tag{30}$$

where λ is the thermal conductivity. The quantities of λ and ρ_0 are assumed to be 0.000233 W/(μm K) and 0.0445 Ω μm [58].

The line shown in Fig. 4 is asymmetrically angled at a higher ratio (25:75) in comparison with a line generally used for acceleration tests [29, 32]. This higher ratio leads to a higher temperature gradient near the corner. Note that here we consider the metal line bent only once between the current input and output. Although heat actually flows from the line to its surroundings [57], this heat flow is not taken into account in Eq. (30); consequently, the highest possible peak temperature is estimated over the limited line length. Thus, the above model can estimate the largest possible temperature gradient not only near the corner, O, but also at the end of the straight part, R.

When input current density j_∞ becomes large, terms related to the current density gradient at point O (e.g.,$X_{\mathrm{j_{xI}}}$ and $X_{\mathrm{j_{yI}}}$), and terms related to the temperature gradient at points R and O (e.g.,$X_{\mathrm{T_I}}$), also become large. When substrate temperature T_s becomes small, terms related to the temperature gradient become large. Accordingly, we assume that j_∞ is larger and T_s is smaller for the boundary conditions of the FE analysis, in order to compare the magnitude of terms under the largest possible estimation. Then, we performed FE analysis under boundary conditions of $j_\infty = 3$ MA/cm^2 and $T_s = 373$ K.

1.6.3 Comparison in Straight Part of Angled Metal Line

Let us derive the *AFD* formula for the straight part in the model discussed in Sect. 1.6.2. Each term in the formula should be estimated to be as large as possible. A numerator of the fraction $X_{\mathrm{T_I}}$ takes the largest value at point R when the direction of Grain boundary-I coincides with longitudinal direction of the line (i.e., $\theta = \pi/2$) because the temperature gradient is greatest along the longitudinal direction of the line (i.e., along the y-axis). Thus, the term for temperature gradient, $X_{\mathrm{T_I}}$, has the following relation:

$$X_{\mathrm{T_I}} = \frac{\frac{\partial T}{\partial x} l \cos\theta + \frac{\partial T}{\partial y} l \sin\theta}{T} \leq \frac{\left|\frac{\partial T}{\partial y}\right| l}{T}. \quad \text{(at point R)} \qquad (31)$$

Then, the quantity of $X_{\mathrm{T_I}}$ is calculated by using the most right-hand side of Eq. (31) to obtain the largest possible estimate. When the grain size d becomes large, the term $X_{\mathrm{T_I}}$ proportionally becomes large because of increasing $l\left[=\sqrt{3}d/6\right]$. The grain size d is 0.4–0.8 μm for general Al lines. We assume that $d = 0.8$ μm and $l = 0.231$ μm to obtain the largest possible estimate of $X_{\mathrm{T_I}}$. The values of each term in the first term of the angle brackets $\langle\rangle$ in Eq. (14) are calculated by using FE analysis results for current density, temperature and temperature gradient (9.05 K/μm) at point R in Fig. 4:

$$X_{\mathrm{T_I}} = 0.00560,$$

$$X_{\mathrm{j_{xI}}} = 0$$

and

$$\frac{Q_{gb}}{kT}X_{T_I} = 0.0988.$$

Considering the above values, the term $X_{T_I}^2$ and higher-order terms included in curly brackets $\{1 - ()_I\}$ in Eq. (14) are much smaller than 1. Here, we neglect the terms smaller than 0.005. The relation $Q_{gb}()_I/(kT) \approx Q_{gb}X_{T_I}/(kT)$ holds, and the term of $1/2\{Q_{gb}()_I/(kT)\}^2$ and higher-order terms are removed from the square brackets [] in Eq. (14). There is no need to consider $X_{j_{xI}}$ in parentheses $(j_x + X_{j_{xI}})$ in Eq. (14) because the current density gradient is zero in the straight part. To preserve an arbitrariness of setting of coordinate system, j_x should remain in the parentheses although the value of J_x vanishes at point R.

After the elimination of negligible terms, the curly brackets {} and square brackets [] in Eq. (14) are expanded. Then, terms smaller than 0.005 are omitted. The negligible term is as follows:

$$\frac{Q_{gb}}{kT}X_{T_I} \times X_{T_I} = 0.000554.$$

In this way, small terms are erased in the first term in angle brackets $\langle\rangle$ in Eq. (14). As for the other terms in the angle brackets$\langle\rangle$, small terms are also eliminated and the curly brackets {} and square brackets [] are expanded in a similar manner. We next obtain the *AFD* formula for the straight part R as follows:

$$\begin{aligned} AFD_{gb\theta} = {} & \frac{C_{gb}\rho\delta}{A}\frac{1}{T}\exp\left(-\frac{Q_{gb}}{kT}\right) \\ & \times \left[j_x(\cos\theta - \cos\varphi_1 - \cos\varphi_2) + j_y(\sin\theta + \sin\varphi_1 - \sin\varphi_2) \right. \\ & + \frac{1}{T}\left(\frac{Q_{gb}}{kT}\right)\left\{\frac{\partial T}{\partial x}j_x l\left(\cos^2\theta + \cos^2\varphi_1 + \cos^2\varphi_2\right)\right. \\ & + \frac{\partial T}{\partial y}j_y l\left(\sin^2\theta + \sin^2\varphi_1 + \sin^2\varphi_2\right) \\ & \left.\left. + \left(\frac{\partial T}{\partial x}j_y + \frac{\partial T}{\partial y}j_x\right)l(\sin\theta\cos\theta - \sin\varphi_1\cos\varphi_1 + \sin\varphi_2\cos\varphi_2)\right\}\right]. \end{aligned} \quad (32)$$

Then, we expand the trigonometric functions including $\varphi_1[= \pi/3 + \Delta\varphi - \theta]$ and $\varphi_2[= \pi/3 + \Delta\varphi + \theta]$with a minimal angle of $\Delta\varphi$ [=−0.0236 rad] [59] by using the Maclaurin series and remove terms smaller than 0.005 in comparison with 1. The *AFD* formula for the straight part is transformed as follows:

$$
\begin{aligned}
AFD_{gb\theta} = {} & C_{gb}\rho\delta\frac{4}{\sqrt{3}d^2}\frac{1}{T}\exp\left(-\frac{Q_{gb}}{kT}\right)\left[\sqrt{3}\Delta\varphi(j_x\cos\theta+j_y\sin\theta)\right. \\
& \left.+\frac{\sqrt{3}d}{4T}\left(\frac{Q_{gb}}{kT}\right)\left(\frac{\partial T}{\partial x}j_x+\frac{\partial T}{\partial y}j_y\right)-\frac{\sqrt{3}d}{4T}\left(\frac{\partial T}{\partial x}j_x+\frac{\partial T}{\partial y}j_y\right)\right],
\end{aligned}
\tag{33}
$$

1.6.4 Comparison Near Corner of Angled Metal Line

The current and temperature distributions near the corner of the angled metal line shown in Fig. 5 are expressed by asymptotic solutions [56]. Cartesian coordinates (x, y) and polar coordinates (r, γ) are shown in Fig. 5. The line width is b, the corner angle is $2\beta_0$ and the input current density is j_∞. Then, the values of current density, its gradient and temperature gradient are obtained by using the following asymptotic solutions accompanied with the results of FE analysis to compare the terms in Eq. (14) near the corner:

$$
j_x = -K_e\left(\frac{r}{b}\right)^{\frac{\pi}{2\beta_0}-1}\sin\left(\frac{\pi}{2\beta_0}-1\right)\gamma, \tag{34}
$$

$$
j_y = -K_e\left(\frac{r}{b}\right)^{\frac{\pi}{2\beta_0}-1}\cos\left(\frac{\pi}{2\beta_0}-1\right)\gamma, \tag{35}
$$

Fig. 5 Infinite metal line

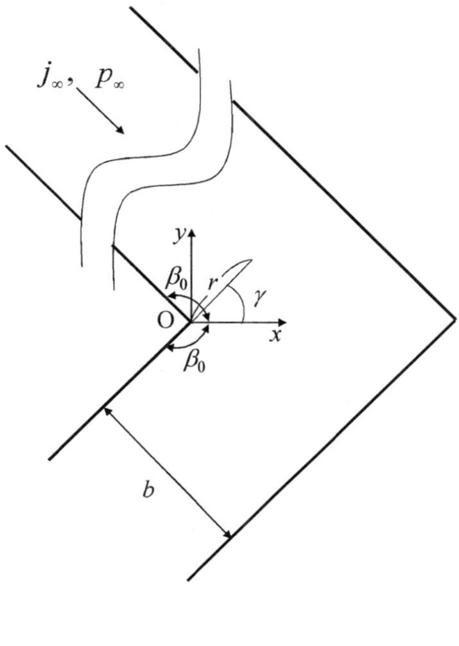

$$\frac{\partial j_x}{\partial x} = -\frac{\partial j_y}{\partial y} = K_e\left(\frac{\pi}{2\beta_0} - 1\right)\frac{1}{b}\left(\frac{r}{b}\right)^{\frac{\pi}{2\beta_0}-2}\sin\left(\frac{\pi}{2\beta_0} - 2\right)\gamma, \tag{36}$$

$$\frac{\partial j_x}{\partial y} = \frac{\partial j_y}{\partial x} = -K_e\left(\frac{\pi}{2\beta_0} - 1\right)\frac{1}{b}\left(\frac{r}{b}\right)^{\frac{\pi}{2\beta_0}-2}\cos\left(\frac{\pi}{2\beta_0} - 2\right)\gamma, \tag{37}$$

$$\frac{\partial T}{\partial x} = \frac{1}{\lambda}K_t\left(\frac{r}{b}\right)^{\frac{\pi}{2\beta_0}-1}\sin\left(\frac{\pi}{2\beta_0} - 1\right)\gamma \tag{38}$$

and

$$\frac{\partial T}{\partial y} = \frac{1}{\lambda}K_t\left(\frac{r}{b}\right)^{\frac{\pi}{2\beta_0}-1}\cos\left(\frac{\pi}{2\beta_0} - 1\right)\gamma, \tag{39}$$

where $K_e = \beta_0^{\pi/2\beta_0-1} j_\infty, K_t = \beta_0^{\pi/2\beta_0-1} p_\infty$, and p_∞ is the heat flux considering Joule heating far from the corner, which is expressed by

$$p_\infty = q_R + \phi_R j_\infty. \tag{40}$$

Here, q_R and ϕ_R are the heat flux and electrical potential at point R far from the corner, as shown in Fig. 4. ϕ_R is taken at point R when the electrical potential is zero at the inner corner.

In essence, p_∞ represents the heat flux induced by the temperature difference between the two points far from the corner, which are symmetric to each other around the bisector of the corner. The temperature difference between the symmetric points originates from the Joule heating generated in the lines around the angle line and/or in the asymmetrically angled line itself. One can directly measure the heat flux q_R, which includes the heat flux induced by the temperature difference between the symmetric points p_∞ and the heat flux due to Joule heating generated symmetrically around the bisector [56]. If the electric current is turned off to eliminate the heat flux by symmetric Joule heating from q_R, the temperature difference between the symmetric points also vanishes because this temperature difference originates from Joule heating. The heat flux p_∞, therefore, cannot be measured separately, but can be extracted from q_R by using Eq. (40). Since the heat flux p_∞ originates from Joule heating, the solutions given by Eqs. (38) and (39) are effective under the flow of electric current.

The current density, its gradient and temperature gradient near the corner in the model described in Sect. 1.6.2 are estimated using Eqs. (34)–(39). The *AFD* formula near the corner is derived using these values.

Now, let us consider the region at $r = d$ as a corner part. The grain size d is 0.4–0.8 μm for a typical Al lines. When the grain size becomes small, the term X_{T_I} near the corner becomes large because of a singularity of temperature gradient at the corner vertex. Each term to be compared should be estimated to be as large as possible. Accordingly, it is assumed that $d = 0.4$ μm and $l = 0.115$ μm. The heat flux q_R and electrical potential ϕ_R at R are calculated by

FE analysis, and thereby we obtain the value of p_∞. Equations (34)–(39) also yield the following relation:

$$|j_x| = \left| -K_e \left(\frac{r}{b}\right)^{\frac{\pi}{2\beta_0}-1} \sin\left(\frac{\pi}{2\beta_0} - 1\right)\gamma \right| \le \left| K_e \left(\frac{r}{b}\right)^{\frac{\pi}{2\beta_0}-1} \right|. \tag{41}$$

Similarly,

$$|j_y| \le \left| K_e \left(\frac{r}{b}\right)^{\frac{\pi}{2\beta_0}-1} \right|, \tag{42}$$

$$\left|\frac{\partial j_x}{\partial x}\right|, \left|\frac{\partial j_y}{\partial y}\right|, \left|\frac{\partial j_x}{\partial y}\right|, \left|\frac{\partial j_y}{\partial x}\right| \le \left| K_e \left(\frac{\pi}{2\beta_0} - 1\right) \frac{1}{b} \left(\frac{r}{b}\right)^{\frac{\pi}{2\beta_0}-2} \right| \tag{43}$$

and

$$\left|\frac{\partial T}{\partial x}\right|, \left|\frac{\partial T}{\partial y}\right| \le \left| \frac{1}{\lambda} K_t \left(\frac{r}{b}\right)^{\frac{\pi}{2\beta_0}-1} \right|. \tag{44}$$

The values of current density, its gradient and temperature gradient can be calculated using the most right-hand sides of Eqs. (41)–(44) to obtain the largest possible estimates.

Regarding the term X_{T_I} in the first term of angle brackets $\langle\rangle$ in Eq. (14), in order to find the largest possible temperature gradient independent of the coordinate setting, $\partial T/\partial x = \partial T/\partial y$ is assumed. By obtaining the extremum of the numerator with respect to θ, the term concerning the temperature gradient, X_{T_I}, has the following relation:

$$X_{T_I} = \frac{\frac{\partial T}{\partial x} l \cos\theta + \frac{\partial T}{\partial y} l \sin\theta}{T} \le \frac{\sqrt{2}\left|\frac{\partial T}{\partial x}\right| l}{T} \quad \text{(at point O)}. \tag{45}$$

Then, the value of X_{T_I} is calculated using the most right-hand side of Eq. (45) to obtain the largest possible estimate. Using Eqs. (44) and (45), the following values are obtained:

$$X_{T_I} = 0.00277$$

and

$$\frac{Q_{gb}}{kT} X_{T_I} = 0.0326,$$

where the angle $\beta_0 = 3\pi/4$ and the values $p_\infty = 0.00101$ W/(μm K) and $T = 559$ K obtained by FE analysis are used.

Thus, the term X_{T_I} and higher-order terms included in curly brackets $\{1 - ()_I\}$ in Eq. (14) are negligible because these terms are smaller than 0.005. Moreover, the relation $Q_{gb}()_I/(kT) \approx Q_{gb}X_{T_I}/(kT)$ holds, and the term $1/2\{Q_{gb}()_I/(kT)\}^2$ and higher-order terms are omitted in the square brackets [] in Eq. (14).

As for the parentheses $(j_x + X_{\mathrm{jxI}})$ in Eq. (14), the current density and its gradient are estimated to be as large as possible using the relations Eqs. (41)–(43). Then, the parentheses $(j_x + X_{\mathrm{jxI}})$ in the first term of the angle brackets $\langle\rangle$ in Eq. (14) is written as

$$(j_x + X_{\mathrm{jxI}}) = j_x\left(1 + \frac{X_{\mathrm{jxI}}}{j_x}\right). \tag{46}$$

If the quantity X_{jxI} is maximized in the evaluation, the following relation is obtained in the same manner as Eq. (45):

$$\frac{X_{\mathrm{j_{xI}}}}{j_x} = \frac{\frac{\partial j_x}{\partial x} l\cos\theta + \frac{\partial j_x}{\partial y} l\sin\theta}{j_x} \leq \frac{\sqrt{2}\left|\frac{\partial j_x}{\partial x}\right| l}{|j_x|} \quad \text{(at point O)}. \tag{47}$$

Note that the value of j_x should be maximized. The quantity X_{jxI}/j_x is calculated using the most right-hand side of Eq. (47). The value is obtained as

$$\frac{X_{\mathrm{j_{xI}}}}{j_x} = \frac{\sqrt{2}l}{r}\left|\frac{\pi}{2\beta_0} - 1\right| = 0.136.$$

This value is greater than 0.005; thus, the term X_{jxI}/j_x remains in the parentheses () on the right-hand side of Eq. (46).

After eliminating negligible terms, the curly brackets {}, square brackets [] and parentheses () concerning current density in Eq. (14) are expanded. Then, terms smaller than 0.005 are omitted. A negligible term is as follows:

$$\frac{Q_{\mathrm{gb}}}{kT} X_{\mathrm{T_I}} \times \frac{X_{\mathrm{j_{xI}}}}{j_x} = 0.00442$$

In the first term in the angle brackets $\langle\rangle$ in Eq. (14), small terms are omitted. As for other terms in angle brackets $\langle\rangle$, negligible terms are also eliminated and the braces are expanded in a similar manner. Then, one can obtain the following *AFD* formula for the corner part, O:

$$\begin{aligned} AFD_{\mathrm{gb}\theta} = {} & \frac{C_{\mathrm{gb}}\rho\delta}{A}\frac{1}{T}\exp\left(-\frac{Q_{\mathrm{gb}}}{kT}\right) \\ & \times \Bigg[j_x(\cos\theta - \cos\varphi_1 - \cos\varphi_2) + j_y(\sin\theta + \sin\varphi_1 - \sin\varphi_2) \\ & + \frac{\partial j_x}{\partial x} l(\cos^2\theta + \cos^2\varphi_1 + \cos^2\varphi_2) + \frac{\partial j_y}{\partial y} l(\sin^2\theta + \sin^2\varphi_1 + \sin^2\varphi_2) \\ & + \left(\frac{\partial j_x}{\partial y} + \frac{\partial j_y}{\partial x}\right) l(\sin\theta\cos\theta - \sin\varphi_1\cos\varphi_1 + \sin\varphi_2\cos\varphi_2) \\ & + \frac{1}{T}\left(\frac{Q_{\mathrm{gb}}}{kT}\right)\Bigg\{\frac{\partial T}{\partial x} j_x l(\cos^2\theta + \cos^2\varphi_1 + \cos^2\varphi_2) \end{aligned}$$

$$+\frac{\partial T}{\partial y} j_y l\left(\sin^2\theta + \sin^2\varphi_1 + \sin^2\varphi_2\right)$$
$$+\left(\frac{\partial T}{\partial x} j_y + \frac{\partial T}{\partial y} j_x\right) l(\sin\theta\cos\theta - \sin\varphi_1\cos\varphi_1 + \sin\varphi_2\cos\varphi_2)\Big\}\Big]. \tag{48}$$

Now, let us expand the trigonometric functions including φ_1 and φ_2 with a minimal angle $\Delta\varphi$ by using the Maclaurin series and omit terms smaller than 0.005. The *AFD* formula for the corner part is transformed as follows:

$$\begin{aligned} AFD_{gb\theta} = {} & C_{gb}\rho\delta\frac{4}{\sqrt{3}d^2}\frac{1}{T}\exp\left(-\frac{Q_{gb}}{kT}\right)\Big[\sqrt{3}\Delta\varphi\left(j_x\cos\theta + j_y\sin\theta\right) \\ & -\frac{d}{2}\Delta\varphi\left\{\left(\frac{\partial j_x}{\partial x} - \frac{\partial j_y}{\partial y}\right)\cos 2\theta + \left(\frac{\partial j_x}{\partial y} + \frac{\partial j_y}{\partial x}\right)\sin 2\theta\right\} \\ & +\frac{\sqrt{3}d}{4T}\left(\frac{Q_{gb}}{kT}\right)\left(\frac{\partial T}{\partial x} j_x + \frac{\partial T}{\partial y} j_y\right)\Big]. \end{aligned} \tag{49}$$

1.6.5 General Expression of *AFD*

To obtain the general expression of *AFD* that is applicable to both the straight part and the corner, the formula should include all terms in Eqs. (33) and (49). Thus, one can obtain a general expression of *AFD* [57] as

$$\begin{aligned} AFD_{gb\theta} = {} & C_{gb}\rho\delta\frac{4}{\sqrt{3}d^2}\frac{1}{T}\exp\left(-\frac{Q_{gb}}{kT}\right)\Big[\sqrt{3}\Delta\varphi\left(j_x\cos\theta + j_y\sin\theta\right) \\ & -\frac{d}{2}\Delta\varphi\left\{\left(\frac{\partial j_x}{\partial x} - \frac{\partial j_y}{\partial y}\right)\cos 2\theta + \left(\frac{\partial j_x}{\partial y} + \frac{\partial j_y}{\partial x}\right)\sin 2\theta\right\} \\ & +\frac{\sqrt{3}d}{4T}\left(\frac{Q_{gb}}{kT} - 1\right)\left(\frac{\partial T}{\partial x} j_x + \frac{\partial T}{\partial y} j_y\right)\Big]. \end{aligned} \tag{50}$$

It can be confirmed that the obtained formula of *AFD* is applicable not only to acceleration conditions but also to conditions with lower current density and higher substrate temperature.

1.6.6 Constant Electrical Resistivity within Rectangular Unit

Temperature may vary slightly within the rectangular unit shown in Fig. 3. Although the electrical resistivity depends on temperature, it was assumed that the electrical resistivity ρ in Eq. (12) was constant within the rectangular unit. Here the effect of the assumption in formulation of *AFD* is discussed. By explicitly expressing the difference in ρ due to temperature distribution in the unit, Eq. (12) is transformed as follows:

$$AFD_{gb\theta}=\frac{(J_{\rm I}+J_{\rm II}+J_{\rm III})\delta}{A}$$
$$=\frac{C_{gb}\delta}{A}\left\{\frac{1}{T_{\rm I}}\exp\left(-\frac{Q_{gb}}{kT_{\rm I}}\right)\rho_{\rm I}j_{\rm I}^{*}+\frac{1}{T_{\rm II}}\exp\left(-\frac{Q_{gb}}{kT_{\rm II}}\right)\rho_{\rm II}j_{\rm II}^{*}+\frac{1}{T_{\rm III}}\exp\left(-\frac{Q_{gb}}{kT_{\rm III}}\right)\rho_{\rm III}j_{\rm III}^{*}\right\}, \tag{51}$$

where the electrical resistivity at the end of Grain boundary-I, -II and -III in Fig. 3 is given by $\rho_i[=\rho(1+\alpha TX_{{\rm T}_i})]$ (i = I, II, III). Considering the derivation of *AFD* described above, one can obtain the following expression:

$$\begin{aligned}AFD_{gb\theta}=\frac{C_{gb}\rho\delta}{A}\Bigg\langle &\frac{1}{T}\exp\left(-\frac{Q_{gb}}{kT}\right)\cos\theta\cdot\{1-()_{\rm I}\}\\
&\times\left[1+\frac{Q_{gb}}{kT}()_{\rm I}+\frac{1}{2}\left\{\frac{Q_{gb}}{kT}()_{\rm I}\right\}^{2}+\cdots+\frac{1}{n!}\left\{\frac{Q_{gb}}{kT}()_{\rm I}\right\}^{n}\right](j_x+X_{{\rm j}_{x{\rm I}}})(1+\alpha TX_{{\rm T}_{\rm I}})\\
&+\frac{1}{T}\exp\left(-\frac{Q_{gb}}{kT}\right)\sin\theta\cdot\{1-()_{\rm I}\}\\
&\times\left[1+\frac{Q_{gb}}{kT}()_{\rm I}+\frac{1}{2}\left\{\frac{Q_{gb}}{kT}()_{\rm I}\right\}^{2}+\cdots+\frac{1}{n!}\left\{\frac{Q_{gb}}{kT}()_{\rm I}\right\}^{n}\right]\left(j_y+X_{{\rm j}_{y{\rm I}}}\right)(1+\alpha TX_{{\rm T}_{\rm I}})\\
&+\frac{1}{T}\exp\left(-\frac{Q_{gb}}{kT}\right)(-\cos\varphi_1)\cdot\{1-()_{\rm II}\}\\
&\times\left[1+\frac{Q_{gb}}{kT}()_{\rm II}+\frac{1}{2}\left\{\frac{Q_{gb}}{kT}()_{\rm II}\right\}^{2}+\cdots+\frac{1}{n!}\left\{\frac{Q_{gb}}{kT}()_{\rm II}\right\}^{n}\right]\left(j_x+X_{{\rm j}_{x{\rm II}}}\right)(1+\alpha TX_{{\rm T}_{\rm II}})\\
&+\frac{1}{T}\exp\left(-\frac{Q_{gb}}{kT}\right)\sin\varphi_1\cdot\{1-()_{\rm II}\}\\
&\times\left[1+\frac{Q_{gb}}{kT}()_{\rm II}+\frac{1}{2}\left\{\frac{Q_{gb}}{kT}()_{\rm II}\right\}^{2}+\cdots+\frac{1}{n!}\left\{\frac{Q_{gb}}{kT}()_{\rm II}\right\}^{n}\right]\left(j_y+X_{{\rm j}_{y{\rm II}}}\right)(1+\alpha TX_{{\rm T}_{\rm II}})\\
&+\frac{1}{T}\exp\left(-\frac{Q_{gb}}{kT}\right)(-\cos\varphi_2)\cdot\{1-()_{\rm III}\}\\
&\times\left[1+\frac{Q_{gb}}{kT}()_{\rm III}+\frac{1}{2}\left\{\frac{Q_{gb}}{kT}()_{\rm III}\right\}^{2}+\cdots+\frac{1}{n!}\left\{\frac{Q_{gb}}{kT}()_{\rm III}\right\}^{n}\right](j_x+X_{{\rm j}_{x{\rm III}}})(1+\alpha TX_{{\rm T}_{\rm III}})\\
&+\frac{1}{T}\exp\left(-\frac{Q_{gb}}{kT}\right)(-\sin\varphi_2)\cdot\{1-()_{\rm III}\}\\
&\times\left[1+\frac{Q_{gb}}{kT}()_{\rm III}+\frac{1}{2}\left\{\frac{Q_{gb}}{kT}()_{\rm III}\right\}^{2}+\cdots+\frac{1}{n!}\left\{\frac{Q_{gb}}{kT}()_{\rm III}\right\}^{n}\right]\left(j_y+X_{{\rm j}_{y{\rm III}}}\right)(1+\alpha TX_{{\rm T}_{\rm III}})\Bigg\rangle.\end{aligned} \tag{52}$$

Here, α of Al takes a value of 0.002 K^{-1} at room temperature for an Al thin film [39, 67]. The value of α for a thin film is smaller than that for the bulk material [53, 67]. The value of αTX_{T_i} (i = I, II, III) is compared with 1 in Eq. (52) using the value of α being 0.002 K^{-1}. When the reference temperature increases as in an acceleration test, α decreases. Thus, if the term αTX_{T_i} can be omitted at room temperature, the term is negligible even at higher temperatures. Because the term concerning the temperature gradient in the straight part X_{Ti} is larger than that near the corner, as described in Sects. 1.6.3 and 1.6.4, the values of αTX_{T_i} are calculated using the values of T and X_{Ti} in the straight part:

$$\alpha TX_{T_i} = 0.00418 < 0.005$$

Thus, the terms αTX_{T_i} can be neglected in Eq. (52), and Eq. (14) is given. Therefore, the effect of temperature distribution in the rectangular unit on the electrical resistivity is unnecessary to consider, and ρ can be regarded constant within the unit in the formulation of *AFD*.

1.6.7 Application of *AFD*

The value of $AFD_{gb\theta}$ changes with the angle θ between the rectangle unit and the x-axis. To extract only the positive values of $AFD_{gb\theta}$ considering void formation, the sum of the value of $AFD_{gb\theta}$ and its absolute value is divided by two, and the expected value from the extracted positive values is obtained by considering the angle θ from 0 to 2π. On the other hand, extracting only the negative values of $AFD_{gb\theta}$ gives the atomic flux divergence AFD_{gen} of hillock formation in the polycrystalline lines:

$$AFD_{gen} = \frac{1}{4\pi}\int_0^{2\pi} \left(AFD_{gb\theta} - \left|AFD_{gb\theta}\right|\right)\mathrm{d}\theta. \tag{53}$$

Equation (53) can be applied to two-dimensional problem. For a one-dimensional problem such as a straight line, the AFD_{gen} of hillock formation is simply formulated by integrating $AFD_{gb\theta}$ with respect to θ [80]:

$$AFD_{gen} = \frac{C_{gb}}{\pi}\frac{\rho\delta}{d^2}\frac{1}{T}\exp\left(-\frac{Q_{gb}}{kT}\right)\left\{\frac{d}{\lambda}\rho j_\infty^3 x\theta_1\frac{1}{T}\left(\frac{Q_{gb}}{kT} - 1\right) - 4\Delta\varphi j_\infty \sin\theta_1\right\}. \tag{54}$$

Here, θ_1 is the critical value of θ and the coordinate x is set along the longitudinal of the line with the origin at the center of the strip.

When the governing parameter for EM damage in the line covered with a passivation layer is considered, the Huntington–Grone equation (10) is replaced with the atomic flux in a passivated metal line, as given by Eq. (11). Then, one can calculate the parameter $AFD^*_{gb\theta}$ in the same manner as for the unpassivated line [62]:

$$
\begin{aligned}
AFD^*_{gb\theta} = {} & C^*_{gb} N \frac{4}{\sqrt{3}d^2}\frac{1}{T}\exp\left\{-\frac{Q_{gb}+\kappa\Omega(N-N_T)/N_0-\sigma_T\Omega}{kT}\right\} \\
& \times\left\langle \sqrt{3}\Delta\varphi\left\{(j_x\cos\theta+j_y\sin\theta)Z^*e\rho-\frac{\kappa\Omega}{N_0}\left(\frac{\partial N}{\partial x}\cos\theta+\frac{\partial N}{\partial y}\sin\theta\right)\right\}\right. \\
& -\frac{d}{2}\Delta\varphi\left\{\left(\frac{\partial j_x}{\partial x}-\frac{\partial j_y}{\partial y}\right)Z^*e\rho\cos 2\theta-\frac{\kappa\Omega}{N_0}\left(\frac{\partial^2 N}{\partial x^2}-\frac{\partial^2 N}{\partial y^2}\right)\cos 2\theta\right. \\
& \left.+\left(\frac{\partial j_x}{\partial y}+\frac{\partial j_y}{\partial x}\right)Z^*e\rho\sin 2\theta-2\frac{\kappa\Omega}{N_0}\frac{\partial^2 N}{\partial x\partial y}\sin 2\theta\right\} \\
& -\frac{\sqrt{3}}{4}d\frac{\kappa\Omega}{N_0}\left(\frac{\partial^2 N}{\partial x^2}+\frac{\partial^2 N}{\partial y^2}\right)-\frac{\kappa\Omega/N_0}{kT} \\
& \times\left[\frac{\sqrt{3}}{4}d\left\{Z^*e\rho\left(j_x\frac{\partial N}{\partial x}+j_y\frac{\partial N}{\partial y}\right)-\frac{\kappa\Omega}{N_0}\left(\frac{\partial N}{\partial x}\frac{\partial N}{\partial x}+\frac{\partial N}{\partial y}\frac{\partial N}{\partial y}\right)\right\}\right. \\
& \left.-\frac{d}{2}\Delta\varphi\left\{Z^*e\rho\left(j_x\frac{\partial N}{\partial y}+j_y\frac{\partial N}{\partial x}\right)-2\frac{\kappa\Omega}{N_0}\frac{\partial N}{\partial x}\frac{\partial N}{\partial y}\right\}\sin 2\theta\right] \\
& +\frac{\sqrt{3}d}{4T}\left\{\frac{Q_{gb}+\kappa\Omega(N-N_T)/N_0-\sigma_T\Omega}{kT}-1\right\} \\
& \left.\times\left\{Z^*e\rho\left(j_x\frac{\partial T}{\partial x}+j_y\frac{\partial T}{\partial y}\right)-\frac{\kappa\Omega}{N_0}\left(\frac{\partial N}{\partial x}\frac{\partial T}{\partial x}+\frac{\partial N}{\partial y}\frac{\partial T}{\partial y}\right)\right\}\right\rangle,
\end{aligned}
\tag{55}
$$

where $C^*_{gb} = D_0\delta/k$.

The governing parameter for EM damage in a passivated polycrystalline line, AFD^*_{gen}, is

$$AFD^*_{gen} = \frac{1}{4\pi}\int_0^{2\pi}\left(AFD^*_{gb\theta}-\left|AFD^*_{gb\theta}\right|\right)d\theta \quad \text{(for hillock formation)} \tag{56}$$

$$AFD^*_{gen} = \frac{1}{4\pi}\int_0^{2\pi}\left(AFD^*_{gb\theta}+\left|AFD^*_{gb\theta}\right|\right)d\theta \quad \text{(for void formation).} \tag{57}$$

The governing parameter *AFD* integrates all the factors that govern the damage, namely, the line structure, film characteristics, atomic density, as well as operating conditions such as current density and temperature. The parameter *AFD* gives the increase or decrease in the number of atoms per unit time and unit volume. By utilizing *AFD,* the distribution of atomic density N within the metal line can be calculated. Then, it is judged whether or not the atomic density is beyond a critical value for beginning void or hillock formation or for seeking the threshold current

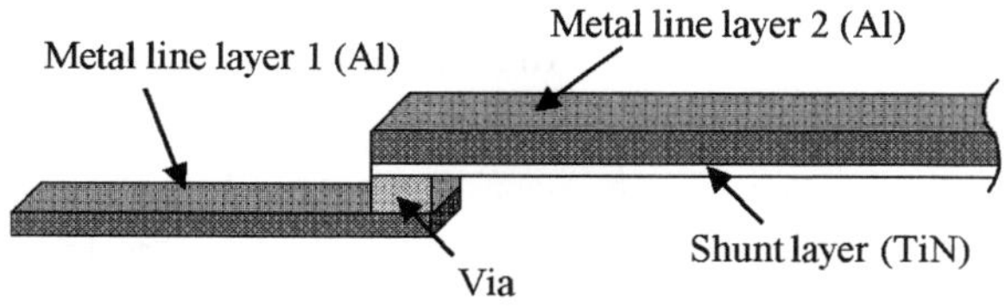

Fig. 6 Metal lines connected by via. Reprinted from Hasegawa et al. [31]. Copyright (2009) with permission from Elsevier

density [1]. An excess of atomic density over the critical value is used for reproducing the damage process until line failure, and lifetime and possible failure site are predicted [62].

1.7 Derivation Method of Characteristic Constants of EM[1]

1.7.1 Failure Mode in Via-Connected Line

The metal lines in IC are often connected by vias and multi-level interconnections are constructed. A schematic diagram of a typical interconnection with a via is shown in Fig. 6. The via for an Al interconnection is made of Al or tungsten (W). The metal lines are often stacked on a shunt layer made of refractory metal such as titanium nitride (TiN), by which the electric current can bypass the void formed in the Al line. In the metal line structure with a via, no Al atoms are supplied to the cathode end by EM because the line is not connected to a reservoir of the atoms such as pads and the atomic flow is intercepted by the via. Therefore, the metal line connected by the via has the failure mode whereby the cathode edge of the line drifts in the direction of electron flow as a result of EM [7]. A governing parameter for EM damage at the ends of a passivated polycrystalline line, $AFD^*_{\mathrm{gen}}|_{\mathrm{end}}$, has been expressed by considering the boundary condition with respect to atomic diffusion [30], where no atoms flow into the cathode end or out of the anode end.

The method for deriving the EM characteristics of a metal line utilizes the EM-induced drift at the line end [31]. In this section the drift velocity induced at the line end by EM is theoretically expressed by using $AFD^*_{\mathrm{gen}}|_{\mathrm{end}}$. By this derivation method, the film's characteristic constants in the formula of the parameter can be determined by measuring drift velocity. Next, the Al line modeled after the via-connected line is considered. The method based on $AFD^*_{\mathrm{gen}}|_{\mathrm{end}}$ is applied to two types of Al polycrystalline lines with different line lengths, and the values of drift velocity in such lines are experimentally measured. By equating the theoretical drift velocity with the experimental value, characteristic constants of the film are obtained. The method based on $AFD^*_{\mathrm{gen}}|_{\mathrm{end}}$ is able to approximate the constants.

---

1 The contents concerning this section have been permitted to reprint from Hasegawa et al. [31], copyright (2009), with permission from Elsevier.

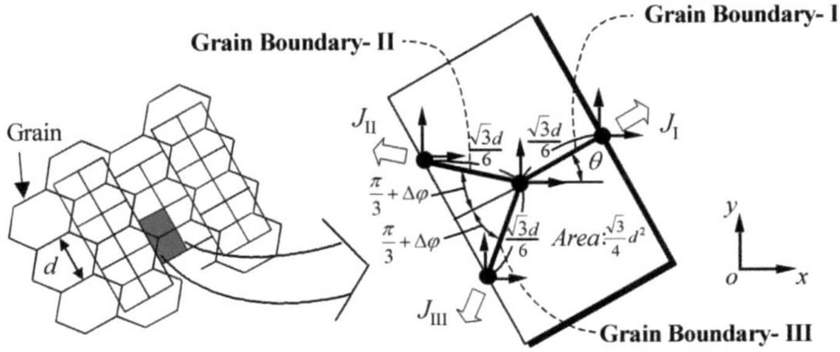

Fig. 7 Model of polycrystalline microstructure. Reprinted from Hasegawa et al. [31]. Copyright (2009) with permission from Elsevier

Table 1 Boundary conditions for atomic diffusion

Range-I	Range-II	Range-III	
$-2\pi/3+\Delta\varphi+\beta<\theta\le-\pi/3-\Delta\varphi+\beta$	$-\pi/3-\Delta\varphi+\beta<\theta\le\beta$	$\beta<\theta\le\pi/3+\Delta\varphi+\beta$	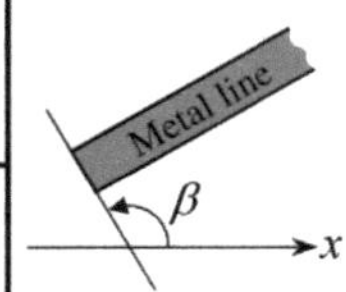
$J_{II}=J_{III}=0$	$J_{II}=0$	$J_{I}=J_{II}=0$	
Range-IV	Range-V	Range-VI	
$\pi/3+\Delta\varphi+\beta<\theta\le2\pi/3-\Delta\varphi+\beta$	$2\pi/3-\Delta\varphi+\beta<\theta\le\pi+\beta$	$\pi+\beta<\theta\le4\pi/3+\Delta\varphi+\beta$	
$J_{I}=0$	$J_{I}=J_{III}=0$	$J_{III}=0$	

Reprinted from Hasegawa et al. [31]. Copyright (2009) with permission from Elsevier

1.7.2 Governing Parameter for EM Damage at Polycrystalline Line Ends

Atom transport in a passivated metal line is assumed to be represented by Eq. (11). Boundary condition concerning the atomic diffusion at the metal line ends is taken into account; that is, there are no incoming and no outgoing flow of atoms at the line end. The boundary condition is set considering the microstructure unit region enclosed by a rectangle in Fig. 7. Let us consider the *AFD* in a unit region that faces the line end. First, the whole range of θ (from 0 to 2π) is divided into six parts (Range-I to -VI), as shown in Table 1; β is defined as the angle between the line edge and the x-axis of the Cartesian coordinates shown in Fig. 7. Then, one or two atomic fluxes in three grain boundaries in the unit region are assumed to be zero for each θ-range: $J_{II} = J_{III} = 0$ in Range-I, $J_{II} = 0$ in Range-II, $J_{I} = J_{II} = 0$ in Range-III, $J_{I} = 0$ in Range-IV, $J_{I} = J_{III}=0$ in Range-V and $J_{III} = 0$ in Range-VI. Thus, the lack of incoming and outgoing atoms at the line ends can be considered by assigning a possible atomic flux of zero within the microstructure unit for each range of θ. The atomic fluxes are summed within the unit. The sum is integrated with respect to θ from 0 to 2π, by distinguishing the six parts. The integrated value is finally multiplied by the cross-sectional area of the grain boundary, δ, and divided by the volume of the unit region, $\sqrt{3}d^2/4$ and by 2π. In this way, the atomic flux divergence at the ends of the passivated polycrystalline line is expressed by [30]

$$
\begin{aligned}
& AFD^{*}_{\mathrm{gen}}\big|_{\mathrm{end}} \\
& = \frac{2}{\sqrt{3}\pi d^2}\frac{C^{*}_{\mathrm{gb}}N}{T}\exp\left(-\frac{Q_{\mathrm{gb}}+\kappa\Omega(N-N_{\mathrm{T}})/N_0-\sigma_{\mathrm{T}}\Omega}{kT}\right)\left\{6D_x\sin\beta-6D_y\cos\beta\right. \\
& +\frac{\sqrt{3}}{4}\pi d\left[-\frac{\kappa\Omega}{N_0}\left(\frac{\partial^2 N}{2}+\frac{\partial^2 N}{\partial y^2}\right)-\frac{\kappa\Omega/N_0}{kT}\left(D_x\frac{\partial N}{\partial x}+D_y\frac{}{\partial y}\right)\right. \\
& \left.\left.+\frac{1}{T}\left(\frac{Q_{\mathrm{gb}}+\kappa\Omega(N-N_{\mathrm{T}})/N_0-\sigma_{\mathrm{T}}\Omega}{kT}-1\right)\left(D_x\frac{\partial T}{\partial x}+D_y\frac{\partial T}{\partial y}\right)\right]\right\},
\end{aligned}
\tag{58}
$$

where $D_x = Z^{*}e\rho j_x - \kappa\Omega/N_0(\partial N/\partial x)$ and $D_y = Z^{*}e\rho j_y - \kappa\Omega/N_0(\partial N/\partial y)$. In Eq. (58), $AFD^{*}_{\mathrm{gen}}|_{\mathrm{end}}$ denotes the governing parameter for EM damage at the ends of a passivated polycrystalline line.

1.7.3 *AFD*-Based Method for Derivation Utilizing Drift Velocity of Line End

The film's characteristic constants included in the *AFD* formula (Eq. (55)), are d, $\Delta\varphi$, Q_{gb}, Z^{*}, C^{*}_{gb} and κ. The average grain size d can be measured by using a focused ion beam (FIB) system. We can obtain $\Delta\varphi$ experimentally by comparing an unpassivated metal line made of the same Al film as a passivated line [57]. The $AFD^{*}_{\mathrm{gen}}|_{\mathrm{end}}$-based method for determining the remaining constants considers a straight metal line. According to Blech [7, 8], the atomic density gradient in a straight metal line is inversely proportional to the length of the line. Stress measurements by Wang et al. [76, 77] indicate that the stress gradient during the initial stage of EM damage can be regarded as linear and independent of the given current density, provided that the input current density is less than several times the threshold current density. The product $\kappa \cdot \partial N/\partial x$, therefore, is considered to be a constant as a first approximation in the derivation method; $\kappa \cdot \partial N/\partial x$ is derived as a characteristic constant depending on the length of the straight line. A current density considerably larger than the threshold current density should not be chosen for the experiment.

The film's characteristic constants Q^{*}_{gb} [$=Q_{\mathrm{gb}} - \sigma_{\mathrm{T}}\Omega$], Z^{*}, C^{*}_{gb} and $\kappa \cdot \partial N/\partial x$ are determined by utilizing the drift velocity at the cathode end of the straight line. First, let us express the drift velocity by using $AFD^{*}_{\mathrm{gen}}|_{\mathrm{end}}$. Here, we introduce the effective length of the microstructure unit, l^{*} [$=0.658d$], which is defined as the square root of the area of the microstructure unit, $A\left[=\sqrt{3}d^2/4\right]$. We shall consider the $l^{*} \times l^{*}$ square region at the cathode end (see Fig. 8). The x-axis is taken from the cathode end along the longitudinal direction of the line as shown in Fig. 8. At $x = l^{*}$, the average atomic flux $J|_{x=l^{*}}$ is obtained by using $AFD^{*}_{\mathrm{gen}}|_{\mathrm{end}}$:

$$
J\big|_{x=l^{*}} = \frac{AFD^{*}_{\mathrm{gen}}\big|_{\mathrm{end}}}{\delta}\frac{\sqrt{3}}{4}d^2. \tag{59}
$$

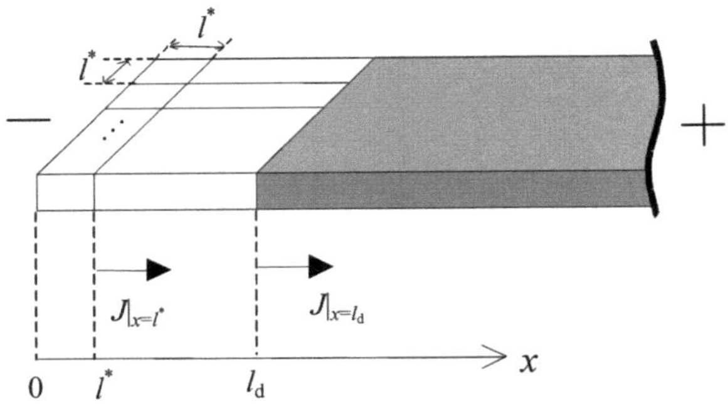

Fig. 8 Schematic diagram of the cathode end of the line. Reprinted from Hasegawa et al. [31]. Copyright (2009) with permission from Elsevier

By using Eq. (59), we find the atomic flux at $x = l_d$, $J|_{x=l_d}$:

$$\begin{aligned} J|_{x=l_d} &= J|_{x=l^*} + \frac{\partial J|_{x=l^*}}{\partial x}(l_d - l^*) \\ &= \frac{\sqrt{3}d^2}{4\delta}\left[AFD^*_{gen}|\text{end} + \frac{\partial AFD^*_{gen}|\text{end}}{\partial x}(l_d - l^*)\right], \end{aligned} \tag{60}$$

where l_d denotes the drift length. It can be assumed that the drift volume disappears through the cross section at $x = l_d$ with the atomic flux $J|_{x=l_d}$. By multiplying the area of grain boundary (*thick*·δ), the number of microstructure units within the line width (w/l^*), the atomic volume (Ω), and the net current application time (t_d) by Eq. (60), the volume of the drift region can be expressed, where w is the line width and *thick* the line thickness. Since the volume of the drift region equals $l_d \cdot w \cdot thick$, the drift velocity of the line end, v_d [$=l_d/t_d$], is given by

$$v_d = \left[AFD^*_{gen}|\text{end} + \frac{\partial AFD^*_{gen}|\text{end}}{\partial x}(l_d - l^*)\right] l^*\Omega. \tag{61}$$

In contrast, accelerated tests are performed for a certain period of time. The metal lines are subjected to a high current density, j_1, at three substrate temperatures: T_{s1}, T_{s2} and T_{s3} ($T_{s1}<T_{s2}<T_{s3}$). In addition, another acceleration test is carried out at T_{s3} at current density j_2, which is smaller than j_1. Let us denote the experimental conditions as follows: j_1 and T_{s1} (Condition 1), j_1 and T_{s2} (Condition 2), j_1 and T_{s3} (Condition 3), and j_2 and T_{s3} (Condition 4). Next, let the temperature at the cathode end of the line be T_1, T_2, T_3 and T_4, respectively, for each testing condition. Voiding area at the line end is measured after current stressing for a certain period of time. The drift length is obtained by dividing the area by the line width. The drift length is then divided by the measured net time of the current application [61], and thereby the velocity of the drift is obtained.

The values of v_d are experimentally obtained under the four testing conditions mentioned above. The unknown constants for the film characteristics in the *AFD* formula can be obtained by using the least-squares method to approximate the measured drift velocity with the theoretical drift velocity in Eq. (61). In particular, the characteristic constants are determined such that the following sum of squares is minimized:

$$F^*_{\mathrm{gb}}|\mathrm{end} = \sum_i \sum_j \left\{ v_{\mathrm{d}}|_{ij} - \left[\left(AFD^*_{\mathrm{gen}}|\mathrm{end} \right)_i + \left(\frac{\partial AFD^*_{\mathrm{gen}}|\mathrm{end}}{\partial x} \right)_i \left(l_{\mathrm{d}}|_{ij} - l^* \right) \right] l^* \Omega \right\}^2. \tag{62}$$

Here, the subscripts i and j represent the condition number and the number of data measured in each experimental condition, respectively. By this method, the film's characteristic constants, namely, Q^*_{gb}, Z^*, C^*_{gb} and $\kappa \cdot \partial N/\partial x$, can be determined as optimized parameters that approximate all experimental data obtained from the measurement of drift velocity.

1.7.4 Experimental Procedure for Derivation

The metal lines used in the experiment were fabricated as shown in Fig. 9. Two specimens of different length, namely, Sample L and Sample S, were prepared. A TiN thin film was reactively sputtered onto a silicon substrate covered with silicon oxide; then, the Al film was continuously deposited on the TiN film by vacuum evaporation. The Al/TiN line specimens were patterned by conventional photolithography and etched by the reactive ion etching (RIE) technique. The small parts of both ends of the Al line were chemically etched and TiN layer was exposed, and thus the Al specimens modeled after the via-connected line were obtained. After that, tetraethyl orthosilicate (TEOS) film was deposited over the specimen's surface by plasma enhanced chemical vapor deposition (PE-CVD). An example of FIB observation of the Al grain microstructure is shown in Fig. 10. It was observed that the average grain size of the line specimen was 0.5 μm.

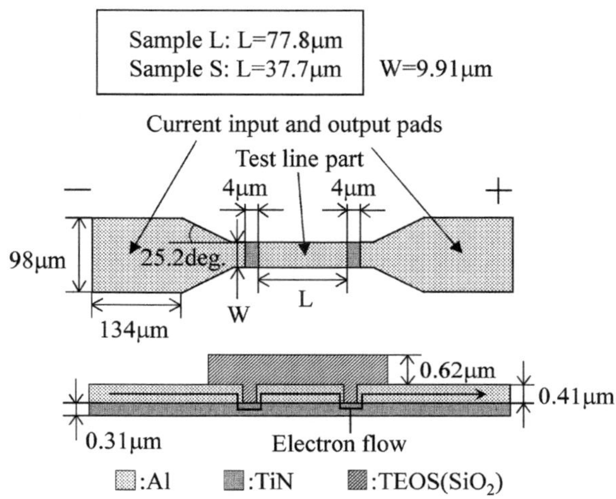

Fig. 9 Metal line specimen used in the experiment. Reprinted from Hasegawa et al. [31]. Copyright (2009) with permission from Elsevier

Fig. 10 FIB observation of Al grain microstructure. Reprinted from Hasegawa et al. [31]. Copyright (2009) with permission from Elsevier

Instead of an actual via-connected line, a metal line modeled after the via-connected line was evaluated in this study. The line was evaluated without the via because of ease of fabrication, and because EM at the cathode end of the line is sufficiently produced in the specimen without using an actual via-connected line. The EM damage in the line specimen appears as the drift of the cathode end and as drift in the via-connected line. Blech [7, 8] and other researchers [48, 63] have often used similar line specimens.

The acceleration tests were performed by using the experimental set-up shown in Fig. 11. To measure the incubation period during which there was no void formation and no drift at the end, the change in the electrical potential drop across the line was monitored. The incubation period was defined from the start of current

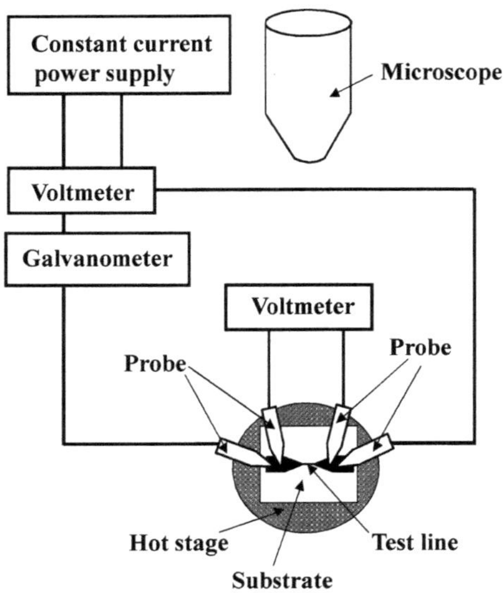

Fig. 11 Experimental setup for derivation. Reprinted from Hasegawa et al. [31]. Copyright (2009) with permission from Elsevier

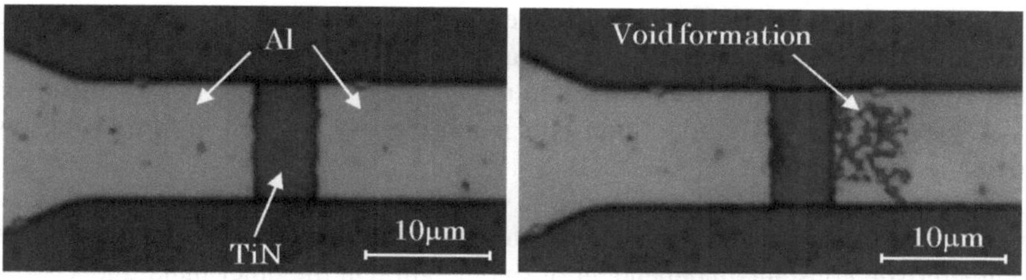

(a) Before current stressing **(b)** After current stressing

Fig. 12 Observation of specimen's cathode end by optical microscopy. Reprinted from Hasegawa et al. [31]. Copyright (2009) with permission from Elsevier

supply to the beginning of the increase in the electrical potential drop in the line as a result of void formation at the end. The net application time of electrical current, t_d, was obtained by subtracting the incubation period from the total application time. Three temperatures, namely, 508, 523 and 538 K, were selected as substrate temperatures. At each temperature, the metal lines of Sample L and Sample S were subjected to direct current with density of 1.5 MA/cm^2 (Conditions 1, 2 and 3). In addition, the test was carried out at a current density of 1.2 MA/cm^2 at substrate temperature of 538 K (Condition 4). Twelve specimens were used for each testing condition. After electric current was supplied until the potential drop increased 20 or 30%, the cathode end of the metal line was observed by optical microscopy, as shown in Fig. 12. The extent of void formation was evaluated by an image processing technique, and the drift length l_d was obtained by dividing the measured area by the line width. The drift length was then divided by the net time of current application. In this way, the drift velocity v_d was obtained from the experiment.

1.7.5 Results and Discussions of Derivation

The experimental data on drift velocity were substituted into Eq. (62) and the film's unknown characteristic constants in $AFD^*_{gen}|_{end}$ were optimized by using the least-squares method. The obtained constants are listed in Table 2. The value of Q^*_{gb} was close to the value of grain boundary diffusion [6, 7]. Furthermore, the value of Z^* appeared to be valid because it was within the range of the previously reported values, −1 to −15 [7, 8, 71, 76]. From a comparison of the values of Q^*_{gb} for Sample L and Sample S, it was found that the Q^*_{gb} values of these samples agreed well; thus, the constant Q^*_{gb} functioned as characteristic constant that was

Table 2 Characteristic constants included in $AFD^*_{gen}|_{end}$

	Q^*_{gb} [eV]	Z^*	C^*_{gb} [Kμm^3/Js]	$\kappa \cdot \partial N/\partial x$ [J/μm^7]
Sample L	0.55	−8.3	2.7×10^{24}	0.36
Sample S	0.54	−8.4	2.4×10^{24}	0.86

Reprinted from Hasegawa et al. [31]. Copyright (2009) with permission from Elsevier independent of the line length and dependent on only film characteristics. The amount of vacancies at the grain boundary depends on the purity of the material, and a higher vacancy concentration directly contributes to a higher diffusion coefficient. Accordingly, mass transport is more easily induced in films with low activation energy [80].

Comparing Z^* and C^*_{gb} of Sample L with those of Sample S, it is found that these values are almost the same. Therefore, the constants Z^* and C^*_{gb} also serve as characteristic constants independent of line length. On the other hand, the $\kappa \cdot \partial N/\partial x$ value of Sample S is more than twice as large as that of Sample L, and only the $\kappa \cdot \partial N/\partial x$ value in Table 2 appeared to depend on line length. According to Blech [7], the atomic density gradient is inversely proportional to line length. Therefore, $\kappa \cdot \partial N/\partial x$ functioned appropriately as a characteristic constant that depends on line length; consequently, the quantity κ was thought to act as a constant independent of the line-length. Hence, it was concluded that the film's characteristic constants were appropriately determined by the $AFD^*_{gen}|_{end}$-based method.

The effective bulk modulus, κ, the critical atomic density for void initiation, N^*_{min}, and the critical atomic density for hillock initiation, N^*_{max}, are obtained by a numerical simulation for the process of building up the atomic density distribution [62]. The values of κ, N^*_{min} and N^*_{max} are determined by simulation of atomic density distribution for the incubation period, during which no EM damage appears [61]. The simulation is carried out during the incubation period measured in the acceleration test. The quantity κ is determined such that the product of κ assumed in the simulation and the atomic density gradient after the simulation agrees with the value of $\kappa \cdot \partial N/\partial x$ obtained from the $AFD^*_{gen}|_{end}$-based method. Then, we obtain the atomic density distribution in the line after the incubation period through the simulation using the determined κ value. The smallest value of atomic density depending on θ [62], N^*, in the line is defined as N^*_{min}, and the largest value as N^*_{max}.

The previous method for deriving the film's characteristic constants based on the parameter AFD^*_{gen} required the observation of void formation by a scanning electron microscopy (SEM) after removal of the passivation layer [61]. In contrast, the present $AFD^*_{gen}|_{end}$ -based method based on drift velocity measurements can be carried out without removal of the passivation layer and requires only the observation of the cathode end by optical microscopy. Thus, the $AFD^*_{gen}|_{end}$-based method presented here is not only accurate but also much easier than the AFD^*_{gen}-based method.

2 Stress Migration

2.1 Introduction

SM is another failure mechanism that is often referred to, together with EM, in the study of reliability issues for interconnection systems. With the development of ultra-large-scale integration for semiconductors, the minimum dimensions for interconnects have been reduced to the sub-micron range. For such small-sized but full-featured devices, multilevel metallization is required. As a result, some unexpected influences have occurred to accompany the introduction of layered conducting lines. Apart from the EM-induced mass transport that occurs due to high density electron flow, residual stress is also generated in thermal processing when there is a difference in thermal expansion coefficients or a chemical reaction between the bonded elements, and so on. At times, these stresses can develop to a level that causes atomic diffusion to the extent where they lead to structural changes, typical known as examples being whisker growth and stress induced voids [11, 13, 15, 18, 19, 54, 70]. If the whiskers that are generated are long enough to connect two conductive layers, or if the voids grow to the dimensions of the line widths, these will result in short circuits or electrical discontinuities, respectively. Here, some typical phenomena relevant to SM will be reviewed in the following sections.

2.2 Historical Review of Typical SM-Induced Phenomena

2.2.1 Spontaneous Sn Whisker Growth

During World War II the electroplated material of choice for electrical components was electroplated cadmium (Cd). Repeated failures of electrical hardware led to a finding that many failures were due to shorting from Cd whiskers [27]. These findings were summarized by Cobb [17], and it was the first report concerning the problems of whisker growth. Starting in 1948, similar failures were experienced by Bell Telephone on the channel filters used for multi-channel transmission lines. Bell Laboratories immediately initiated the use of pure Sn electroplating to replace Cd, but they quickly found that pure Sn also had whisker problems similar to those experienced with Cd plating [18]. Since then, Sn whisker research has attracted considerable attention over 50 years due to the Sn and Sn-based alloys that have been widely used in the electronic industry until now. Figure 13 shows a typical example of a Sn whisker-induced reliability issue. As shown in the figure, many whiskers grew on a eutectic SnCu finish on a leadframe. One of the whiskers is so long that it has bridged a pair of the legs.

It is striking to note that essentially all of the fundamental concepts still debated today were initially established during the 1950s. Many of these 1950s era proposals were perhaps little more than well-informed speculation at the time, but they were based on sound principles of materials science and they formed a basis for all current discussion relevant to whisker formation [27]. Some highlights of this research are worthy of further review, as listed below:

(1) As reported in Bell Laboratories' work published in 1951, whisker growth was recognized as a spontaneous process, not only on Cd, Zn and Sn electroplating, but also on Al casting alloys and Ag electroplating exposed to an atmosphere of hydrogen sulfide [18].

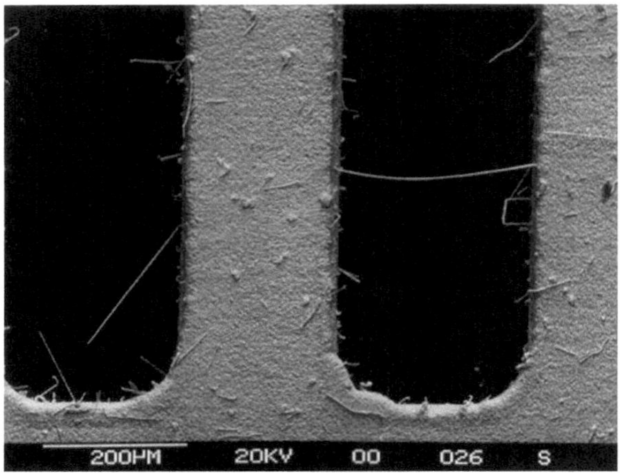

Fig. 13 SEM image of Sn whiskers grown on eutectic SnCu finish on a leadframe. One of the whiskers shorts two of the leadframe legs [78]

(2) Herring and Galt [34] inferred that Sn whiskers were single crystals without lattice defects by investigating their elastic and plastic properties.
(3) By showing electron micrographs of the different stages in the growth of Sn whiskers, Koonce and Arnold [41] established the fact that Sn whisker growth took place as a result of the addition of material at the base, rather than at the tip.
(4) Fisher et al. [24] established that compressive stress gradients were the driving force for whisker growth. This knowledge has made such a great contribution to the subsequent proposed whisker growth models.

Going through several decades of research assisted by experimental tools that have been developed over the years, Sn whisker growth has been confirmed as a spontaneous process, driven by a compressive stress gradient, growing from the base, and readily occurring at room temperature. However, the mechanism is still unclear to date. Several models have been constructed to describe the mechanism of Sn whisker growth. Dislocation-based theories have been proposed independently by Eshelby [22], Frank [25], Amelinckx et al. [3] and Lee and Lee [43]. Recrystallization-based theories were proposed by Ellis et al. [21], and subsequently developed by Furuta and Hamamura [26]. In 1973, Tu [70] published his paper on Sn whisker growth. As reported in his paper, Sn whiskers were observed at room temperature growing from the Sn surfaces of Cu–Sn bimetallic films, but not at Sn films without a Cu under-layer. This was attributed to a driving force generated by the formation of Cu_6Sn_5 intermetallic compounds (IMCs) in the Cu–Sn films [70]. Later, Tu and co-authors have demonstrated the existence of IMC by utilizing new analytical techniques such as TEM and FIB examinations. In addition, Tu [72] proposed that the oxide layer on the Sn film played a dominant role in affecting Sn whisker growth. Weak spots in the oxide layer are important in enabling local stress relaxation to form Sn whiskers. Without the surface oxide, a homogeneous relaxation occurs over the entire film [72]. This theory can be well understood from a comparison of the illustrations shown in Figs. 14 and 15. Besides the models mentioned above, a great deal of important data has been

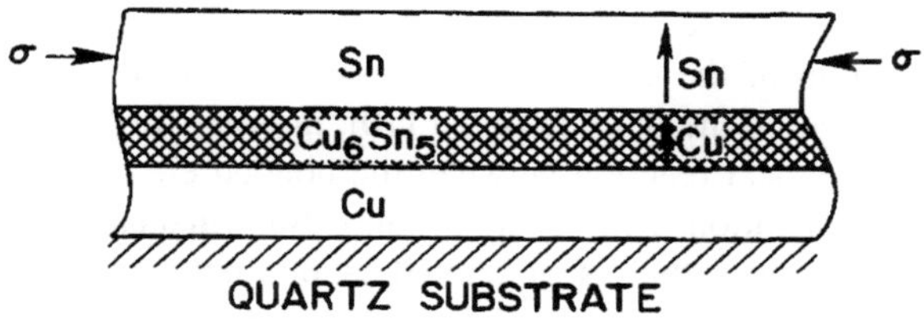

Fig. 14 A sketch of the cross section of the bimetallic Cu–Sn thin films forming the compound Cu_6Sn_5. The surface of Sn is assumed to be free of oxide. The *arrows* indicate the fluxes in the Sn and the compound. Reprinted with permission from Tu [71]. Copyright 1994 by the American Physical Society

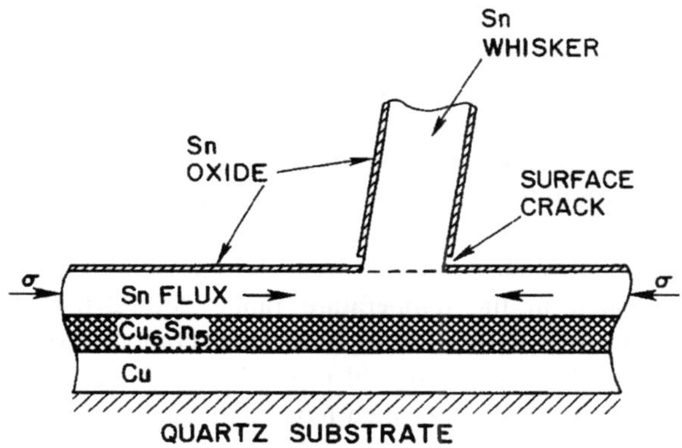

Fig. 15 A sketch of the cross section of the bimetallic Cu–Sn thin films forming the compound Cu_6Sn_5 and a whisker. The surfaces of the Sn and the whisker are oxidized except the base of the whisker where the oxide is broken. A lateral flux of Sn is indicated by arrows in the Sn film. Reprinted with permission from Tu [72]. Copyright 1994 by the American Physical Society

reported, as represented by Zhang et al. [79], Choi et al. [16], Barsoum et al. [4] and Galyon and Palmer [28].

The SM-induced reliability issue has been the main driver for these studies ever since the whisker problem was discovered. With the ongoing reduction of circuit feature sizes, it is increasingly relevant to the electronic packaging industries. Most of the correlative works focus on suppressing whisker growth as far as possible. It is noted that one may consider ways such as the creation of an environment in which no natural oxide layer is formed on the surface of the material, or covering the surface of the material with a tough artificial layer [69], for the suppression of whisker formation.

2.2.2 Fabrication of Nanowires by Utilizing Controllable SM

With the newly reported works on various nanowires fabricated by utilizing SM [15, 35, 51, 54, 64, 66], the field of SM research has entered a new phase, not only in terms of its suppression, but also in terms of its applications. This is because one-dimensional nano-structures have attracted considerable attention due to their unique mechanical, electrical, and magnetic properties, and their fundamental importance to MEMS/NEMS in recent years.

It has been demonstrated that whisker growth is driven by compressive stress gradients. The origin of the compressive stress can be mechanical, thermal, and chemical [78]. Based on the opinion of Tu [70], in order for Sn whiskers to grow, there must be a chemical reaction between the bonded elements to guarantee the necessary stress generation. Moreover, because Sn whisker growth is a spontaneous process, the stress that is generated is internal and so the geometrical properties of whiskers are uncontrollable. All of the characteristics described above have certain limitations that must be taken into account in consideration of nanowire formation by utilizing SM, such as the choice of source material, the growth rate, the generation of the driving force and controllability.

In contrast to 'traditional' Sn whisker growth, alternative approaches have been developed in which an external applied stress is used to induce atomic diffusion. Almost all of these fabrication techniques aimed at utilizing the thermal stresses that result from a mismatch in thermal expansion coefficients in bilayer/multilayer structures. Let us refer to the work by Shim et al. [66] to give a schematic representation of the nanowire growth mechanism. Taking Bi nanowire growth as an example, and as illustrated in Fig. 16a, a trilayer structure with an oxidized Si substrate followed by a Bi layer is used. It should be mentioned that there is a large difference between the thermal expansion coefficients of Bi (13.4 $\mu m\ m^{-1}\ K^{-1}$) and SiO_2/Si (0.5 $\mu m\ m^{-1}\ K^{-1}$/2.4 $\mu m\ m^{-1}\ K^{-1}$). The Bi film expands while it is annealed in the temperature range 260–270°C, while the substrate restricts expansion, putting the Bi film under compressive stress [66]. By making use of stress relief and atomic diffusion, Bi nanowires can be fabricated. The mass of Bi nanowires that are formed have high aspect ratios (length/diameter), as shown in Fig. 16b. Figure 16c shows TEM analysis of a formed Bi nanowire. The nanowire was found to be uniform in diameter and to have formed a 10 nm thick Bi oxide layer on its outer surface [66]. The presence of the Bi oxide layer seems to agree well with the oxide layer theory proposed by Tu [72]. The effect of the oxide layer on stress-induced extrusion was also studied by other researchers [14, 15, 38].

It is noteworthy that the growth of SM-induced nanowires is governed by temperature, film thickness, grain size and the time that the film is subjected to stress during the process [54]. Therefore, by adjusting these parameters, the growth of nanowires can be controlled. This seems to be especially important in order to achieve higher aspect ratios, or the rapid and mass growth of these nanowires. Although the current technique for the fabrication of nanowires by utilizing SM is imperfect, we can expect it to be applied to mass production process in industry in the near future.

2.3 Summary

Since the atomic diffusion induced by SM is a stress relief phenomenon, it must relate to the stress gradient. To be precise, it occurs due to a gradient of compressive hydrostatic stress. Denoting the hydrostatic stress by σ and considering a

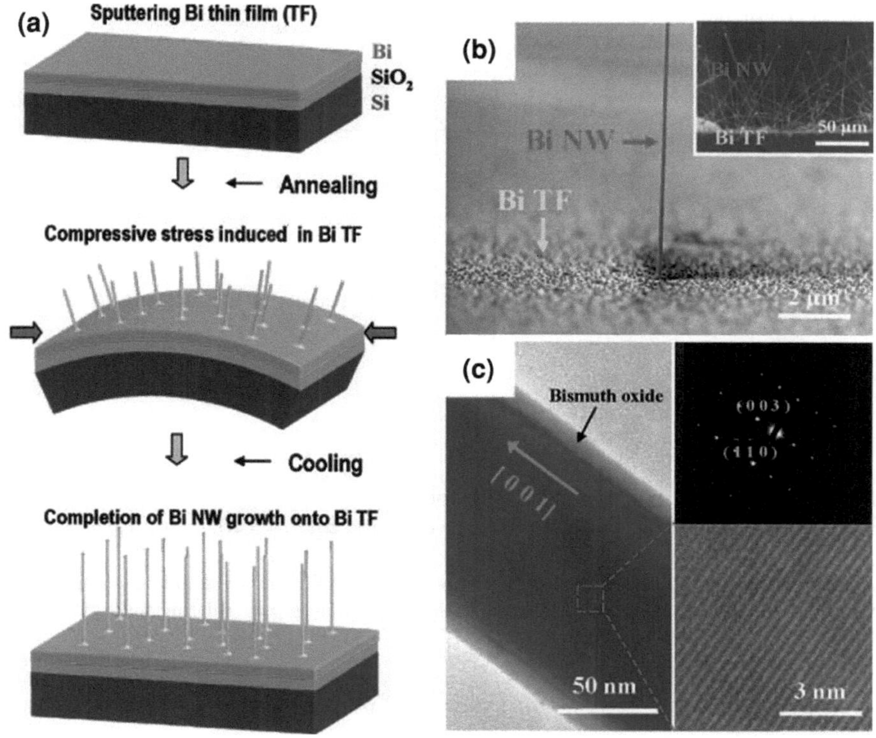

Fig. 16 Growth mechanism and structural characteristic of the single-crystalline Bi nanowires. **a** a schematic representation of the growth of Bi nanowires by on-film formation of nanowires, **b** a SEM image of a Bi nanowire grown on a Bi thin film, and **c** a low-magnification TEM image of a Bi nanowire [66]

material with a distribution of compressive stress as shown in Fig. 17, the atoms diffuse from position A with more-negative stress (higher compressive stress) towards position B with less-negative stress (lower compressive stress). As a result, a local atomic accumulation is caused at position B. The hydrostatic stress is expressed as $\sigma = (\sigma_x + \sigma_y + \sigma_z)/3$, where σ_x, σ_y and σ_z are the corresponding normal stresses in the Cartesian coordinates system (x, y, z). In most of these cases, the surface of a material subjected to SM is covered by an oxide layer or by a passivation layer. In those cases, the normal stresses σ_x, σ_y and σ_z caused by the accumulation of atoms are the same as each other, $\sigma_x = \sigma_y = \sigma_z$, and hence σ is equal to σ_x, where x is usually taken in the longitudinal direction of the tested material and z is in the normal direction to the surface of the material. The SM-induced atomic flux, $\mathbf{J}_\mathrm{S}$ is given by [33, 42]

$$\mathbf{J}_\mathrm{S} = \frac{N\Omega D_0}{kT} \exp\left(-\frac{Q - \Omega\sigma}{kT}\right) \mathrm{grad}\sigma \tag{63}$$

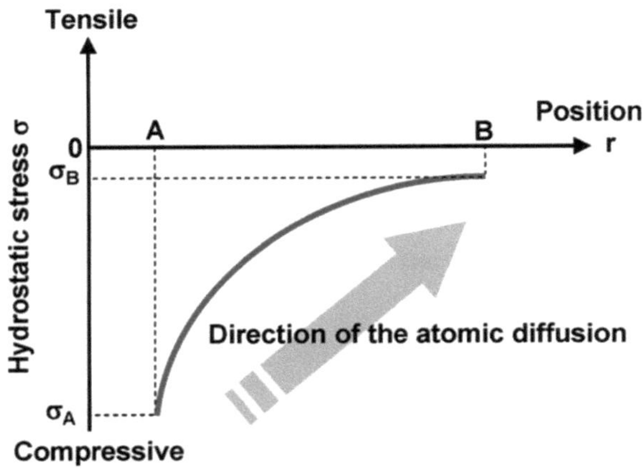

Fig. 17 Illustration for the phenomenon of SM in a material with a distribution of compressive stress. Reprinted with permission from Saka et al. [55]. © 2008 IEEE

Here, the gradient of σ is the driving force for atomic diffusion, and this differs with the electron flow, which is used to describe that of EM.

Acknowledgments K. S. acknowledges T. Abo for his help in preparing the manuscript. M. S. wishes to express his thanks to X. Zhao for his kind help in preparing the manuscript.

References

1. Abé, H., Sasagawa, K., Saka, M.: Electromigration failure of metal lines. Int. J. Fract. **138**, 219–240 (2006)
2. Ainslie, N.G., d'Heurle, F.M., Wells, O.C.: Coating, mechanical constraints, and pressure effects on electromigration. Appl. Phys. Lett. **20**, 173–174 (1972)
3. Amelinckx, S., Bontinck, W., Dekeyser, W., Seitz, F.: On the formation and properties of helical dislocations. Phil. Mag. **2**, 355–377 (1957)
4. Barsoum, M.W., Hoffman, E.N., Doherty, R.D., Gupta, S., Zavaliangos, A.: Driving force and mechanism for spontaneous metal whisker formation. Phys. Rev. Lett. **93**, 206104 (1-4) (2004)
5. Black, J.R.: Electromigration—a brief survey and some recent results. IEEE Trans. Electron Devices **ED-16**, 338–347 (1969)
6. Black, J.R.: Electromigration failure modes in aluminum metallization for semiconductor devices. Proc. IEEE **57**, 1587–1593 (1969)
7. Blech, I.A.: Electromigration in thin aluminum films on titanium nitride. J. Appl. Phys. **47**, 1203–1208 (1976)
8. Blech, I.A.: Diffusional back flows during electromigration. Acta Mater. **46**, 3717–3723 (1998)
9. Blech, I.A., Herring, C.: Stress generation by electromigration. Appl. Phys. Lett. **29**, 131–133 (1976)
10. Blech, I.A., Meieran, E.S.: Electromigration in thin Al films. J. Appl. Phys. **40**, 485–491 (1969)
11. Blech, I.A., Petroff, P.M., Tai, K.L., Kumar, V.: Whisker growth in Al thin films. J. Cryst. Growth. **32**, 161–169 (1975)
12. Böhm, J., Volkert, C.A., Mönig, R., Balk, T.J., Arzt, E.: Electromigration-induced damage in bamboo Al interconnects. J. Electron. Mater. **31**, 45–49 (2002)

13. Børgesen, P., Lee, J.K., Gleixner, R., Li, C.-Y.: Thermal-stress-induced voiding in narrow, passivated Cu lines. Appl. Phys. Lett. **60**, 1706–1708 (1992)
14. Chang, C.Y., Vook, R.W.: The effect of surface aluminum oxide films on thermally induced hillock formation. Thin Solid Films **228**, 205–209 (1993)
15. Cheng, Y.-T., Weiner, A.M., Wong, C.A., Balogh, M.P.: Stress-induced growth of bismuth nanowires. Appl. Phys. Lett. **81**, 3248–3250 (2002)
16. Choi, W.J., Lee, T.Y., Tu, K.N., Tamura, N., Celestre, R.S., McDowell, A.A., Bong, Y.Y., Liu, N.: Tin whiskers studied by synchrotron radiation scanning X-ray micro-diffraction. Acta Mater. **51**, 6253–6262 (2003)
17. Cobb, H.L.: Cadmium whiskers. Mon. Rev. Am. Electroplaters Soc. **33**, 28–30 (1946)
18. Compton, K.G., Mendizza, A., Arnold, S.M.: Filamentary growths on metal surfaces-Whiskers. Corrosion **7**, 327–334 (1951)
19. Curry, J., Fitzgibbon, G., Guan, Y., Muollo, R., Nelson, G., Thomas, A.: New failure mechanisms in sputtered aluminum-silicon films. IEEE Proc. Int. Reliab. Phys. Symp. **22**, 6–8 (1984)
20. d'Heurle, F., Ames, I.: Electromigration in single crystal aluminum films. Appl. Phys. Lett. **16**, 80–81 (1970)
21. Ellis, W.C., Gibbons, D.F., Treuting, R.C.: Growth of Metal Whiskers from the Solid. Growth and Perfection of Crystals, pp. 102–120. Wiley, New York (1958)
22. Eshelby, J.D.: A tentative theory of metallic whisker growth. Phys. Rev. **91**, 755–756 (1953)
23. Fiks, V.B.: On the mechanism of the mobility of ions in metals. Sov. Phys. Solid State **1**, 14–28 (1959)
24. Fisher, R.M., Darken, L.S., Carroll, K.G.: Accelerated growth of tin whiskers. Acta Metall. **2**, 368–373 (1954)
25. Frank, F.C.: On tin whiskers. Phil. Mag. **44**, 854–860 (1953)
26. Furuta, N., Hamamura, K.: Growth mechanism of proper tin-whisker. Jpn. J. Appl. Phys. **8**, 1404–1410 (1969)
27. Galyon, G.T.: A History of Tin Whisker Theory: 1946 to 2004. SMTA International Conference, Chicago (2004)
28. Galyon, G.T., Palmer, L.: An integrated theory of whisker formation: the physical metallurgy of whisker formation and the role of internal stresses. IEEE Trans. Elect. Packag. Manuf. **28**, 17–30 (2005)
29. Gonzalez, J.L., Rubio, A.: Shape effect on electromigration in VLSI interconnects. Microelectron Reliab. **37**, 1073–1078 (1997)
30. Hasegawa, M., Sasagawa, K., Saka, M., Abé, H.: Expression of a governing parameter for electromigration damage on metal line ends. In: Proceedings of the ASME InterPACK '03(CD-ROM): InterPack 2003-35064 (2003)
31. Hasegawa, M., Sasagawa, K., Uno, S., Saka, M., Abé, H.: Derivation of film characteristic constants of polycrystalline line for reliability evaluation against electromigration failure. Mech. Mater. **41**, 1090–1095 (2009) (See corrigendum to this article, Doi: 10.1016/j.mechmat.2010.09.004, for Table 2)
32. Hau-Riege, S.P., Thompson, C.V.: Experimental characterization and modeling of the reliability of interconnect trees. J. Appl. Phys. **89**, 601–609 (2001)
33. Herring, C.: Diffusional viscosity of a polycrystalline solid. J. Appl. Phys. **21**, 437–445 (1950)
34. Herring, C., Galt, J.K.: Elastic and plastic properties of very small metal specimens. Phys. Rev. **85**, 1060–1061 (1952)
35. Hinode, K., Homma, Y., Sasaki, Y.: Whiskers grown on aluminum thin films during heat treatments. J. Vac. Sci. Technol. A **14**, 2570–2576 (1996)
36. Hu, C.-K., Small, M.B., Rodbell, K.P., Stanis, C., Blauner, P., Ho, P.S.: Electromigration failure due to interfacial diffusion in fine Al alloy lines. Appl. Phys. Lett. **62**, 1023–1025 (1993)
37. Huntington, H.B., Grone, A.R.: Current-induced marker motion in gold wires. J. Phys. Chem. Solids **20**, 76–87 (1961)
38. Iwamura, E., Takagi, K., Ohnishi, T.: Effect of aluminum oxide caps on hillock formation in aluminum alloy films. Thin Solid Films **349**, 191–198 (1999)

39. Kawamura, M., Mashima, T., Abe, Y., Sasaki, K.: Formation of ultra-thin continous Pt and Al film by RF sputtering. Thin Solid Films **377–378**, 537–542 (2000)
40. Kirchheim, R., Kaeber, U.: Atomistic and computer modeling of metallization failure of integrated circuits by electromigration. J. Appl. Phys. **70**, 172–181 (1991)
41. Koonce, S.E., Arnold, S.M.: Growth of metal whiskers. J. Appl. Phys. **24**, 365–366 (1953)
42. Korhonen, M.A., Børgesen, P., Tu, K.N., Li, C.-Y.: Stress evolution due to electromigration in confined metal lines. J. Appl. Phys. **73**, 3790–3799 (1993)
43. Lee, B.-Z., Lee, D.N.: Spontaneous growth mechanism of tin whiskers. Acta Mater. **46**, 3701–3714 (1998)
44. Lloyd, J.R., Smith, P.M.: The effect of passivation thickness on the electromigration lifetime of Al/Cu thin film conductors. J. Vac. Sci. Technol. A **1**, 455–458 (1983)
45. Marcoux, P.J., Merchant, P.P., Naroditsky, V., Rehder, W.D.: New 2D simulation model of electromigration. Hewlett-Packard J. 79–84 (June 1989)
46. Nikawa, K.: Monte Carlo calculations based on the generalized electromigration failure model. Proc. 19th IEEE Int. Reliab. Phys. Symp. 175–181 (1981)
47. Oates, A.S.: Electromigration in multilayer metallization: drift-controlled degradation and the electromigration threshold of Al–Si–Cu/TiN_XO_Y/$TiSi_2$ contacts. J. Appl. Phys. **70**, 5369–5373 (1991)
48. Oates, A.S.: Electromigration transport mechanisms in Al thin-film conductors. J. Appl. Phys. **79**, 163–169 (1996)
49. Oates, A.S., Barr, D.L.: Lattice electromigration in narrow Al alloy thin-film conductors at low temperatures. J. Electron. Mater. **23**, 63–66 (1994)
50. Park, Y.J., Thompson, C.V.: The effects of the stress dependence of atomic diffusivity on stress evolution due to electromigration. J. Appl. Phys. **82**, 4277–4281 (1997)
51. Prokes, S.M., Arnold, S.: Stress-driven formation of Si nanowires. Appl. Phys. Lett. **86**, 193105 (1–3) (2005)
52. Proost, J., Maex, K., Delaey, L.: Electromigration-induced drift in damascene and plasma-etched Al(Cu). II. Mass transport mechanisms in bamboo interconnects. J. Appl. Phys. **87**, 99–109 (2000)
53. Renucci, P., Gaudart, L., Petrakian, J.P., Roux, D.: New model for the temperature coefficient of resistivity of polycrystalline films: application to alkaline-earth metals. J. Appl. Phys. **54**, 6497–6501 (1983)
54. Saka, M., Yamaya, F., Tohmyoh, H.: Rapid and mass growth of stress-induced nanowhiskers on the surfaces of evaporated polycrystalline Cu films. Scr. Mater. **56**, 1031–1034 (2007)
55. Saka M, Yasuda M, Tohmyoh H, Settsu N, Fabrication of Ag micromaterials by utilizing stress-induced migration. In: Proc. 2nd Electronics Systemintegration Technology Conference © 2008 IEEE (2008)
56. Sasagawa, K., Saka, M., Abé, H.: Current density and temperature distributions near the corner of angled metal line. Mech. Res. Commun. **22**, 473–483 (1995)
57. Sasagawa, K., Nakamura, N., Saka, M., Abé, H.: A new approach to calculate atomic flux divergence by electromigration. Trans. ASME J. Electron. Packag. **120**, 360–366 (1998)
58. Sasagawa, K., Naito, K., Saka, M., Abé, H.: A method to predict electromigration failure of metal lines. J. Appl. Phys. **86**, 6043–6051 (1999)
59. Sasagawa, K., Naito, K., Kimura, H., Saka, M., Abé, H.: Experimental verification of prediction method for electromigration failure of polycrystalline lines. J. Appl. Phys. **87**, 2785–2791 (2000)
60. Sasagawa, K., Hasegawa, M., Saka, M., Abé, H.: A governing parameter for electromigration damage in passivated polycrystalline line and its verification. In: Baker, S.P. (ed.) Stress-induced Phenomena in Metallization 612. AIP, Melville (2001)
61. Sasagawa, K., Hasegawa, M., Saka, M., Abé, H.: Governing parameter for electromigration damage in the polycrystalline line covered with a passivation layar. J. Appl. Phys. **91**, 1882–1890 (2002)
62. Sasagawa, K., Hasegawa, M., Saka, M., Abé, H.: Prediction of electromigration failure in passivated polycrystalline line. J. Appl. Phys. **91**, 9005–9014 (2002)

63. Schreiber, H.-U.: Electromigration mechanisms in aluminum lines. Solid-State Electron **28**, 1153–1163 (1985)
64. Settsu, N., Saka, M., Yamaya, F.: Fabrication of Cu nanowires at predetermined positions by utilizing stress migration. Strain **44**, 201–208 (2008)
65. Shewmon, P.: Diffusion in Solids. Minerals Metals and Materials Society, Warrendale (1989)
66. Shim, W., Ham, J., Lee, K., Jeung, W.Y., Johnson, M., Lee, W.: On-film formation of Bi nanowires with extraordinary electron mobility. Nano Lett. **9**, 18–22 (2009)
67. Shin, W.C., Besser, R.S.: A micromachined thin-film gas flow sensor for microchemical reactors. J. Micromech. Microeng. **16**, 731–741 (2006)
68. Shingubara, S., Nakasaki, Y., Kaneko, H.: Electromigration in a single crystalline submicron width aluminum interconnection. Appl. Phys. Lett. **58**, 42–44 (1991)
69. Tohmyoh, H., Yasuda, M., Saka, M.: Controlling Ag whisker growth using very thin metallic films. Scr. Mater. **63**, 289–292 (2010)
70. Tu, K.N.: Interdiffusion and reaction in bimetallic Cu-Sn thin films. Acta Metall. **21**, 347–354 (1973)
71. Tu, K.N.: Electromigration in stressed thin films. Phys. Rev. B **45**, 1409–1413 (1992)
72. Tu, K.N.: Irreversible processes of spontaneous whisker growth in bimetallic Cu–Sn thin-film reactions. Phys. Rev. B **49**, 2030–2034 (1994)
73. Vaidya, S., Sheng, T.T., Sinha, K.: Linewidth dependence of electromigration in evaporated Al-0.5%Cu. Appl. Phys. Lett. **36**, 464–466 (1980)
74. Villars, P.: Pearson's Handbook Desk Edition—Crystallographic Data for Intermetallic Phases, 1. ASM International, Material Park (1997)
75. Walton, D.T., Frost, H.J., Thompson, C.V.: Development of near-bamboo and bamboo microstructures in thin-film strips. Appl. Phys. Lett. **61**, 40–42 (1992)
76. Wang, P.-C., Cargill III, G.S., Noyan, I.C., Hu, C.-K.: Electromigration-induced stress in aluminum conductor lines measured by X-ray microdiffraction. Appl. Phys. Lett. **72**, 1296–1298 (1998)
77. Wang, P.-C., Noyan, I.C., Kaldor, S.K., Jordan-Sweet, J.L., Liniger, E.G., Hu, C.-K.: Topographic measurement of electromigration-induced stress gradients in aluminum conductor lines. Appl. Phys. Lett. **76**, 3726–3728 (2000)
78. Zeng, K., Tu, K.N.: Six cases of reliability study of Pb-free solder joints in electronic packaging technology. Mater. Sci. Eng. R **38**, 55–105 (2002)
79. Zhang, Y., Xu, C., Fan, C., Vysotskaya, A., Abys, J.: Understanding whisker phenomenon-Part I: Growth rate. In: Proc. of the 2001 AESF SUR/FIN Conf.: NA (2001)
80. Zhao, X., Saka, M., Yamashita, M., Togoh, F.: Evaluation of the dominant factor for electromigration in sputtered high purity Al films. Trans. ASME J. Electron. Packag. **132**, 021003 (1–9) (2010)

Fabrication of Micro and Nano Metallic Materials

Masumi Saka and Kazuhiko Sasagawa

Abstract Convenience of application and controllability are obvious requirements for engineering methodologies that are intended for use in the fabrication of micro and nano metallic materials. In order to meet these requirements, fabrication techniques based on atomic diffusion that utilize protective layers over the specimens have been trialed. Various EM-based approaches are introduced first. Attempts to collect metallic atoms by the utilization of temperature gradients in passivated specimens and by improvements to the structures of the specimens are discussed, and the fabrication of Al wires, spheres and belts with micro/nano sized dimensions are shown. In addition, the effect of the purity of the source materials on the fabrication and the effect of temperature on the selective production of various shapes of micro/nano materials are mentioned. Numerical simulations of formation of micro/nano materials are explained and are verified as being useful tools when investigating conditions for efficient fabrication. Next, some approaches are introduced that involve SM-based methods. The rapid and mass formation of Cu nanowires and the fabrication of Ag micro materials, which can be realized by covering the specimen with a protective layer, are described.

M. Saka (✉)
Department of Nanomechanics, Tohoku University,
Aoba 6-6-01, Aramaki, Aoba-ku, Sendai 980-8579, Japan
e-mail: saka@ism.mech.tohoku.ac.jp

K. Sasagawa
Department of Intelligent Machines and System Engineering,
Hirosaki University,
3 Bunkyo-cho, Hirosaki 036-8561, Japan
e-mail: sasagawa@cc.hirosaki-u.ac.jp

M. Saka (ed.), *Metallic Micro and Nano Materials*, Engineering Materials,
DOI: 10.1007/978-3-642-15411-9_3,

1 Introduction

It is well known that whisker growth is occasionally generated as a result of a compressive stress release phenomenon [11, 18, 48]. Generally speaking, this phenomenon requires atomic diffusion and a protective surface oxide, such as in the case when Sn whiskers are formed [46]. When the protective surface oxide breaks at a certain point, a whisker consisting of diffused atoms grows to release the compressive stress at that point. This mechanism has been a great inspiration in research into the formation of micro and nano metallic materials. Here two subjects arise. The first is how to supply the driving force for atomic diffusion and the protective layer. The other is how to treat the two key points: atomic diffusion and the protective layer.

First, as stated in chapter titled Introduction, convenience of application and a controllable nature are our most significant requirements for processes involved in the formation of micro and nano metallic materials. This attracted our attention to the topics of EM and SM, which we have introduced in chapter titled Basis of Atomic Diffusion. Unlike complex chemical methods, EM and SM are physical phenomena involving atomic diffusion, although they differ in the driving force that brings this about.

On the other hand, in order to realize for the role of the protective layer, we shall refer to the work that has been undertaken on Al hillock growth. It has been reported by Chang and Vook [6] that in an ultra-high vacuum, no hillocks were found on Al surfaces even under compression. Al hillocks only grew when the Al surface was oxidized, and this result demonstrated that the surface oxide acted as a protective layer. Without the presence of a protective layer in an ultra-high vacuum environment, a free Al surface is a good site for the release of compressive stress. In these circumstances, the compressive stress is relieved uniformly over the surface. To effectively apply a suitable protective layer to form micro and nano metallic materials, we first need to consider how such a layer can be formed. Not only native oxide layers, but also synthetic passivation layers can be used as the protective layer. By fabricating an artificial layer we may be able to control predetermined positions where stress release can occur to form the desired materials, or, we may be able to control the point size for stress release to control the shape and size of the formed materials.

Based on the considerations above, a lot of work has been undertaken to investigate the formation of micro and nano metallic materials by utilizing the protective layer in EM/SM. In the next part, some of these achievements will be introduced.

2 Fabrication of Micro and Nano Metallic Materials by Utilizing EM

2.1 Effect of Temperature Gradient on Atomic Diffusion

Basically, EM is a phenomenon involving atomic diffusion induced by a high density of electrons flowing in a metal line. The number of atoms that are transported depends on the current density, the temperature and the microstructure of the metal line. At a location where depletion of metallic atoms is caused by EM, the voids are formed. Contrarily, at a location where accumulation of atoms occurs, the hillocks are formed.

In the earlier work [29], in order to apply EM to the formation of micro and nano metallic materials, first of all some parameters that would be necessary for the process were studied. By considering these parameters to be 'adjustable variables', they were considered to play a significant role in determining the final result.

One set of parameters that can be used to control EM are the substrate temperature [19], the current density [19] and the sample line length [4]. These cause atomic diffusion, resulting in an accumulation of atoms at the anode side of the line, and thus produce higher compressive stress at the anode end than that of the cathode end.

In addition to providing the necessary electron flows as the driving force for EM, the current also plays an important role in atomic accumulation. Known as 'the electro-thermal' problem, when current passes through a conductive material, Joule heating occurs and thus a resulting temperature distribution is generated. A temperature gradient is hence formed that depends on the line length. This is believed to be the cause of atomic flux divergence and of the formation of hillocks and voids [33, 44]. Considering a metal line under a constant direct current and the resulting effect of Joule heating, the diffusivity of the atoms in a high temperature region is higher than that of the atoms in a relatively lower temperature region [4, 14]. Consider a very small region where EM occurs, and assume that atoms flow through the region from a hot space to a cold one. Then, the number of atoms entering the region is more than the number leaving the region due to the different diffusivities presented above. Therefore, hillocks are formed in the region due to atomic accumulation. Contrarily, voids can be formed by related processes due to atomic depletion. Figure 1 shows an example of the temperature distribution in a line obtained by using a finite element analysis for the experimental conditions used in the work [29]. There is a sharp negative temperature gradient at the anode side, as shown in Fig. 1, which is due to Joule heating in the sample line. The area in which the temperature gradient has a considerable negative value is practically coincident with the area in which the sites of accumulated atoms were distributed in the experiment [29].

The other parameter concerns the line structure, which serves to accumulate atoms at a predetermined position, while the shape of the small slot in the protective layer through which the accumulated atoms are discharged clearly also needs to be controlled. The mechanism for generating micro and nano metallic materials by utilizing EM and the influence of the parameters mentioned above can be described as follows. A high density of flowing electrons, which is related to the current density in the line, causes atomic diffusion and atoms accumulate at the anode side of the line. These are held in place due to the contribution of the passivation layer. As a result, a high compressive stress is generated in the thin-film line. When a patterned slot is present in the passivation layer, the compressive

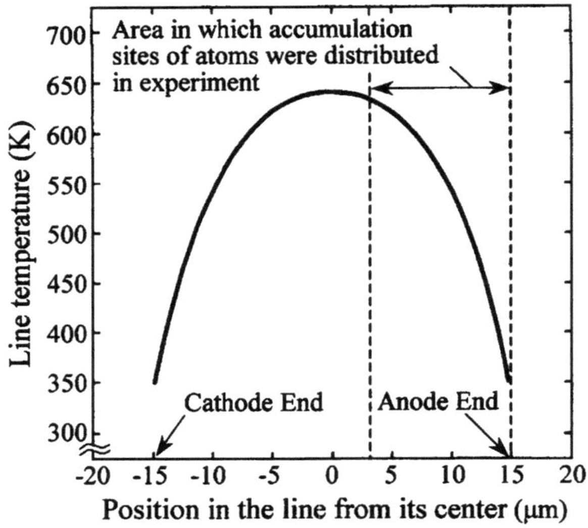

Fig. 1 Temperature distribution obtained by finite element analysis of Al bamboo line with a current density of 28.5 MA/cm^2 and substrate temperature of 298 K. The area between the two *broken lines* shows the region in which the distribution of the accumulation sites of the atoms was observed in the experiment, and this area almost coincides with the region having a sharp negative temperature gradient [29]

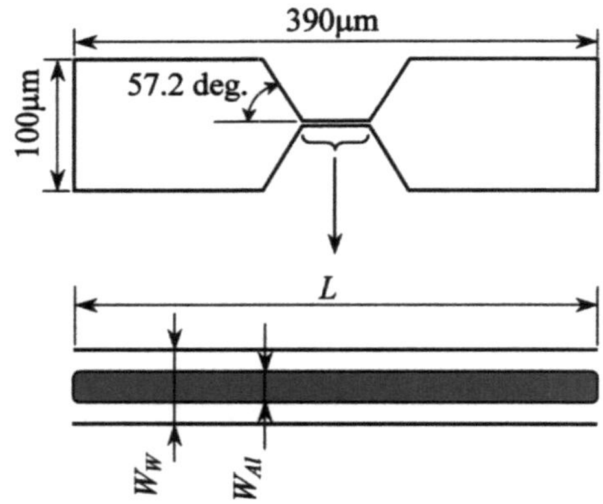

Fig. 2 Shape and dimensions of the specimens used in the experiment. The width of the W line W_W is 7.1 μm, and that of the embedded Al line W_{Al} is 3.1 μm. The thickness of the Al line is 0.29 μm and that of tungsten line is 0.45 μm [29]

stress is released as the atoms are discharged. Therefore, shaped metallic material can be generated.

Figure 2 shows the detailed shape and dimensions of the specimens used in the earlier work [29]. Taking into consideration commonly-used interconnection structures, a Si wafer was used as the substrate. A tungsten (W) film incorporating a groove filled with Al through the film thickness was formed by sputtering onto a Si substrate, and was then connected up via a set of input and output pads to allow current to pass through. The resistivity of Al is approximately half of that of W, so, for the specimen illustrated in Fig. 2, in which the cross-sectional area of the Al line is half that of the W line, the current flowing in each metal is almost identical. Consequently the current density in the Al line is higher than in the later, which makes it easy to generate EM. A tetraethyl orthosilicate (TEOS) oxide film was deposited as a passivation layer covering the layer composed of W and Al, and the

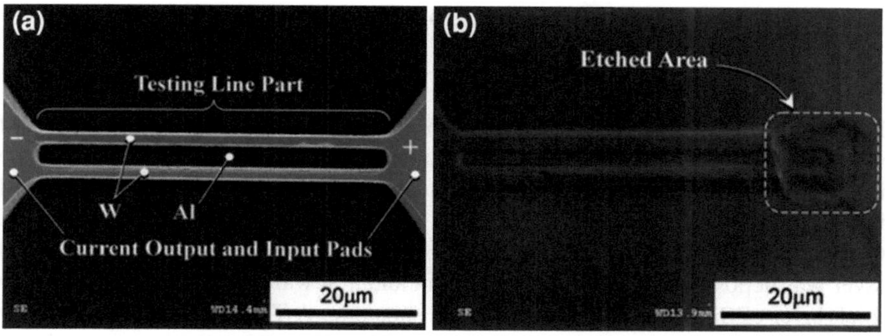

Fig. 3 FE-SEM image of fabricated Al line buried in a W line. **a** Unpassivated line, in which the *left-hand side* is the cathode and the *right-hand side* is the anode so that diffused atoms will accumulate at the *right-hand end*, and **b** Al line buried in a W line covered with a 3.5 μm thick passivation layer where the thickness of the passivation layer at the anode end has been reduced to 2.0 μm by wet etching [29]

region near the anode end of the line was patterned by photolithography and wet etching to achieve an ultrathin TEOS film. Figure 3 shows an image of a typical well-defined specimen obtained using field emission scanning electron microscopy (FE-SEM). Based on the specimen illustrated above, very small slots were finally patterned in the passivation layer by using a FIB system.

An example of one of the Al nanowires that were generated is shown in Fig. 4a. It has a diameter of 900 nm, and was derived from a 0.5 μm square slot penetrating the passivation layer. Another example with a diameter of 250 nm is shown in Fig. 4b, which was generated from a 0.2 μm square slot that did not completely penetrate the passivation layer. The aspect ratio of the nanowire shown in Fig. 4b is about 30.

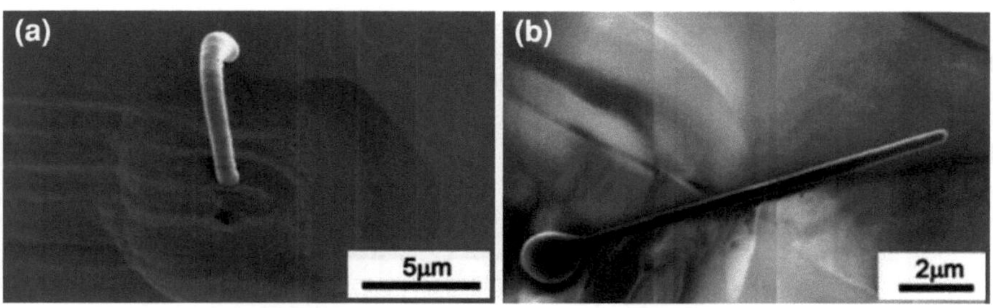

Fig. 4 The examples of the generated nanowires. **a** Al nanowire from a 0.5 μm square slot penetrating the passivation layer, and **b** Al nanowire from a 0.2 μm square slot not penetrating the passivation layer [29]

2.2 *Effective Atomic Accumulation Due to an Introduced Artificial Slit*

In 2006, an effective new method of forming metallic thin wires by utilizing EM was reported [28]. The most important contribution of this method is due to the improved sample structure, and this is illustrated in Fig. 5 showing an FE-SEM image, and Fig. 6, which explains the cross-sectional structure, respectively. Compared with the line structure used in the previous work [29], a 200 nm TiN layer was deposited between an Al thin film and a Si wafer covered with a 300 nm thick layer of SiO_2, which was used as the substrate. A TEOS film was used as the passivation layer for the Al film. On the other hand, an artificial slit was introduced into the Al layer by etching just the Al at the anode end of the line structure, which finally turned out to be very effective for atomic accumulation. To form a source for discharging the accumulated atoms, a circle-like hole with a diameter of approximately 900 nm was introduced into the TEOS layer at the anode end of the lines by using a FIB system.

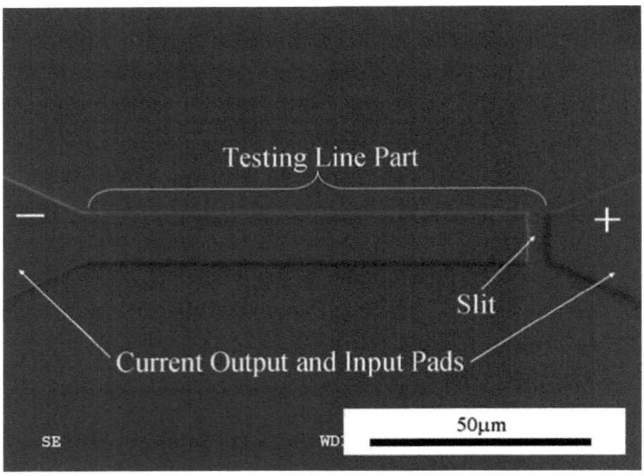

Fig. 5 FE-SEM image of a passivated Al thin film line that has a slit at the anode end [28]

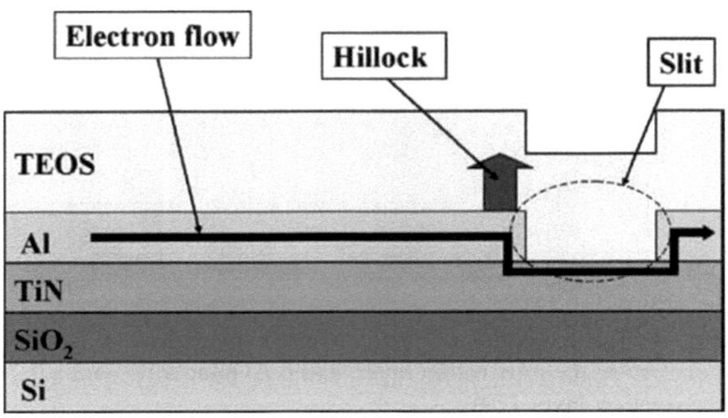

Fig. 6 Illustration of the electron flow in the test sample (cross-sectional view) [28]

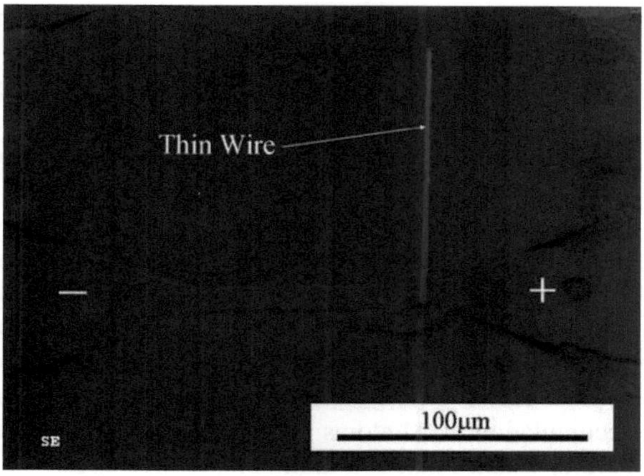

Fig. 7 Result of applied current test on a passivated Al line that had a slit at the anode end. It is confirmed that there is a wire product from the introduced hole [28]

The electrical resistance of TiN is 20 or more times that of Al, so, as illustrated in Fig. 6, when a current is supplied to the line, the electron flow will pass along the inside of the Al (except the slit portion) and along the inside of the TiN in the slit portion. However, the Al atoms that are transported due to the high density electron flow cannot pass through the TiN layer. Therefore, the introduction of the slit makes it possible to collect large numbers of Al atoms in a specific area at the anode end of the lines. As a result, a high compressive stress is generated in these areas. The compressive stress is released as the atoms are discharged from the introduced hole.

FE-SEM observations confirmed the growth of a 'wire-like' product from the introduced hole when the current was applied, as shown in Fig. 7. The formed wire has the appearance of a straight column with a diameter of approximately 1 μm. The aspect ratio of the wire exceeded 100. This value is three times larger than that achieved in the previous work [29].

2.3 Fabrication of Al Micro Spheres by Utilizing EM

Functional nano- and micro-scaled spheres are of both scientific and technological interest. It has already been proposed that they may have extensive applications in catalysis [42], the leather industry [45], electron-detectors [41], luminescence technology [22], and so on. Up to now, a great many nano- and micro-scaled spheres, e.g., carbon [49], sulfide [40], metal [3], alloy [27], and so on, have been synthesized, and many methods for the synthesis of spheres have been reported. Template-based methods, including the bio-organic template method [9] and the inorganic template method [50] are well known "bottom-up" approaches for fabricating micro spheres (MSs). Template-free methods such as the arc plasma method [26], the solvo-thermal method [51] and the chemical beam epitaxy method [2] have also been used for the fabrication of MSs. The above mentioned methods are based on chemical reactions. Unlike the techniques mentioned above, a typical physical method based on EM does not depend on the chemical reactions of the materials. Therefore, this approach is expected to provide more selectivity for the diffused elements by controlling the electron flow and therefore more controllability for the fabrication process. In Sects. 2.1 and 2.2, we described that EM has been successfully employed to fabricate micro and nano Al wires [28, 29]. In this section, the fabrication of Al MSs by utilizing EM [43] will be presented.

The sample structure used to fabricate Al MSs is almost same as that used in the previous study [28]. When an electric current is applied to the sample, electrons will flow inside the Al line, except for the slit portion, where the TiN layer is a bypass. However, Al atoms cannot pass the slit portion and hence an effective accumulation of Al atoms is formed near the anode end of the line, as mentioned before, which induces a high compressive stress.

A ceramic heater was positioned beneath the sample, and the substrate temperature was kept constant. It is known that the temperature at the anode end achieves the maximum value in the Al line when a current is flowing, as shown later. The maximum temperature, T_{max}, may be a governing parameter for the formation of MSs, and it controls the number of accumulated atoms.

The values of T_{max} are determined from the current density and the substrate temperature. Figure 8 shows an example of the temperature distribution in the Al line obtained by finite element analysis (see the work by Sasagawa et al. [34] for details of the analysis of this electro-thermal problem). The conditions are the absolute value of applied current density ($j[=|\mathbf{j}|]$) of 2.9 MA/cm^2 and substrate temperature (T_{sub}) of 613 K. The slit length is 3 μm. It is shown that the temperature increases along the line from the cathode end to the anode end. The melting point of Al is 933 K, and part of the line near to the anode end melts under these conditions, while the temperature at the cathode end is much lower than the melting point. Experiments were performed under the following conditions: [j (MA/cm^2), T_{sub} (K)] = [2.4, 653], [2.6, 653], [2.6, 613] and [2.9, 613], respectively. The experimental conditions listed above give higher values of T_{max} when compared with the conditions used for fabricating the Al lines [28].

Figure 9a and b shows FE-SEM micrographs of an example of an Al MS formed by utilizing EM. A fine MS could be successfully obtained under certain experimental conditions [43], and the diameter of the MS is 4.3 μm. Cross-sectional micrographs of the Al MS, which was cut open utilizing a FIB technique, are shown in Fig. 9c and d, and they clearly show that the MS is a solid structure. Figure 9c shows that the pre-introduced hole was filled with Al, and it clarifies that the Al MS is formed by atoms that are ejected from the hole to the outside.

Now considering the mechanism for the formation of Al MSs by utilizing EM, the Al atoms are accumulated at the anode end under EM conditions due to the existence of the slit. Because the anode end achieves a higher temperature, as shown in Fig. 8, the accumulated atoms in that region may melt. That is to say, the

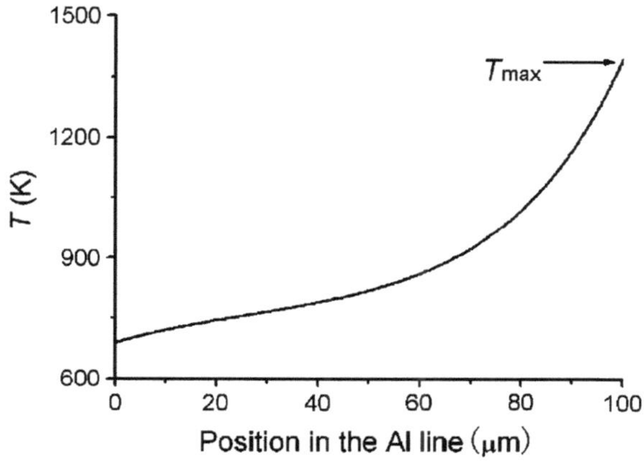

Fig. 8 An example of temperature distribution plotted against the position in the Al line [43]

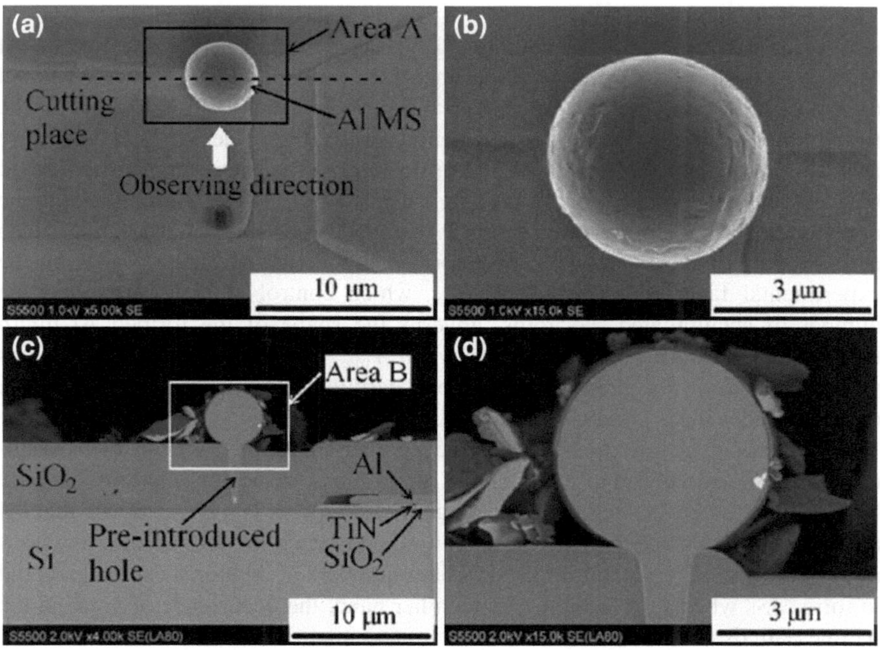

Fig. 9 **a** FE-SEM micrograph of a MS formed by utilizing EM, **b** the magnification view of *area A*, the diameter of the MS was about 4.3 μm, **c** FE-SEM micrograph of the cross sectional view of the MS after being cut off, the Al atoms were pushed out from the pre-introduced hole, and **d** the magnification view of *areaB*, the fragments were resist used for covering the MS during the cutting process [43]

Al line contains solid and liquid phases. When the melted atoms are ejected from the holes, liquid spheres are initially formed under atmospheric conditions due to the effects of surface tension. The spheres then cool and solidify in the atmosphere,

Fig. 10 Relationship between d and T_{max} [43]

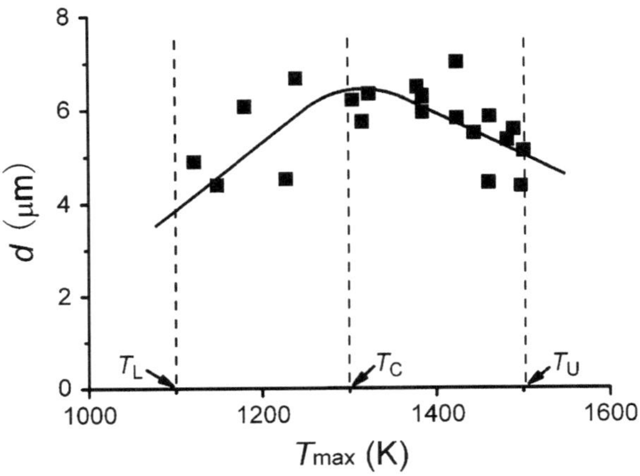

and so the Al MSs are formed. Figure 10 shows the diameters of the formed Al MSs, d, against T_{max}. During the experiments, it was found that MSs were formed over a specific temperature range, $T_L < T_{max} < T_U$, and that T_{max} governs the formation of Al MSs. From the experimental results, the values of T_L and T_U were found to be about 1,100 and 1,500 K, respectively. The diameters of the formed MSs were in the range of 4–7 μm. In particular, larger MSs are likely to be obtained under the temperature regime at the anode end; T_C ($\cong$ 1,300 K). It is also confirmed that the Al nanowires were formed in the case of $T_{max} < T_L$. From the experimental facts, it is clear that T_{max}, which controls the viscosity and the number of accumulated Al atoms, governs the shape of the formed Al micro features. On the other hand, in the case where $T_{max} > T_U$, although some Al atoms were ejected from the pre-introduced hole, hardly any regular-shaped MSs were obtained under these conditions.

In order to generate Al MSs, a large number of Al atoms need to be accumulated at the anode end of the line. We suppose that sufficient Al atoms can only be accumulated in the temperature range $T_L < T_{max} < T_U$. It is known that Al atoms move faster under higher temperature, and hence the atomic flux increases as T_{max} increases. It is natural that sufficient numbers of Al atoms can be gathered to form MSs when T_{max} is high. On the other hand, the occurrence of voids in the Al line will also be facilitated under high temperature, which may lead to a shorter lifetime for the line. As a result, less Al atoms insufficient for MS formation are accumulated.

We know that the formation of Al MSs is governed by the maximum temperature of the Al line, and is successful at a specific range of temperatures. Since temperature is a very important factor of the EM characteristic, we should ask what we can obtain by using other temperature ranges and why this should suffice. We will clarify the mechanism for this technique later, by which selectively forming metallic micro/nano structures with different shapes at predetermined positions can be realized.

2.4 Fabrication of Al Micro Belts by Utilizing EM

As the other typical material, the focus of interest on belt-shaped micro and nano materials has increased because of their unique properties in terms of their absence of dislocations and other line defects [24, 52]. The belt-shaped morphology is distinct from wire-shaped micro and nano materials since it has perfect crystallinity with a well-defined geometry, and is an ideal material that can be used to gain a theoretical and experimental understanding of dimensionally-confined transport phenomena in functional oxides and for producing functional devices. Thus far, ZnO, SnO_2, In_2O_3, CdO and β-Ga_2O_3 micro/nano belts have been synthesized successfully via thermal evaporation or chemical vapor deposition [5, 24]. In this section, we turn our attention to the fabrication of Al micro belts by utilizing EM.

The test structure that was used to fabricate Al micro belts is similar to that used in the previous work [28], which has been illustrated in Sect. 2.2. A sample consisting of a passivated polycrystalline Al line into which a slit has been introduced at the anode end is used. The line structure brings together a large number of diffused atoms at one position. The slit plays an important role in effectively collecting the diffused atoms in a specific area, and the passivation layer is used to release atoms at a certain position. Something different from the previous work is that a rectangular hole was introduced into the passivation layer by using FIB instead of the circular holes that were used in the past [21, 28]. Figure 11 shows the structure of the test sample.

For the experiment, the samples were placed on a ceramic heater under atmospheric conditions, and subjected to a constant direct current by placing a pair of probes in contact with the input and output pads of the line. All of the experimental conditions are listed in Table 1, where T_H is the temperature of the ceramic heater, j is the absolute value of applied current density, t is the current-stressing time, 2 and 4 N are the purities of the Al materials, corresponding to 99 and 99.99%, respectively, and the sectional dimensions of the belts almost correspond to the sizes of the rectangular holes that were introduced when seen in plan view. The generation

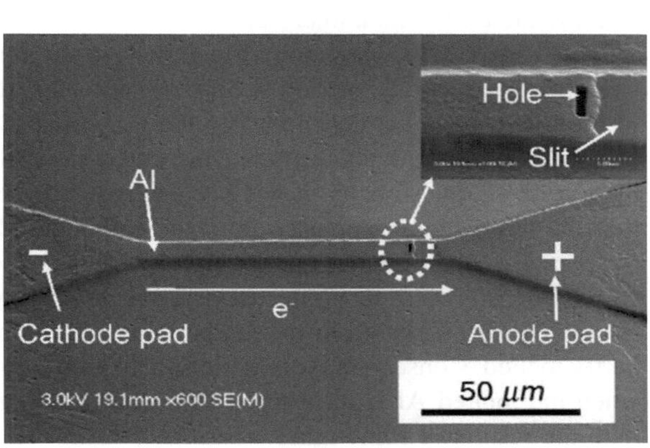

Fig. 11 Top view of the sample structure with a slit and a rectangular hole at the anode end of the line [21]

Table 1 Experimental conditions and results [21]

Sample	Purity	T_H (K)	j (MA/cm^2)	t (s)	Formed structure	Diameter of spheres/length of belts (μm)	Sectional size of belts (thickness × width) (μm)
a	2 N	613	2.4	600	Micro-belt	45	0.8 × 3
b	2 N	613	2.4	300	Micro-belt	12	0.8 × 3
c	2 N	613	2.4	90	Micro-belt	5	0.8 × 3
d	2 N	613	2.8	300	Micro-sphere	3.9	0.8 × 3
e	4 N	613	2.4	300	Micro-particle	–	0.8 × 3
f	4 N	613	2.4	300	Micro-belt	7.5	0.8 × 6

Fig. 12 FE-SEM images of samples **a–f** after current stressing under different experimental conditions [21]

results under each of these conditions are also shown in Table 1. FE-SEM images of test lines a–f are shown in Fig. 12, which show that micro belts are formed in lines a–c and f, and a MS and a micro particle were formed in lines d and e.

The micro belts shown in Fig. 12a–c were formed after different current stressing times, corresponding to 600, 300 and 90 s, respectively. Thus, the lengths of the micro belts are dependent on the current-stressing time.

Figure 12b and d suggests that the current density is a key factor in the formation of the micro belts. It has been reported, based on finite element analysis, that the temperature increases from the cathode end to the anode end [43]. The temperature at the anode end is higher than the melting point of Al when the current density and substrate temperature are 2.8 MA/cm^2 and 613 K, respectively. Therefore, the Al in the region where the atoms have accumulated may be partly melted. Consequently, a liquid sphere would form in atmospheric conditions when the melted Al is discharged from the rectangular hole, which is consistent with the results of some other works [31, 43].

A micro belt and a micro particle were formed in lines b and e, respectively, as shown in Fig. 12, revealing that the purity of Al material also has an effect on the growth properties.

In this section, the micro belts were clarified to be formed by introducing rectangular holes at prearranged positions through FIB etching. Additionally, the widths and thicknesses of the Al micro belts could be readily controlled, and their lengths could be determined by the current-stressing time. Potential applications for Al micro belts are as interconnections, light-weight materials, micro-structured experimental composites, etc. On the other hand, it should be noted that, as well as the length being dependent on the current stressing time, the shapes of the fabricated material are related to the purity of the source material and the temperature. In the next section, we will discuss the effects of the purity of the source material and the temperature on the fabrication of micro materials by utilizing EM.

2.5 Effect of Purity of Source Material on Fabrication of Al Micro/Thin Materials

In this section, the effect of the purity of the source material on the fabrication of Al micro/thin materials (MTMs) by utilizing EM will be described. Different high-purity Al materials [Al purity 99% (2 N) and 99.99% (4 N)] were used. The test structure that was used is similar to that described in the previous work [28]. The samples were placed on a ceramic heater under atmospheric conditions and were subjected to a constant direct-current stressing. All of the experimental conditions and results are listed in Table 2, where l is the length of the slit, n is the number of holes introduced by the FIB etching, and n_p is the number of practical thin wires and spheres that were formed. FE-SEM images of samples a–h are shown in

Table 2 Experimental conditions and results [20]

Sample	Purity	T_H (K)	j (MA/cm^2)	l (μm)	n	n_p	Formed structure	Diameter of spheres/length of wires (μm)
a	2 N	613	2.6	11.5	2	2	Large sphere/thin wire	4.8/6.3
b	4 N	613	2.6	11.4	2	1	Large sphere/irregular micromaterial	4.5/–
c	2 N	613	2.4	10.2	2	2	Thin wire	80.1/8.2
d	4 N	613	2.4	12.2	2	2	Thin wire/small sphere	17.3/2.5
e	2 N	613	2.3	12.1	1	1	Thin wire	12.1
f	4 N	613	2.3	12.2	1	1	Thin wire	9.8
g	2 N	613	2.3	13.1	1	1	Thin wire	14.5
h	4 N	613	2.3	12.3	1	1	Thin wire	4.5

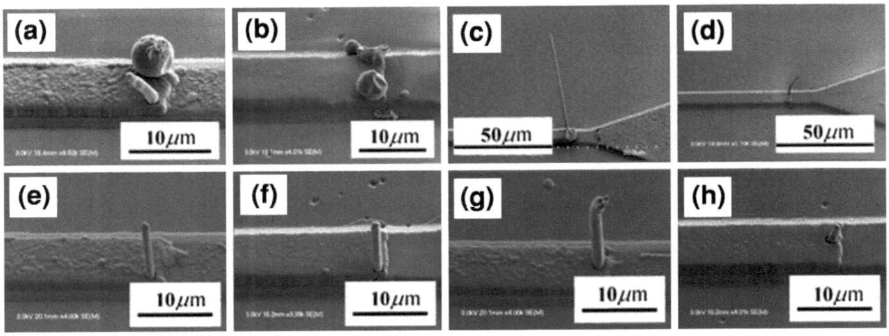

Fig. 13 FE-SEM images of samples **a–h** after current stressing. **a** A large sphere and a short wire were formed in sample a, **b** a large sphere and irregular micro material were formed in sample b, **c** two long wires were formed in sample c (see, Table 2 for details), **d** a long wire and a small sphere were formed in sample d, **e** a long wire was formed in sample e, **f** a long wire was formed in sample f, **g** a long wire was formed in sample g, and **h** a short wire was formed in sample h [20]

Fig. 13, in which we can see that large spheres were formed in samples a and b, and thin wires were formed in samples c–h.

As has been shown above, decreasing the purity of the Al material can lead to an enhancement in the discharge of Al atoms from the holes at the anode end of the samples. It has been revealed that the 2 N Al sample has more prominent EM properties than the 4 N Al sample. The significant increase in the atomic flux of the 2 N purity Al material due to EM is evidence that sample purity has a significant effect on the experimental results. Considering the work that has been published previously [14], we can clearly identify the reason for this effect by using the absolute value of the atomic flux in an Al line, $J[=|\mathbf{J}|]$. The quantity J is related to the activation energy Q_{gb} and the absolute temperature T as

$$J \propto \frac{1}{T}\exp\left(-\frac{Q_{gb}}{kT}\right). \tag{1}$$

In a polycrystalline Al film, the diffusivity of atoms along a grain boundary is significantly larger than that in lattice, indicating that we can ignore the latter [47]. The density of defects increases with decreasing purity in Al materials, and the structures of the grain boundaries also deteriorate, causing an increase in the coefficient for grain boundary diffusion [15]. It is consistent with some previous reports that an atomic flux divergence due to EM arises at microstructural inhomogeneities [1]. Consequently, the drift velocity along the grain boundaries is higher in a 2 N film than that of a 4 N film, based on a decrease in the activation energy of atomic diffusion along the grain boundary, Q_{gb}, in the former. Therefore, the atomic flux will increase as an exponential function with decreasing purity (from Eq. (1)), which will enhance the ease of fabrication of Al MTMs due to EM.

It has been noted that the temperature T also has a significant effect on J, which will increase following an exponential function based on $-(1/T)$. Accordingly, the effects of Joule heating should receive careful attention. When the electrons travel from the cathode to the anode, they will take a detour along the TiN layer during their passage through the slit, where the Al has been removed by wet etching. Since the resistivity of TiN is significantly greater than that of Al, and since the TiN region is close to the holes, the Joule heating due to TiN will impact Al atoms discharging from the holes while current stressing is taking place. From this viewpoint, the slits in the samples should be etched to a controlled length and should correspondingly show a controlled level of Joule heating.

2.6 Effect of Temperature on Controlling EM to Selectively Form Micro and Nano Materials

In the preceding sections, we have introduced wire/sphere/belt-shaped micro and nano metallic materials fabricated by utilizing EM. These formed materials depend on the shape and size of the hole through which the accumulated atoms are discharged [21, 29, 43]. On the other hand, different end products can also be formed even in cases where the hole is the same shape and size. Figure 14 shows an example of different shaped Al materials formed at the anode ends of lines containing holes of the same shape and same size. It has been mentioned that a very important factor of the EM characteristic is temperature, which significantly affects the shape of the formed materials. How does the temperature effect work in these cases? In this section, we will clarify the mechanism of the effect of temperature on controlling EM to selectively form micro and nano materials.

Now let us consider a model for classifying different shaped materials by assuming a steady state (see the work by Saka et al. [31] for details of the verification of this condition and the schematic of the line structure), in which Al atoms are continuously discharged from a small hole, as illustrated in Fig. 15. The pressure at point K within region R is considered to be constant with respect to time, *t*, and the pressure is proportional to the atomic density [16]. Hence, under these conditions, the atomic density in R is constant with respect to time. Thus, the number of atoms entering R is considered to be equal to the number discharged through the small holes in unit time.

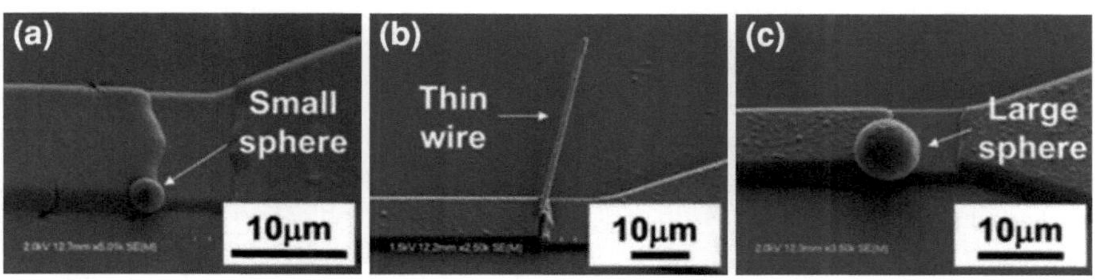

Fig. 14 FE-SEM images of typical shapes formed at the anode end of the lines with the hole in the same shape and same size for discharging atoms. **a** A small sphere, **b** a thin wire, and **c** a large sphere [31]

Fig. 15 An illustration of the assumed steady state in which Al atoms are continuously discharged from the small holes [31]

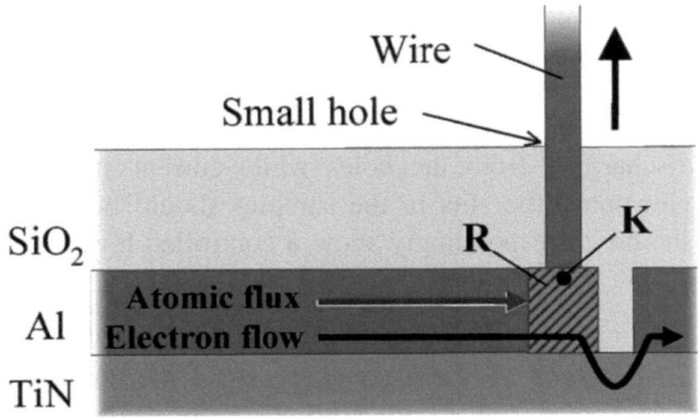

By using the absolute value of the atomic flux in a metal line, J, the cross section of the line, S, and the atomic volume, Ω, the volume of atoms entering R in unit time is given by $JS\Omega$, and that of atoms discharged through the small hole is the same. Therefore, the speed, u, at which the wire is discharged from the hole is given by $JS\Omega/(nS')$, where n is the number of structures fabricated simultaneously on the line and S' is the cross-sectional area of the wire. The diameter of the wire $[= 2\sqrt{S'/\pi}]$ is considered to be equivalent to the side length of the hole.

During the discharge of portion V, which is initially concealed at the root of the wire, as shown in Fig. 16, the material in V has to be cooled by the surrounding ambient so that the atoms that are discharged at high temperature and with low viscosity can remain wire-like. V is assumed to have a characteristic length, Δl, which, for simplicity, is independent of the wire diameter. The temperature of V after discharge, T^*, is governed by its temperature at the start of the discharge, T, and the loss of thermal energy from V during the discharge. So, let us consider the loss of energy required for the temperature to fall to T^* and discuss the shapes of the features that are fabricated. In the following discussion, thin wires are always assumed to be formed for the range of T that is being considered.

The interval of time during which V is being discharged is given by $nS'\Delta l/(JS\Omega)$. The energy, ΔE, lost in time period dt from the surface of V is given by

$$\Delta E = (T - T_0)h_m A(t)dt \qquad (2)$$

Fig. 16 An illustration of the portion V which is ejected at speed u and which loses thermal energy as it is discharged [31]

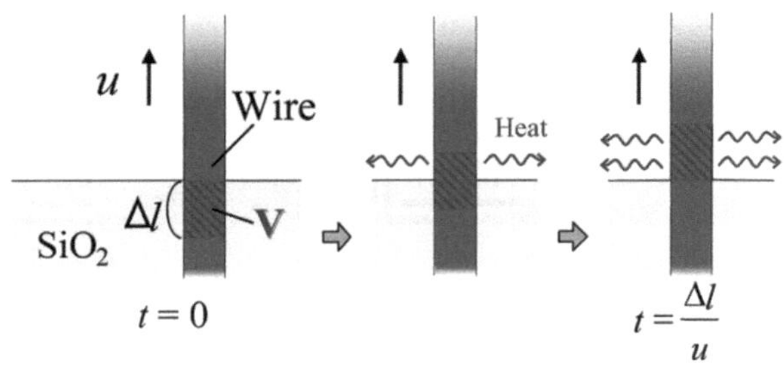

where T_0 is the ambient air temperature, $A(t)[= 2\pi JS\Omega t\sqrt{S'/\pi}/(nS')]$ is the area of the surface concerned, and h_m is the average heat-transfer coefficient. Therefore, the thermal energy, *E*, lost from V in the time interval from the commencement to the end of the discharge is found by integrating Eq. (2) with respect to *t* within limits given by $0 < t < nS'\Delta l/(JS\Omega)$:

$$E = (T - 300)h_m\sqrt{\pi S'^3}\frac{n}{JS\Omega}(\Delta l)^2 \tag{3}$$

where T_0 is assumed to be 300 K. The temperature, *T*, of V as the discharge commences is considered to be the same as that of the Al at the anode end of the line, and is a result of Joule heating. The atomic flux *J* is given by Eq. (10) in Basis of Atomic Diffusion. For Eq. (10) in Basis of Atomic Diffusion, the temperature-dependent resistivity, ρ, can be expressed as $\rho = \rho_0[1 + \alpha(T - T_s)]$, where ρ_0 denotes the resistivity at temperature T_s, and α is the temperature coefficient of resistivity. By substituting Eq. (10) in Basis of Atomic Diffusion into Eq. (3), we get

$$E = (T - 300)\frac{T}{\exp(-\frac{Q}{kT})j\rho_0[1 + \alpha(T - T_s)]}\frac{n\sqrt{S'^3}B}{S} \tag{4}$$

where $B = \sqrt{\pi}h_m k(\Delta l)^2/(\Omega N D_0|Z^*|e)$. Although the atomic flux actually varies due to grain boundaries with various relative angles that exist in a polycrystalline structured material, for simplicity Eq. (4) is obtained by just considering grain boundaries which are parallel with the current density passing through the metal line. Figure 17 shows the relationship between $E/\left(n\sqrt{S'^3}B/S\right)$ and *T*, based on Eq. (4), with $j = 2.0, 2.4, 2.8$, and 3.0 MA/cm^2. While the value of *T* is essentially dependent on *j* due to Joule heating, *T* and *j* are regarded as independent quantities in Fig. 17 based on Eq. (4). The value of *Q* for the passivated polycrystalline Al line is assumed to be 0.74 eV [34], which was derived simply based on Eq. (10) in Basis of Atomic Diffusion for a value of the activation energy seemingly affected by

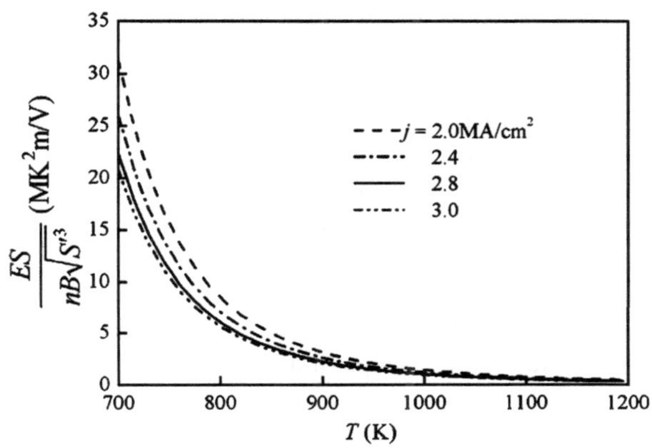

Fig. 17 Relation between $E/(n\sqrt{S'^3}B/S)$ and *T* based on Eq. (4), with $j = 2.0$, 2.4, 2.8, and 3.0 MA/cm^2 [31]

the back flow. The following assumptions are also made: $T_s = 621$ K, which is the average of the values used in the experiment [31], $\rho_0 = 8.01 \times 10^{-2}$ Ω μm, and $\alpha = 1.61 \times 10^{-3}$ K^{-1}. As clearly described in Fig. 17, the current density j has little effect on $E/\left(n\sqrt{S'^3}B/S\right)$ because the range of applied values of j is narrow. Additionally, $E/\left(n\sqrt{S'^3}B/S\right)$ decreases monotonically with T in the range of T shown in Fig. 17. This is because atoms are discharged at a faster rate with higher values of T, and thus the integration is over a shorter time interval for a given value of Δl.

We denote the range of T^* within which wires are formed as

$$T_l < T^* < T_h \tag{5}$$

where T_h is the maximum temperature at which the discharged portion V maintains the shape of a wire, and T_l is the minimum temperature at which V forms a wire without fouling the hole. In addition, T^* is given by

$$T^* = T - C^{-1}E \tag{6}$$

where C is the heat capacity of V. The term $C^{-1}E$ in Eq. (6) should be a more complex term obtained from an analysis of the heat-transfer problem, considering not only the thermal energy lost from the sides of V, but also heat conduction through the sections above and below V; however, such an analysis is not easy. On the other hand, it is apparent that the relevant temperature drop is affected by the thermal energy lost from the sides of V. With these thoughts in mind, the temperature drop is approximated by $C^{-1}E$ in this case. From Eqs. (5) and (6), we find

$$\frac{CT}{B} - \frac{CT_h}{B} < \frac{E}{B} < \frac{CT}{B} - \frac{CT_l}{B}. \tag{7}$$

The heat capacity C depends on S'. Equation (7) is satisfied in zone D as shown in Fig. 18, which is a plot of E/B against T given by Eq. (4). We can use this zone to classify the structures formed for fixed values of S' and T_s. Thus, the micro structures formed in zone D are wires.

Thus far, only the formation of wires has been considered. Actually, in the region of low T and large E/B, i.e., low T^*, a small MS would be fabricated because atoms cannot be continuously discharged and the thermal energy lost from the surface of the sphere, which is spherical because of surface tension, satisfies the requirement for maintaining a spherical shape. If a large quantity of thermal energy is lost from V, its temperature decreases to such an extent that atoms foul the small hole, thus preventing further discharge, and so the sphere remains small. In contrast, in the realm of high T and small E/B, i.e., high T^*, discharged atoms with low viscosity would form a MS that would grow to a certain size [43], and the thermal energy lost from the surface of the sphere would satisfy the requirement for maintaining the shape of a large MS. It is inferred that the thermal energy that is lost from the surface is a contributory factor in the retention of the shape of the small and large spheres as well as the wires.

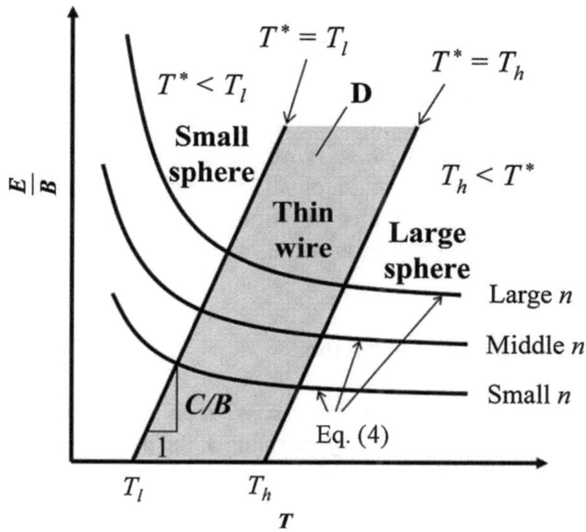

Fig. 18 An illustration showing zone D, where Eq. (7) is satisfied, in a plot of E/B against T [31]

Based on the theory outlined above, for conditions on the parts of the curves in zone D as shown in Fig. 18, a wire is formed. In the case of $T_{\mathrm{h}} < T^*$, on the right-hand side of zone D, a large sphere is formed because of insufficient loss of energy and in the case of $T^* < T_{\mathrm{l}}$, on the left-hand side of zone D, a small sphere is formed because of the higher energy loss.

Now, let us speculate whether it is possible to classify the structures that are formed on the basis of T only. It is imagined that the atoms in V, which has a high value for T, are discharged at different speeds; some extremely fast, and others extremely slow. It would be difficult for V to become subject to clotting due to cooling in the former case, but it would be easier in the latter case. Therefore, one would expect that the structures that are formed would be affected by the difference in the discharge speed. Therefore, it seems logical that the structures that are formed cannot be classified by T alone.

By using the values of T, j, S, S' and n as used in the experiments, E/B is calculated from Eq. (4), and the relationship between E/B and T is analyzed and plotted (see the work by Saka et al. [31] for details). As a result, it has been shown that the structures that are formed fall into distinct groups. It is noted that large MSs are fabricated under conditions with high T and small E/B, i.e., high T^*; thin wires with intermediate T and E/B, i.e., intermediate T^*; and small MSs under conditions of low T and large E/B, i.e., low T^*.

In this way, the structures that are formed can be classified by simply considering T^*. Finally, it should be noted that a phenomenological classification based on T is valid because T is the main variable affecting E as far as j, of which range is narrow as shown here, is treated, and hence T^* is a single-valued function of T. In actuality, it is a difficult task to evaluate the value of T^* itself, and one can use T for classifying the structures that are formed under the conditions examined.

2.7 Numerical Simulation of Fabrication of Micro/Nano Materials

In this section, a method is introduced for numerically simulating the fabrication of micro/nano materials such as thin wires utilizing EM [37]. The simulation method was developed by modifying the numerical simulation for EM failure, considering a collection of metal atoms [35]. The formation under various conditions is simulated and the results are compared with the experimental results to discuss the usefulness of the simulation.

2.7.1 Simulation Method

The atomic flux **J** in a passivated metal line is expressed by Eq. (11) in Basis of Atomic Diffusion. Using **J**, the governing parameter for EM damage, *AFD*, is formulated as described in Sect. 1.6 in Basis of Atomic Diffusion. The value of the parameter, AFD^*_{gen}, is given by Eqs. (55), (57) in Basis of Atomic Diffusion. The parameter at the line end is especially defined as $AFD^*_{\mathrm{gen}}|_{\mathrm{end}}$ by Eq. (58) in Basis of Atomic Diffusion given in Sect. 1.7 in Basis of Atomic Diffusion. The formation of micro/nano materials utilizing EM is simulated using an AFD^*_{gen}-based numerical method, covering the concentration and overflow of the atoms, as well as initial void formation in the specimen.

Figure 19a shows a schematic diagram of the shape and size of the simulated specimen [28]. Figure 19b shows the cross section of the specimen used in experiment. The atoms are discharged from a hole and a micro/nano material is formed. In the simulation, the specimen is divided into elements in two dimensions as shown in Fig. 19c. The atomic density of each element is calculated based on the value of AFD^*_{gen}, and the atomic density distribution in the specimen changes with time. Using AFD^*_{gen} depending on atomic density distribution, the change in the atomic density distribution is calculated iteratively until the beginning of specimen damage. The formation of micro/nano materials is simulated by calculating the volume of the atoms that flow out of the specimen.

Figure 20 shows the flowchart of the computation procedure [37]. First, the distributions of current density and temperature in the specimen are obtained by two-dimensional FE analysis. The current density is obtained as the solution of the Laplace equation regarding electrical potential, and the temperature distribution is obtained as the solution of the heat conduction problem considering Joule heating [35]. The values of parameters AFD^*_{gen} and $AFD^*_{\mathrm{gen}}|_{\mathrm{end}}$ are calculated based on the results of the FE analysis and the film's characteristic constants, which were obtained in advance by the $AFD^*_{\mathrm{gen}}|_{\mathrm{end}}$-based method described in Sect. 1.7 in Basis of Atomic Diffusion [13] or AFD^*_{gen}-based method [34]. The atomic density of an element depends largely on the orientation of the microstructure, θ, in the *AFD* formula [34]. Thus, the atomic density taken in relation to θ is denoted as N^*, and all values of N^* for the whole range of θ (i.e., from 0 to 2π) are calculated. The atomic density N^* is represented by $N^* = N_0 - \int_0^t AFD^*_{\mathrm{gb}\theta}\mathrm{d}t$,

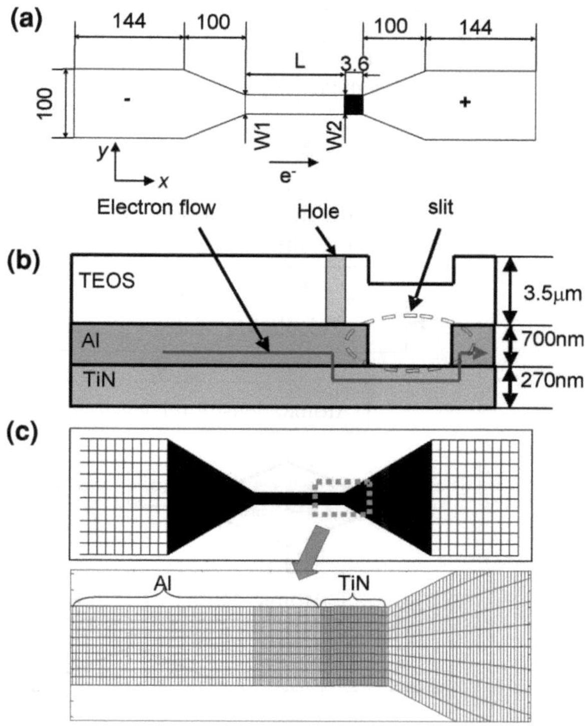

Fig. 19 Al line specimen in numerical simulation [38]. **a** Schematic diagrams of the shape and size of specimen to be simulated. Dimensions are in micrometers. **b** Schematic diagram of specimen's cross section and electron flow in the specimen. **c** Mesh generation in the simulation

where t is time. Then, the atomic density of the element, N, is obtained by averaging all values of N^* for the element. Assuming a critical atomic density for void initiation, $N^*_{\min}$, and for hillock initiation, $N^*_{\max}$. The material constants $N^*_{\min}$ and $N^*_{\max}$ can be obtained by a simple experiment based on AFD^*_{gen} [35]. In the simulation of the formation of micro/nano materials, it is judged whether or not N^* reaches $N^*_{\max}$. When $N^* \geq N^*_{\max}$, an excess atomic density of N^*_{over} is assumed to occur, which is calculated as $N^*_{over} = N^* - N^*_{\max}$. Then, the value of $N^*_{\max}$ is substituted for N^*. All the excess atoms accumulate at the anode end of the line part, as a result of the structure of the specimen. Thus, we can calculate the total volume of these atoms using N^*_{over}, which is assumed to be the volume of the fabricated material. The iterative calculation is carried out until the beginning of specimen damage, which is defined as the state where any N^* in the specimen reaches $N^*_{\min}$ (i.e., void initiation), as the first approximation. Initial void formation is correlated with the time for failure [35], and micro/nano materials are considered to grow in proportion to the time of EM-induced production [31]. Therefore, the simulation is developed to discuss efficient conditions for the production of micro/nano materials by taking into account the time of initial void formation.

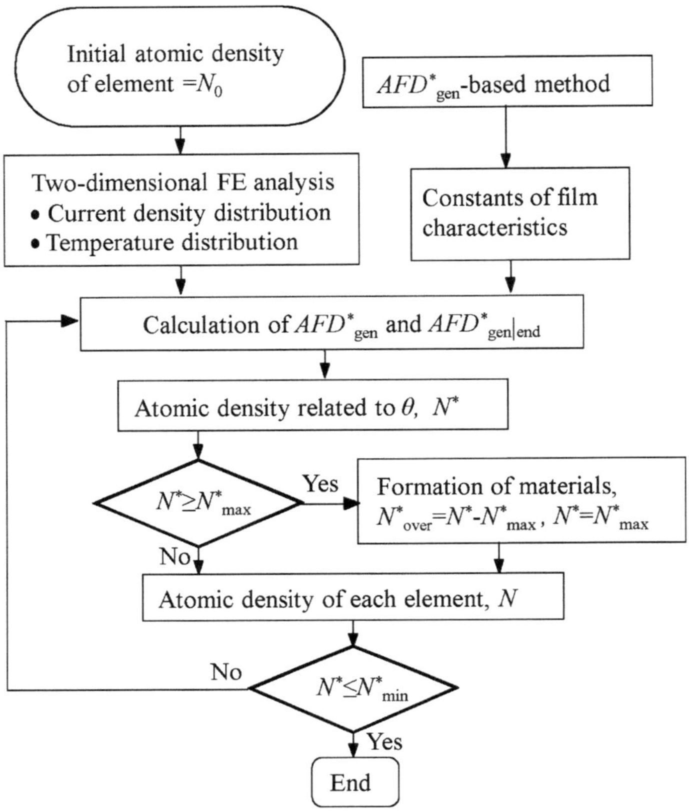

Fig. 20 Simulation flowchart. With kind permission from Sasagawa et al. [37]

2.7.2 Effect of Product Condition on Fabrication[1]

For the Al polycrystalline line shown in Fig. 19a, the volume of micro/nano material and the location of void formation in the Al specimen were predicted by the simulation. The Al line had width of $W1 = W2 = 10.8$ μm, thickness of 700 nm, and length of $L = 101.1$ μm. The slit length was 3.6 μm, and the thickness of the TiN layer was 270 nm. The thickness of the TEOS layer was 3.5 μm. As described later, the specimen fabricated on a silicon substrate was stable at high temperatures. It is known that the EM-induced formation of micro/nano materials varies with the input current density in the line, j, as well as with the substrate temperature, T_s. Three combinations of j and T_s, as shown in Table 3, were selected as simulation conditions. The current density, temperature and

---

[1] With kind permission from Sasagawa et al. [37].

Table 3 Simulation conditions and results

Condition	j (MA/cm^2)	T_s (K)	T_P(K)	Fabricated volume (μm^3)
1	2.4	653	1,140	9.5
2	2.6	613	1,200	7.5
3	2.8	553	1,190	6.2

With kind permission from Sasagawa et al. [37]

atomic density in the line are distributed differently under the three conditions. The differences in these distributions are expected to affect the formation of micro/nano materials, as well as the damage to the line.

In setting the simulation conditions, Condition 2 was taken as the standard condition, and T_s in Conditions 1 and 3 was varied for lower and higher values of j than that under Condition 2, considering standardization of formation speed. It is known that the formation speed of micro/nano materials depends on j and T_s. Moreover, the peak temperature T_p in the temperature distribution along the Al line part depends on j and T_s, and T_p might affect the formation speed. Accordingly, when j was changed, we selected values of T_s such that T_p coincided with the values for Condition 2 in the FE analysis [35]. The value of T_p under the three conditions was verified by FE analysis, as shown in Table 3. It was found that the values of T_p were almost the same under Conditions 2 and 3, but lower under Condition 1. If T_s in Condition 1 were higher, T_p would be almost the same in all conditions. But we could not raise T_s above 653 K, owing to the limitation of the hot stage used in the experimental verification. The maximum difference of T_p between the three conditions was about 60 K, which corresponds to only 5% of T_p in kelvin.

Sasagawa et al. [34, 35] and Hasegawa et al. [13] have given the values of the film's characteristic constants. In this simulation, the following values were used: $Q_{gb} = 0.548$ eV, $Z^* = -8.26$ and $C^*_{gb} = 1.15 \times 10^{24}$ K μm^3/(J s). The value of initial atomic density was $N_0 = 6.024 \times 10^{10}$ μm^{-3}. The values of N^*_{min}, N^*_{max} and κ depend on the thickness of the TEOS film [36]. For the present specimen, the values were set as follows: $N^*_{min} = 5.894 \times 10^{10}$ μm^{-3}, $N^*_{max} = 6.148 \times 10^{10}$ μm^{-3} and $\kappa = 44.8$ GPa.

The last column in Table 3 shows the predicted volume of the fabricated materials. The volume was the largest under Condition 1 and smallest under Condition 3 due to difference of lifetime. Figure 21 shows the distribution of normalized atomic density, N/N_0, at the completion of the simulation. The abscissa represents the location along the longitudinal axis of the line. The valley of N/N_0 corresponds to the location of void formation. The damage occurred near the cathode end under Condition 1, near the anode end under Condition 3, and between the two under Condition 2. It was shown that the failure site moved to the anode side as j increased. Note that the simulation results are based on the failure criterion of initial void formation. If we consider the specimen's open-circuit failure instead of the initial void formation, the predicted volume becomes larger and the failure site moves closer to cathode side in comparison with the present results.

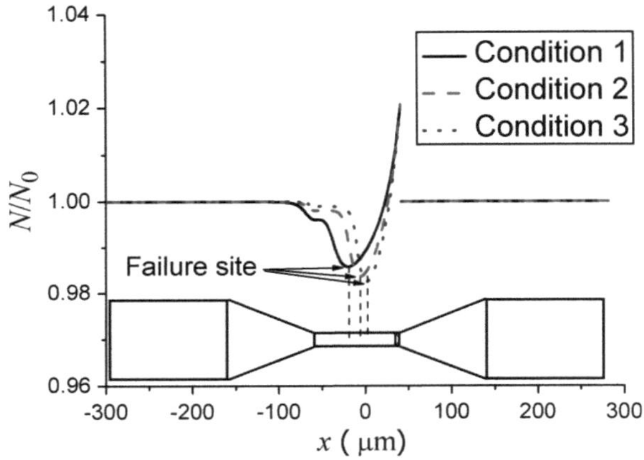

Fig. 21 Distribution of N/N_0 along the longitudinal axis and the predicted failure sites in the specimen (with kind permission from Sasagawa et al. [37])

In order to verify the results of the simulation, experiments were performed using specimens with the same line shape and under the same conditions as those in the simulation. The specimens were fabricated as follows. A TiN film was deposited by sputtering and then an Al film was deposited by vacuum evaporation. Al and TiN layers were patterned into the shape of the line by wet etching and fast-atom-beam etching. Then, the Al layer was etched only at the anode end of the line by wet etching to introduce a slit. TEOS film was deposited by plasma-enhanced chemical vapor deposition. Finally, three square holes with side length of about 800 nm were introduced through the TEOS film at the anode end of the line by a focused ion beam technique, as shown in Fig. 19b.

The specimen was positioned on a ceramic hot stage kept at a constant temperature, and was subjected to a constant direct current. When the specimen was subjected to EM, not only micro/nano materials but also voids were formed and grew in the specimen. During this process, the potential drop across the line increased. A direct current was supplied to the Al line until the circuit through specimen opened and current supply was stopped automatically. Most experiments took <10 min.

Under each condition, experiments were performed using six specimens and wire-shaped materials were formed for five of them. The volume of the formed thin wires under each condition was measured using two-directional micrographs taken by FE-SEM. First, the length of a wire was measured in three dimensions from FE-SEM micrographs taken from two orthogonal directions. Next, the length was multiplied by the diameter of the wire measured from the micrograph, and thereby the volume of the wire was obtained. The results are shown in Fig. 22. The inset photo in Fig. 22 shows the FE-SEM micrograph of wires formed under Condition 2. It was found that the volume of micro/nano materials under Condition 1 was largest, while that under Condition 3 was smallest. These results exhibited a trend similar to the simulation results shown in Table 3, although the magnitude of the experimental results is much greater than that of the simulation results because the simulation is terminated at the beginning of the failure process

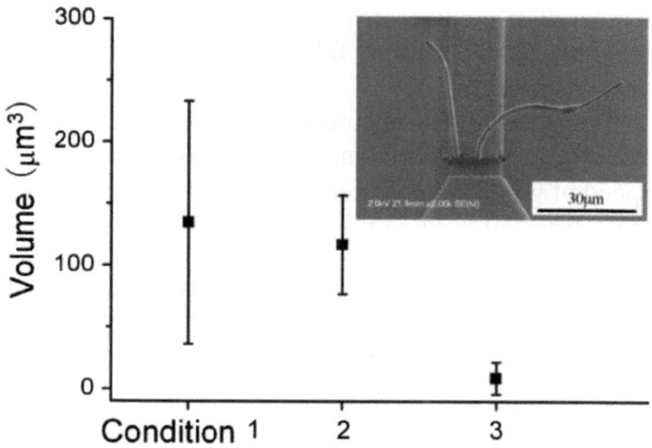

Fig. 22 Volumes of experimentally formed wires under the three conditions. Average values and standard deviations are shown. Inset photo: wires formed under Condition 2 (with kind permission from Sasagawa et al. [37])

in actual specimen. The agreement between the calculated and experimental results supports the methodology of the simulation.

After current was supplied, an optical micrograph was also taken and the failure site where the specimen opened was identified. Figure 23 shows the histogram of the failure site. The failure sites were mainly distributed near the cathode end and at the center of the line under each condition. The inset photo in Fig. 23 shows the optical micrograph of the failure site under Condition 2 where failure occurred at cathode end and at the center as a special case. It seems that the ratio of the number of failures at the center to the number of failures near the cathode increases with increasing current density, from Condition 1 to Condition 3. It was shown that the tendency of failure sites was similar to the predicted one in Fig. 21. Now, let us discuss the difference of failure modes in the formation process of the micro/nano materials. When the specimen is subjected to EM, there are two failure modes in the specimen. One is failure near the cathode end, which is caused by the current density gradient at the connection part of the line and the large contact pad (Mode I), and the other is failure near the anode end, which is caused by the steep

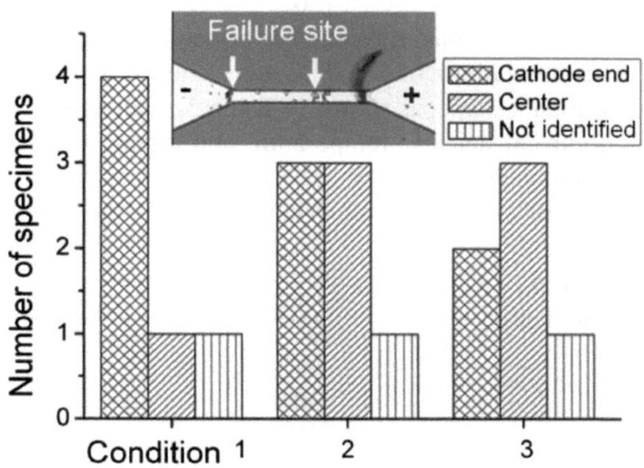

Fig. 23 Histograms of failure sites in the experiments under the three conditions. Inset photo: failures at cathode end and at the center as a special case in Condition 2 (with kind permission from Sasagawa et al. [37])

temperature gradient near the hot spot at the slit (Mode II). Basically, Mode I is inherent in the present fabrication method, while Mode II appears as current density increases.

Considering the theoretical atomic flux due to EM given by Eq. (11) in Basis of Atomic Diffusion, the specimen's lifetime would increase but the formation speed of wire would decrease with decreasing j at a given T_p, and thus the change in the fabricated volume would be negligible. However, the change in failure mode observed in the simulation showed an extension of lifetime under Condition 1. Thus, the change in the dominant failure mode depending on the condition made affected the formation of micro/nano materials, more specifically, the fabricated volume.

2.7.3 Effect of Specimen's Shape on Fabrication

It was shown in Sect. 2.7.2 that the AFD^*_{gen}-based numerical method reproduces well the effect of product condition on the fabrication of micro/nano materials [37]. Now, assuming passivated polycrystalline lines with differing line shape, the volume of the micro/nano materials fabricated until damage initiation for the Al specimen was predicted [38]. Figure 19a schematically shows the shape and size of the simulated specimen with Table 4. Shape I was 10.8 μm wide and 101.1 μm long, that is, long and straight, Shape II was 10.8 μm wide and 50.5 μm long, that is, short and straight, Shape III was 11.8 μm wide at the cathode end, 10.8 μm wide at the anode end and 101.1 μm long, that is, long and tapered, and Shape IV was 10.8 μm wide at the cathode end, 11.8 μm wide at the anode end and 101.1 μm long, that is, long and inversely tapered. The Al layer was 700 nm thick, the TiN layer was 270 nm thick and the TEOS layer was 3.5 μm thick, as shown in Fig. 19b.

The following characteristic constants were used in the simulation: $Q_{gb} = 0.548$ eV, $Z^* = -8.26$ and $C^*_{gb} = 1.15 \times 10^{24}$ K μm^3/(J s). The value of initial atomic density was $N_0 = 6.024 \times 10^{10}$ μm^{-3}. The values of N^*_{min}, N^*_{max} and κ were chosen as follows: $N^*_{min} = 5.894 \times 10^{10}$ μm^{-3}, $N^*_{max} = 6.148 \times 10^{10}$ μm^{-3} and $\kappa = 44.8$ GPa [36].

Micro/nano materials can be formed by using EM under high values of j and T_s. Six combinations of j and T_s were chosen as the simulation conditions for each specimen's shape, as shown in Table 5, such that T_p was the same in all conditions. The current density, temperature and atomic density were distributed differently depending on line shape. The differences in these distributions affected the formation of micro/nano materials as well as the failure of the specimen.

Table 4 Four line shapes used in simulation [38]

Shape	L (μm)	$W1$ (μm)	$W2$ (μm)
I	100	10.8	10.8
II	50	10.8	10.8
III	100	11.8	10.8
IV	100	10.8	11.8

Table 5 Six combinations of j and T_s for each shape [38]

j (MA/cm^2)	2.0	2.2	2.4	2.6	2.8	3.0
T_s in Shape I	818	748	678	618	557	503
T_s in Shape II	818	748	680	618	558	505
T_s in Shape III	818	748	683	618	560	507
T_s in Shape IV	873	808	743	680	621	568

Table 6 Calculation condition and results [38]

Shape	j (MA/cm^2)	T_s (K)	T_p (K)	NS volume (μm^3)	Lifetime (s)
I	2.0	818	1,200	12.3	49.5
II	2.0	818	1,198	10.9	45.3
III	2.0	818	1,196	13.7	56.6
IV	2.2	808	1,199	12.0	44.9

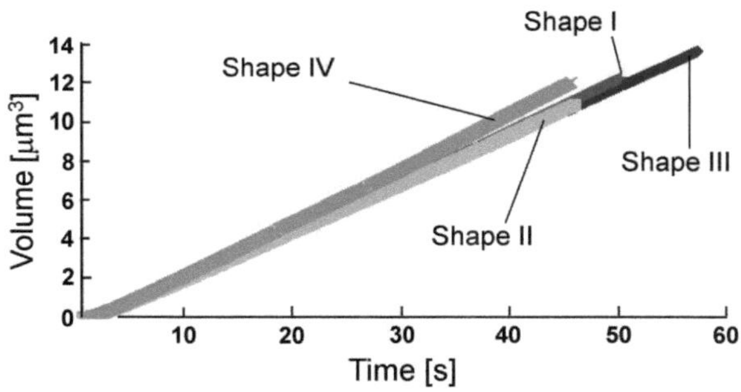

Fig. 24 Simulation results of the predicted volume of micro/nano materials [38]

Table 6 and Fig. 24 show the calculation results where the largest volume was obtained in each specimen's shape, respectively. The current density that resulted in the largest volume was 2.0 MA/cm^2 for Shape I, Shape II and Shape III, and 2.2 MA/cm^2 for Shape IV. The abscissa of Fig. 24 is time, and the ordinate is the predicted volume of the fabricated micro/nano materials. The volume of micro/nano materials formed until damage initiation was the largest for Shape III. The volume of fabricated material decreased as follows, from highest to lowest: Shape III, Shape I, Shape IV and Shape II. Though the formation velocity of Shape I and Shape III did not differ greatly, the lifetime of Shape III was longer than that of Shape I.

Figure 25 shows the distribution of normalized atomic density, N/N_0, at the completion of the simulation selected in Table 6. The abscissa represents location along the longitudinal axis of the line. The location of the valley in the plot of N/N_0 corresponds to the location of void formation in the specimen. The damage occurred near the cathode end in all shapes. Figures 26 and 27 show the current density distribution and temperature distribution in the specimen under the condition where the largest volume was obtained for each shape. Notably, the distributions of atomic density, current density and temperature near the anode end,

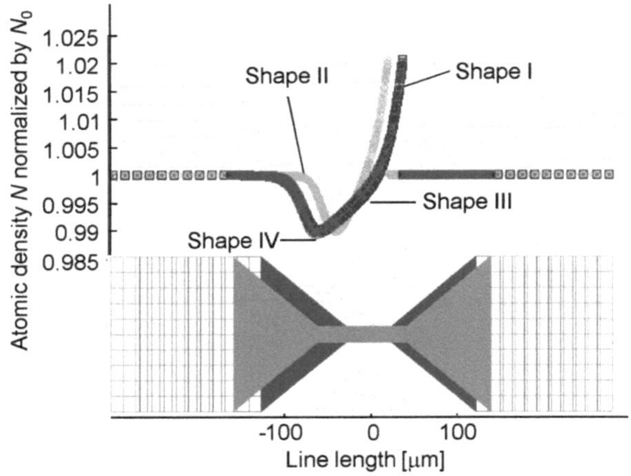

Fig. 25 Distributions of atomic density along the longitudinal axis of the specimen [38]

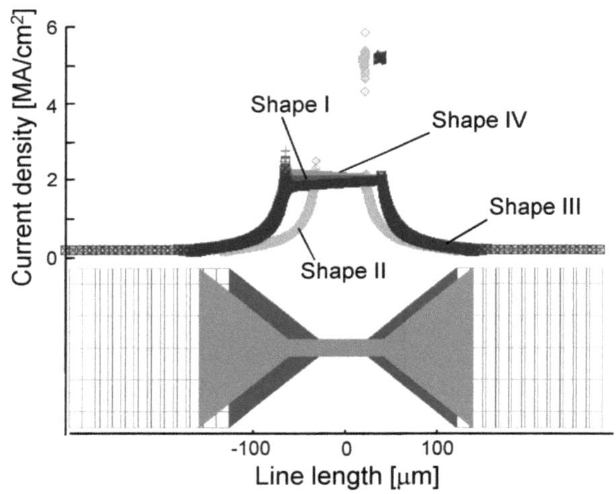

Fig. 26 Current density distributions in the specimen [38]

which influences the formation speed, did not appreciably differ between Shape I and Shape III. Therefore, these distributions did not lead to a difference between these two shapes in formation speed. On the other hand, looking at these distributions near the cathode end, which influences the lifetime, the current density and temperature in Shape III were smaller than those in Shape I. Accordingly, less EM damage was induced in Shape III in comparison with Shape I, and the lifetime of Shape III was longer than that of Shape I. Thus, it is conceivable that the specimen's shape made a difference in predicted volume between Shape I and III.

A numerical simulation method of the formation of micro/nano materials utilizing EM was applied to Al specimens with various shapes. It was shown that the volume of micro/nano materials that form prior to specimen damage varied depending on the specimen's shape, owing to the shape-dependent distributions of atomic density, current density and temperature, which influenced EM damage.

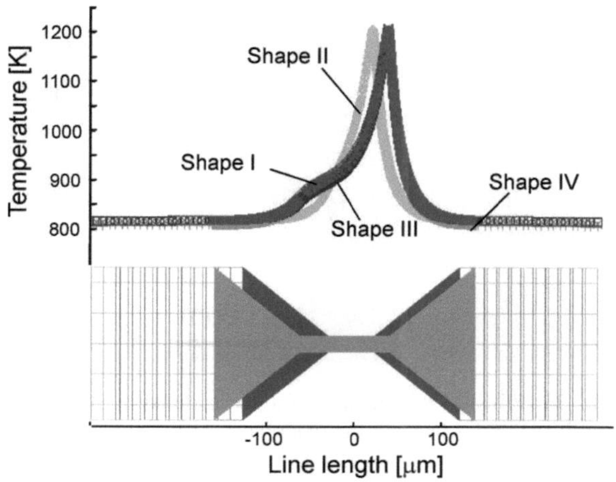

Fig. 27 Temperature distributions in the specimen [38]

Furthermore, it was clarified that this simulation method contributes to seeking the optimal specimen shape for efficient fabrication.

3 Fabrication of Micro and Nano Metallic Materials by Utilizing SM

Some typical SM-induced phenomena have been reviewed in chapter titled Basis of Atomic Diffusion. Ways in which we can apply SM to the fabrication of micro and nano metallic materials will become the topic of this section.

3.1 Rapid and Mass Formation of Cu Nanowires by Utilizing SM

Since Cu has better electrical and thermal conductivities than the traditional Al material, and because this is particularly important when we consider the trend for the miniaturization of electronic devices today, we first focus on Cu as a preferred material for rapid and mass formation of nanowires by utilizing SM.

3.1.1 Stress Concentration in Polycrystalline Cu Films

In chpater titled Basis of Atomic Diffusion, it was mentioned that SM in most bilayer/multilayer structures is caused by thermal stress resulting from a mismatch

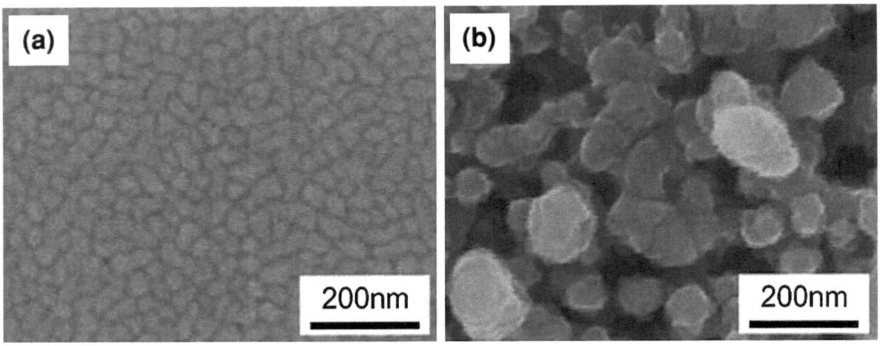

Fig. 28 FE-SEM micrographs of **a** the oxide layer covering a 416 nm thick evaporated Cu film, and **b** the grains beneath the oxide layer [30]

of the thermal expansion coefficients. The gradient of the hydrostatic stress is the driving force for atomic diffusion. It is of interest to ask how the gradient is generated and how it drives atomic diffusion.

Taking Cu as an example, it is well known that Cu is easily oxidized in the atmosphere and is covered with an oxide layer (Cu_2O). As an example, Fig. 28a shows an oxide layer covering the surface of a 416 nm thick evaporated Cu film. By removing the oxide layer using 5% HF, the Cu film was found to show a polycrystalline structure, as shown in Fig. 28b, and it was noted that Cu grain boundaries were randomly located within the film. On the other hand, for such a Cu film with an oxide layer in a Cu/Ta/SiO_2/Si multilayer structure (see the work of [30] for the details of this structure) in which the layer ordering is from top to bottom, hydrostatic stress will be generated when it is subjected to heating.

In the Cu/Ta/SiO_2/Si multilayer structure mentioned above, the distribution of hydrostatic stress was found by elastic finite element analysis to be uniform on a macroscopic level over all of the Cu film. However, if we take into account the anisotropy of the grains in the polycrystalline Cu films, stress concentrations can occur at the grain boundaries due to material singularities, i.e., sudden changes in the material properties across the grain boundaries. Also, these stress concentrations due to material singularities are further enhanced by geometrical singularities at triple points composed of three grains. In fact, stress concentrations around grain boundaries have already been reported [23, 25]. In this case, the stresses take extreme values around the grain boundaries, and hence local gradients of hydrostatic stress are generated in Cu grains. This drives the diffusion of Cu atoms from the grain boundaries into the centers of the Cu grains.

3.1.2 Process of SM-Induced Cu Nanowire Growth

Regarding the fabrication of Cu nanowires by utilizing SM, Cu/Ta/SiO_2/Si structures incorporating Cu films of different thickness (79, 416 and 2,013 nm)

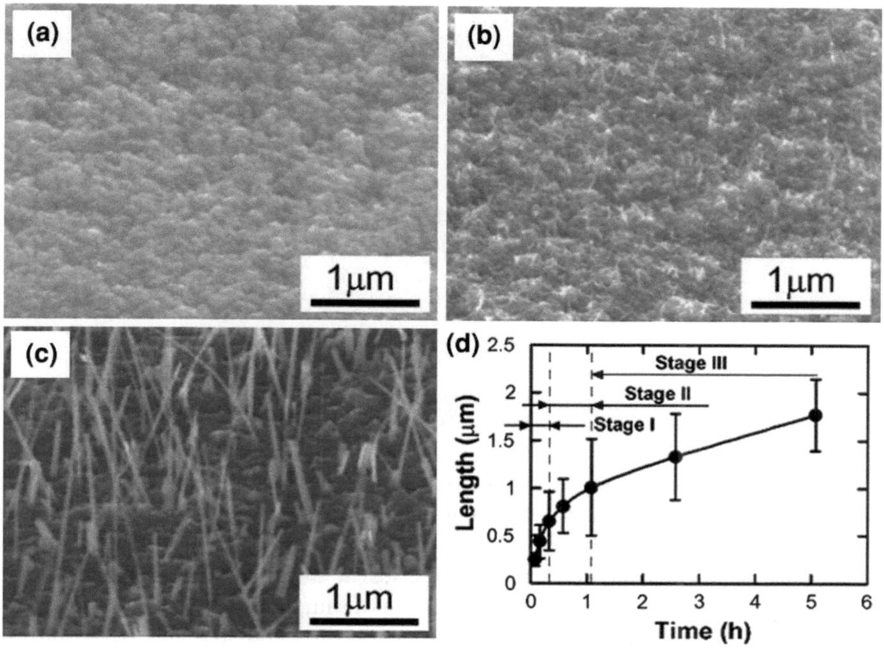

Fig. 29 Growth of Cu nanowires on sample B at 340°C observed by FE-SEM. The micrographs show the examined surfaces **a** shortly after starting heating (180 s), **b** after 270 s, and **c** after 5 h. **d** The relationship between wire length and the time after starting heating. Note that the growth curve is divided into three stages [30]

were used and labeled as samples A, B and C, respectively [30]. The corresponding average grain sizes of samples A, B and C are 59, 83 and 208 nm. The experiments were performed under atmospheric conditions, and a ceramic heater positioned beneath the samples heated the samples uniformly. As an example of the experimental results, FE-SEM micrographs of sample B after when it was maintained at 340°C are shown in Fig. 29a–c. Immediately after starting the heating, uneven patterns appeared on the surface of sample B (Fig. 29a). Subsequently, many point-sources for the initiation for Cu nanowires were observed (Fig. 29b). After they had been initiated, the Cu nanowires then proceeded to grow to about 1.8 μm in length during 5 h of heating (Fig. 29c).

The role of the oxide layer must be important in nanowire generation, as stated in chapter titled Basis of Atomic Diffusion. As far as Cu nanowires are concerned, if the Cu film is not fully covered by an oxide layer, diffused Cu atoms will never form nanowires. The uneven pattern shown in Fig. 29a may be due to local accumulations of Cu atoms at the interface between the oxide layer and the Cu film, i.e., Cu_2O/Cu. If the compressive hydrostatic stress induced by the accumulation of Cu atoms at the interface attains a critical value, the accumulated Cu atoms start to penetrate the oxide via any weak spots in the layer, after which Cu

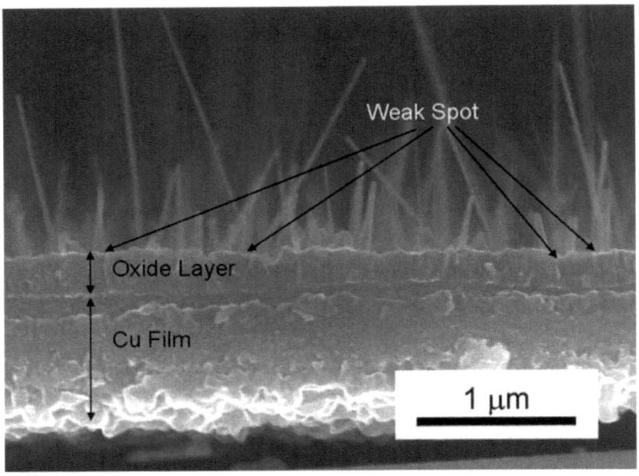

Fig. 30 FE-SEM micrograph of the cross-sectional observation of sample B which has experienced heating for 5 h at 340°C

nanowires are nucleated on its surface (see Fig. 29b). The weak spots in the oxide layer form pathways through which Cu atoms can migrate externally due to the lower compressive stress that operates at the interface of Cu_2O/Cu, and such weak spots may be located at the grain boundaries in the oxide layer. As shown in Fig. 30, a FE-SEM cross-sectional observation of sample B, which has experienced heating for 5 h at 340°C, was undertaken to demonstrate the hypothesis. According to the wire growth mechanism described above, the diameters of the Cu nanowires correspond with the sizes of the weak spots in the oxide layer, and these can be several tens of nanometers. The sizes of the weak spots in the oxide layer are much smaller than those suggested by other groups [17, 39].

The growth curve, i.e., the relationship between the wire length and the time after starting the heating, shows that wire growth can be roughly divided into three stages (see Fig. 29d). The growth rate of Cu nanowires may essentially be governed by the gradient of the hydrostatic stress between the top surface of the Cu_2O layer and the Cu_2O/Cu interface. The hydrostatic stress at the top surface of the Cu_2O layer is practically independent of time, and it takes an almost constant value during the whole period of the nanowire growth. On the other hand, the hydrostatic stress at the Cu_2O/Cu interface is caused by the accumulation of Cu atoms due to surface diffusion; hence the growth rate is affected by the gradient of the hydrostatic stress and is directly determined by the surface diffusivity of the Cu atoms. In addition, surface diffusion and grain boundary diffusion in the direction of the film thickness occur simultaneously, and the atoms that migrate due to grain boundary diffusion contribute to the subsequent wire growth after they experience surface diffusion. It is well known that surface diffusion can occur more readily than grain boundary diffusion for Cu films [47]. In stage I, the Cu nanowires are initiated due to surface diffusion and then grow mainly by further surface diffusion. However, the continuous surface diffusion leads to a lack of Cu atoms around the grain boundaries and a local accumulation of Cu atoms in the grains, and, as a result, the gradients of the local stresses in the Cu grains will be reduced. In stage II, both surface diffusion and grain boundary diffusion in the direction of the film thickness contribute to wire growth. Cu atoms are supplied via grain boundaries for the continuous surface diffusion. In stage III, the growth of Cu nanowires is governed by grain boundary diffusion. It is natural that the Cu nanowires keep on growing as long as grain boundary diffusion processes.

3.1.3 Governing Parameters for SM-Induced Cu Nanowire Growth

It is noteworthy that thinner whiskers were generated under lower temperature conditions, as shown in Fig. 31a. Figure 31b shows the average diameters of the Cu nanowires after 5 h of heating in relation to temperature. In the case of sample A at 340°C, we hardly observed any Cu nanowires on the surface after 5 h of heating. On the other hand, Cu nanowires were generated at 240°C on sample B and at 220°C on sample C, and a clear correlation between wire generation and the thickness of the Cu films was observed. It was found by FE-SEM observations that the average grain sizes of Cu became smaller with decreasing film thickness. Usually, small-grained materials tend to be isotropic. In the case of sample A, local

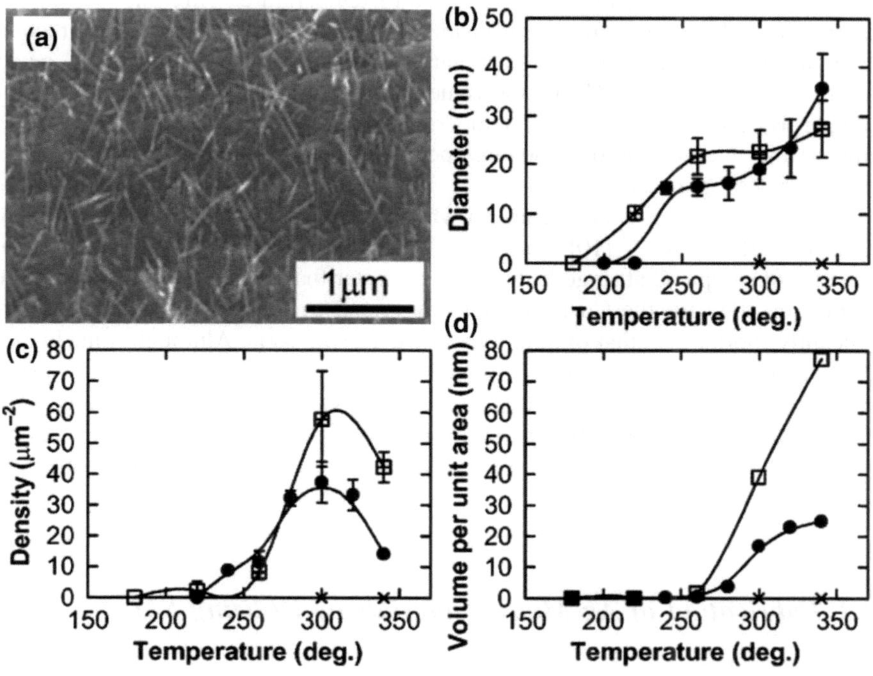

Fig. 31 **a** FE-SEM micrograph of Cu nanowires on sample B grown at 280°C after starting heating (5 h), **b** wire diameter after 5 h versus temperature, **c** wire density after 5 h versus temperature, and **d** volume of nanowires per unit area after 5 h in relation to temperature. The *cross* is for sample A, the *solid circle* is for sample B and the *square* is for sample C, respectively [30]

stresses that were sufficient to induce either surface diffusion or grain boundary diffusion were not evident.

In the case of samples B and C, provided that they grew, the nanowires that were generated became thicker under higher temperature. The Cu atoms could diffuse more actively at higher temperature. Therefore, a greater number of Cu atoms are accumulated at the interface of Cu_2O/Cu under higher temperature, and this causes higher values of compressive hydrostatic stress at the interface. Since the higher compressive hydrostatic stress expands the sizes of the weak spots in the oxide layers, the diameters of the nanowires become larger at higher temperatures. In this case, the coalescence of extra weak spots occurs, and the wire density, i.e., the number of wires per unit area, decreases with increasing temperature (Fig. 31c). It was found that the densities of the wires on samples B and C took their maximum values at about 300°C. This may be indicative of the coalescence of weak spots in the oxide layers.

Figure 31d shows the relationship between temperature and the volumes of nanowires per unit area after 5 h of heating, v [$=0.25\pi d^2 l\rho$], where d is the average diameter of the Cu nanowires, l is the average length and ρ is the wire density, respectively. The values of v increased with increasing time. Note that the values of v for sample B saturated with respect to temperature. This may be due to saturation of the grain boundary diffusion of atoms in the film-thickness direction at higher temperature. On the other hand, the values of v for sample C never saturated right up to 340°C, and the values were significantly larger than those for sample B, especially at higher temperature. These experimental observations suggest that the volume of nanowires per unit area is governed by temperature, film thickness, grain size and time.

The growth rate of spontaneous Sn whiskers has been summarized in a typical range of 0.01–0.1 Å s^{-1} [10]. That of the Cu nanowires was about 1.5–10 Å s^{-1} in the case of sample C at stage I; this is much faster than the former. Moreover, the incubation time for the Cu nanowires was found to be <270 s, which was significantly shorter than that of other types of wires/whiskers. Although some higher growth rates have been reported for Sn whiskers by using accelerated whisker growth methods [7, 8, 12], their diameters, on the scale of microns, prevented their application as nanomaterials. The rapid and mass formation of Cu nanowires by the growth mechanism presented here offers us a new direction for using metallic wires as nanomaterials.

3.2 Fabrication of Ag Micro Materials by Utilizing SM[2]

Ag micro materials are highly regarded as prospective materials for the modern electronics industry, especially in terms of their high electrical conductivities.

[2] The contents concerning this section have been permitted to reprint from Saka et al. [32].

Based on the successes shown in Sect. 3.1 in the rapid and mass formation of Cu nanowires by utilizing SM, attempts were made to fabricate Ag micro materials in a similar way. However, something different appeared, and a new challenge to the application of SM to these processes emerged.

In Sect. 3.1, the oxide layer was recognized as playing a significant role in rearranging atoms that were discharged due to SM into wire shapes. Unlike Cu, which is easily oxidized and which forms an oxide layer that can be maintained during the process, the oxide layer associated with Ag, Ag_2O, is well known to gradually decompose at temperatures higher than 160°C. For an Ag-included multilayer that is subjected to heating at such a temperature range, there is no path for the generation of Ag micro/nano materials in those cases where we cannot utilize restriction due to the oxide layer. With this in mind, the use of a synthetic passivation layer of SiO_2 was considered in order to provide a suitable restriction.

3.2.1 Effect of Temperature on the Formation of SM-Induced Ag Micro Materials

In studying the effect of temperature on the formation of SM-induced Ag micro materials, the usefulness of the synthetic passivation layer was verified.

A 100 nm thick SiO_2 layer covering the Ag film was used as the passivation layer in a SiO_2/Ag/Ti/SiO_2/Si multilayer structure, of which the layer ordering is from top to bottom (see the work by Saka et al. [32] for details of the sample preparation). At the same time, Ag/Ti/SiO_2/Si-structured samples without a SiO_2 passivation layer were also prepared. Both kinds of sample were heated under different temperatures of 260, 420 and 600°C for 3 h under atmospheric conditions.

Regarding the experimental results, first, the significance of the SiO_2 passivation layer was verified. All of the samples without the passivation layer showed no formation of micro/nano materials. This is attributed to the decomposition of the Ag_2O layer, as mentioned above. One of the samples that had undergone heating at 260°C for 3 h is shown in Fig. 32a. On the other hand, the samples with a 100 nm thick SiO_2 layer were found to show formations of nanowires and hillocks. Now, let us discuss the effect of temperature on the formation of SM-induced Ag micro materials. The SiO_2 passivation layers contain some relatively weak connections of grain boundaries and triple points, from which the accumulated Ag atoms can be discharged. When a sample was heated at relatively low temperature, i.e., 260°C, Ag nanowires of 50–100 nm in diameter and 3–5 μm in length were formed, as shown in Fig. 32b. In this case, lower compressive stress occurred at the SiO_2/Ag interface. The numbers of Ag atoms that were discharged from these paths were very low. Since the diameters of the paths were small, the discharged atoms were rearranged into thin wire shapes. Figure 32c shows examples of the round hillocks that were formed at an intermediate temperature, 420°C. Compared with the case of heating at 260°C, higher compressive stress is generated and hence induces an increase in the number of accumulated atoms for

Fig. 32 **a** FE-SEM micrograph of a sample without a SiO_2 layer after heating at 260°C for 3 h. FE-SEM micrographs of the samples with 100 nm thick SiO_2 layers after heating at different temperatures of **b** 260°C, **c** 420°C, and **d** 600°C, for 3 h (reprinted with permission from Saka et al. [32])

discharging. This resulted in the expansion of the paths in the SiO_2 to form large hillocks. At a high temperature of 600°C, larger hillocks were formed, as shown in Fig. 32d. In this case, the Ag atoms became more active and a large number of Ag atoms accumulated at the SiO_2/Ag interface to form higher compressive stress. Therefore, the paths were more extended to form larger hillocks.

3.2.2 Effect of Thickness of Passivation Layer on the Formation of SM-Induced Ag Micro Materials

Another set of experiments was carried out to study the effect of the thickness of the passivation layer on the formation of SM-induced Ag micro materials. A set of SiO_2/Ag/Ti/SiO_2/Si multilayer structures with SiO_2 passivation layers with respective thicknesses of 20, 100, 300 and 500 nm were prepared. All of the samples were heated at a constant temperature of 600°C under atmospheric conditions. Figure 33a–d shows the materials that were formed for the respective cases. Large hillocks were found to be formed on the sample with the 20 nm thick SiO_2 layer, as shown in Fig. 33a. Accumulated atoms were easily discharged from

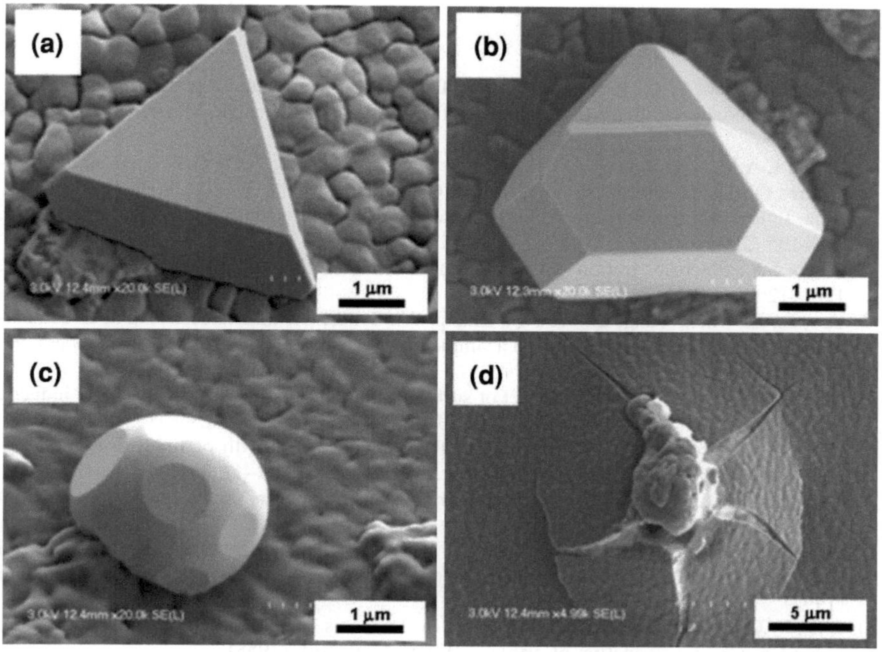

Fig. 33 FE-SEM micrographs of the samples with SiO_2 layers of different thickness after heating at 600°C, **a** 20 nm thick SiO_2 layer, **b** 100 nm thick SiO_2 layer, **c** 300 nm thick SiO_2 layer, and **d** 500 nm thick SiO_2 layer (reprinted with permission from Saka et al. [32])

the thin passivation layer. It was noted that the hillocks that were formed exhibit a clear triangular shape. This phenomenon seems to be caused by the recrystallization of hillocks at a high temperature. Samples with 100 nm thick SiO_2 layers showed a similar tendency, as shown in Fig. 33b. On the other hand, on samples with a 300 nm thick SiO_2 layer, the number of hillocks decreased and the shape of the hillocks tended to be rounded, as shown in Fig. 33c. Few hillocks were formed on samples with a 500 nm thick SiO_2 layer, and the SiO_2 layer was ruptured in places, as shown in Fig. 33d. It is recognized that as the SiO_2 layer becomes thicker, it is more difficult for the accumulated Ag atoms to discharge through the paths in the SiO_2 layer, and therefore the accumulated atoms broke through the SiO_2 layer.

On the other hand, when hillocks were formed all over the surface and recrystallization into clear shaped hillocks occurred, no more atoms were accumulated into the hillocks because large numbers of Ag atoms were discharged and the hydrostatic stress distribution disappeared. Therefore, the shapes of the hillocks did not change any further.

Acknowledgments M. S. wishes to express his thanks to X. Zhao for his kind help in preparing the manuscript.

References

1. Baker, S.P., Joo, Y.-C., Knauβ, M.P., Arzt, E.: Electromigration damage in mechanically deformed Al conductor lines: dislocations as fast diffusion paths. Acta. Mater. **48**, 2199–2208 (2000)
2. Barea, L.A.M., von Zuben, A.A.G., Márquez, A.Z., Frateschi, N.C.: GaN nano- and microspheres fabricated selectively on silicon. J. Cryst. Growth **308**, 37–40 (2007)
3. Bica, I.: Pore formation in iron micro-spheres by plasma procedure. Mater. Sci. Eng A **393**, 191–195 (2005)
4. Blech, I.A.: Electromigration in thin aluminum films on titanium nitride. J. Appl. Phys. **47**, 1203–1208 (1976)
5. Cao, L., Li, M.K., Yang, Z., Wei, Q., Zhang, W.: Synthesis and characterization of dentate-shaped β-Ga_2O_3 nano/microbelts via a simple method. Appl. Phys. A Mater. Sci. Process. **91**, 415–419 (2008)
6. Chang, C.Y., Vook, R.W.: The effect of surface aluminum oxide films on thermally induced hillock formation. Thin Solid Films **228**, 205–209 (1993)
7. Chen, K., Wilcox, G.D.: Observations of the spontaneous growth of tin whiskers on tin-manganese alloy electrodeposits. Phys. Rev. Lett. **94**, 066104(1–4) (2005)
8. Chuang, T.H., Lin, H.J., Chi, C.C.: Rapid growth of tin whiskers on the surface of Sn-6.6Lu alloy. Scr. Mater. **56**, 45–48 (2007)
9. Coffer, J.L., Bigham, S.R., Pinizzotto, R.F., Yang, H.: Characterization of quantum-confined CdS nanocrystallites stabilized by deoxyribonucleic acid (DNA). Nanotechnology **3**, 69–76 (1992)
10. Ellis, W.C., Gibbons, D.F., Treuting, R.C.: In: Doremus, R.H., Roberts, B.W., Turnbull, D. (eds.) Growth and perfection of crystals. Wiley, New York (1958)
11. Eshelby, J.D.: A tentative theory of metallic whisker growth. Phys. Rev. **91**, 755–756 (1953)
12. Furuta, N., Hamamura, K.: Growth mechanism of proper tin-whisker. Jpn. J. Appl. Phys. **8**, 1404–1410 (1969)
13. Hasegawa, M., Sasagawa, K., Uno, S., Saka, M., Abé, H.: Derivation of film characteristic constants of polycrystalline line for reliability evaluation against electromigration failure. Mech. Mater. **41**, 1090–1095 (2009)
14. Huntington, H.B., Grone, A.R.: Current-induced marker motion in gold wires. J. Phys. Chem. Solids **20**, 76–87 (1961)
15. Kononenko, O.V., Ivanov, E.D., Matveev, V.N., Khodos, I.I.: Electromigration activation energy in pure aluminum films deposited by partially ionized beam technique. Scr. Metall. Mater. **33**, 1981–1986 (1995)
16. Korhonen, M.A., Børgesen, P., Tu, K.N., Li, C.-Y.: Stress evolution due to electromigration in confined metal lines. J. Appl. Phys. **73**, 3790–3799 (1993)
17. Lau, J.H., Pan, S.H.: 3D nonlinear stress analysis of tin whisker initiation on lead-free components. Trans. ASME J. Electron. Packag. **125**, 621–629 (2003)
18. Lee, B.-Z., Lee, D.N.: Spontaneous growth mechanism of tin whiskers. Acta Mater. **46**, 3701–3714 (1998)
19. Lloyd, J.R., Smith, P.M., Prokop, G.S.: The role of metal and passivation defects in electromigration-induced damage in thin film conductors. Thin Solid Films **93**, 385–395 (1982)
20. Lu, Y., Saka, M.: Effect of purity on the fabrication of Al micro/thin-materials by utilizing electromigration. Mater. Lett. **63**, 2294–2296 (2009)
21. Lu, Y., Saka, M.: Fabrication of Al micro-belts by utilizing electromigration. Mater. Lett. **63**, 2227–2229 (2009)
22. Mayr, T., Moser, C., Klimant, I.: Luminescence decay time encoding of magnetic micro spheres for multiplexed analysis. Anal. Chim. Acta **597**, 137–144 (2007)
23. Miyamoto, H., Shiratori, M., Miyoshi, T., Oto, M.: Interpretation of mechanical properties of metals in terms of microstructure. Bull. JSME **14**, 893–900 (1971)

24. Pan, Z.W., Dai, Z.R., Wang, Z.L.: Nanobelts of semiconducting oxides. Science **291**, 1947–1949 (2001)
25. Park, Y.J., Thompson, C.V.: The effects of the stress dependence of atomic diffusivity on stress evolution due to electromigration. J. Appl. Phys. **82**, 4277–4281 (1997)
26. Qiu, J., Li, Y., Wang, Y., Liang, C., Wang, T., Wang, D.: A novel form of carbon micro-balls from coal. Carbon **41**, 767–772 (2003)
27. Sacilotti, M., Cheyssac, P., Patriarche, G., Decobert, J., Chiaramonte, Th., Cardoso, L.P., Pillis, M.F., Brasil, M.J., Iikawa, F., Nakaema, M., Lacroute, Y., Vial, J.C., Donatini, F.: Organometallic precursors as catalyst to grow three-dimensional micro/nanostructures: spheres, clusters & wires. Surf. Coat. Technol. **201**, 9104–9108 (2007)
28. Saka, M., Nakanishi, R.: Fabrication of Al thin wire by utilizing controlled accumulation of atoms due to electromigration. Mater. Lett. **60**, 2129–2131 (2006)
29. Saka, M., Ueda, R.: Formation of metallic nanowires by utilizing electromigration. J. Mater. Res. **20**, 2712–2718 (2005)
30. Saka, M., Yamaya, F., Tohmyoh, H.: Rapid and mass growth of stress-induced nanowhiskers on the surfaces of evaporated polycrystalline Cu films. Scr. Mater. **56**, 1031–1034 (2007)
31. Saka, M., Kato, K., Tohmyoh, H., Sun, Y.: Controlling electromigration to selectively form thin metal wires and metal microspheres. J. Mater. Res. **23**, 3122–3128 (2008)
32. Saka, M., Yasuda, M., Tohmyoh, H., Settsu, N.: Fabrication of Ag micromaterials by utilizing stress-induced migration. In: Proceedings of the 2nd Electronics Systemintegration Technology Conference, vol. 1, pp. 507–510 (2008)
33. Sasagawa, K., Nakamura, N., Saka, M., Abé, H.: A new approach to calculate atomic flux divergence by electromigration. Trans. ASME J. Electron. Packag. **120**, 360–366 (1998)
34. Sasagawa, K., Hasegawa, M., Saka, M., Abé, H.: Governing parameter for electromigration damage in the polycrystalline line covered with a passivation layer. J. Appl. Phys. **91**, 1882–1890 (2002)
35. Sasagawa, K., Hasegawa, M., Saka, M., Abé, H.: Prediction of electromigration failure in passivated polycrystalline line. J. Appl. Phys. **91**, 9005–9014 (2002)
36. Sasagawa, K., Hasegawa, M., Yoshida, N., Saka, M., Abé, H.: Prediction of electromigration failure in passivated polycrystalline line considering passivation thickness. In: Proceedings of the ASME InterPACK '03(CD-ROM): InterPack2003-35065 (2003)
37. Sasagawa, K., Fukushi, S., Sun, Y., Saka, M.: A numerical simulation of nanostructure formation utilizing electromigration. J. Electron. Mater. **38**, 2201–2206 (2009)
38. Sasagawa, K., Kirita, A., Fukushi, S., Saka, M.: Simulation of nanostructure production by electromigration considering specimen's shape. J. Nanosci. Nanotechnol. **10**, 6036–6040 (2010)
39. Sheng, G.T.T., Hu, C.F., Choi, W.J., Tu, K.N., Bong, Y.Y., Nguyen, L.: Tin whiskers studied by focused ion beam imaging and transmission electron microscopy. J. Appl. Phys. **92**, 64–69 (2002)
40. Shi, Y., Chen, J., Shen, P.: ZnS micro-spheres and flowers: chemically controlled synthesis and template use in fabricating MS(shell)/ZnS(core) and MS (M = Pb, Cu) hollow microspheres. J. Alloys Compd. **441**, 337–343 (2007)
41. Słówko, W., Prasol, H.: Micro-sphere plate as an electron detector at low vacuum. Vacuum **67**, 191–198 (2002)
42. Song, D., Yi, W.B.: Polymethyl methacrylate micro-spheres supported palladium: a new catalyst for Heck and Suzuki reactions. J. Mol. Catal. A Chem. **280**, 20–23 (2008)
43. Sun, Y., Tohmyoh, H., Saka, M.: Fabrication of Al microspheres by utilizing electromigration. J. Nanosci. Nanotechnol. **9**, 1972–1975 (2009)
44. Tan, C.M., Roy, A.: Investigation of the effect of temperature and stress gradients on accelerated EM test for Cu narrow interconnects. Thin Solid Films **504**, 288–293 (2006)
45. Tegtmeyer, D., Vorlaender, O., Zeyen, W., Tysoe, C., Silberkuhl, F.: New micro-sphere application in the leather industry. J. Am. Leather Chem. Assoc. **102**, 288–292 (2007)
46. Tu, K.N.: Irreversible processes of spontaneous whisker growth in bimetallic Cu-Sn thin-film reactions. Phys. Rev. B **49**, 2030–2034 (1994)

47. Tu, K.N.: Recent advances on electromigration in very-large-scale-integration of interconnects. J. Appl. Phys. **94**, 5451–5473 (2003)
48. Tu, K.N., Li, J.C.M.: Spontaneous whisker growth on lead-free solder finishes. Mater. Sci. Eng. A **409**, 131–139 (2005)
49. Wang, Q., Cao, F., Chen, Q., Chen, C.: Preparation of carbon micro-spheres by hydrothermal treatment of methylcellulose sol. Mater. Lett. **59**, 3738–3741 (2005)
50. Yang, J., Lin, H., He, Q., Ling, L., Zhu, C., Bai, F.: Composition of hyperbranched conjugated polymers with nanosized cadmium sulfide particles. Langmuir **17**, 5978–5983 (2001)
51. Zhang, X., Chen, Y., Jia, C., Zhou, Q., Su, Y., Peng, B., Yin, S., Xin, M.: Two-step solvothermal synthesis of α-MnS spheres: growth mechanism and characterization. Mater. Lett. **62**, 125–127 (2008)
52. Zhang, Z.Q., Jiang, C.B., Li, S.X., Mao, S.X.: Nucleation and growth of ZnO micro- and nano-belts during thermal evaporation. J. Cryst. Growth **277**, 321–329 (2005)

Evaluation of Mechanical Properties

Mikio Muraoka and Hironori Tohmyoh

Abstract Mechanical properties such as elastic modulus, fracture stress, and yield stress of nano/micromaterials are fundamental data for practical design of nano/micromaterial-based devices. These properties generally differ from those of bulk material because of size effects. This chapter is devoted to an introduction of some techniques for evaluating the mechanical properties of nanowires and thin wires. In order to clarify the advantages of the techniques that we introduce, the first section gives an overview of typical techniques reported so far. In the subsequent sections, we take up atomic force acoustic microscopy using a concentrated-mass cantilever and a bending method based on the geometrically nonlinear problem on the bent shape, i.e., *elastica*, for evaluating elastic modulus and bending strength of brittle nanowires. Finally, evaluation of elastic–plastic properties of metallic thin wires is demonstrated by means of unsymmetrical, small-span bending test.

1 Overview of Mechanical Tests

With miniaturization of a solid down to nanometer scale, surface and interfacial effects become dominant. Such effects have led to amazing changes of mechanical properties of a solid, such as elastic modulus and strength, as well as other physical and chemical properties. Increasing interests have also been paid to solids of

---

M. Muraoka (✉)
Department of Mechanical Engineering, Akita University, 1-1 Tegatagakuen-machi, Akita 010-8502, Japan
e-mail: muraoka@ipc.akita-u.ac.jp

H. Tohmyoh
Department of Nanomechanics, Tohoku University, Aoba 6-6-01, Aramaki, Aoba-ku, Sendai 980-8579, Japan
e-mail: tohmyoh@ism.mech.tohoku.ac.jp

M. Saka (ed.), *Metallic Micro and Nano Materials*, Engineering Materials,
DOI: 10.1007/978-3-642-15411-9_4,

sub-micrometer scale such as ultrathin wires of several hundreds of nanometers in diameter because they can be a key element that bridges from nanomaterial to micro/macroscopic systems. Their mechanical properties are important as design parameters, not only for mechanical applications such as nano-springs and nano-actuators, but also for the reliability of electrically and optically functional structures. In addition, long-term mechanical reliability may become a significant concern in the coming stages of practical nano/micromaterial-based engineering.

Several methods have been reported for the mechanical testing of nanowires (NWs), including nanotubes (NTs) and nanobelts (NBs), see Table 1 and Fig. 1. Most of these methods rely on the use of transmission electron microscopy (TEM), scanning electron microscopy (SEM), or atomic force microscopy (AFM). According to the testing principle, these methods are grouped into mechanical resonance, tensile load (TL) (pulling), bending (cantilever bending and three-point bending), string deflection, and nanoindentation. The mechanical resonance (MR) method, reported with excitation techniques such as electric-field-induced resonance and contact resonance, measures the resonant frequency, and thus a value of density is required to determine the Young's modulus. The MR method could be extended to a fatigue test, but no strength test has been reported so far.

Other methods using a static load, which enables measurement of the strength in addition to the elastic modulus, employ an AFM cantilever as a load sensor [78] successfully performed tensile testing of an individual multi-walled carbon nanotube (MWCNT) mounted between two opposing AFM tips under the observation by SEM (Fig. 2). In the case of static testing in SEM, an AFM cantilever is installed on a nano-manipulator integrated in the SEM, and the cantilever deflection is measured from the SEM image. For the TL and string deflection (SD) methods, where an NW is subject to large tension, mechanical fixing of the NW to the AFM tip or a substrate is a critical technique, achieved by means of electron-beam-induced deposition or shadow-mask deposition of joining materials. The cantilever bend (CB) method, which normally uses an NW vertically grown or laterally pinned on a substrate, acquires force–deflection data to determine Young's modulus. For the laterally pinned NWs, friction forces between the NWs and the substrate have to be considered [73]. The bending strength is evaluated from the curvature of the NW at fracture. In the three-point bend (TPB) test, a conventional AFM provides surface imaging and acquisition of force–deflection data are applied to an NW suspended on a pore or a trench patterned in a substrate. Fracture testing by TPB using a lateral load requires the fixing of both ends to avoid sample movement [74].

All of the CB and TPB methods restrict the deflection to small values, because they rely on the linear theory of beam bending [62], see the linear relations between the load P and deflection δ in Fig. 1b–d. Therefore, it is difficult to apply these methods to NWs thinner than approximately 100 nm. Significant reduction in the flexural (bending) rigidity, proportional to a biquadrate of the diameter, lowers the allowable load for the linear theory to the critical level of AFM performance. For the TPB method, the frictional sliding of the supported end or reactive tension may be encountered. In addition, attractive interactions between

Table 1 Mechanical tests of inorganic nanowires for Young's modulus E and fracture strength σ_s

References	View	Method	Material	E (GPa)	σ_s (GPa)
[68]	TEM	TV	MWCNT (ϕ 6–25 nm)	400–3,700	–
[44]	TEM	MR	MWCNT (ϕ 8–40 nm)	100–1,000[b]	–
[14]	TEM	MR	Bumboo CNT (ϕ 33–64 nm)	23–32	–
[4]	TEM	MR	ZnO[0001] NB (t 20–40 nm)	40–65	–
[40]	TEM	MR	GaN[120] (ϕ 36–84 nm)	230-300[c]	–
[12]	SEM	MR	ZnO[0001] (ϕ 17–550 nm)	140–220[b]	–
78]	SEM	TL	MWCNT (ϕ 13–36 nm)	270–950	11–63
[2]	SEM	CB, BK	MWCNT (ϕ 26 nm)	460–500	–
[17]	SEM	CB	Si[111] (ϕ 100–200 nm)	-	12
[43]	SEM	CB	a-C (ϕ 180–540 nm)	17–84	–
[18]	SEM	CB, TL	ZnO[0001] (ϕ 100–140 nm)	72–130	3.7–5.5
[73]	AFM	CB[a]	SiC[111] (ϕ 21.5–23 nm)	610–660	53.4
			MWCNT (ϕ 26–76 nm)	1,280 ± 590	14.2 ± 8.0
[42]	AFM	CB	Cr beam (t 50–83 nm)	40–75[c]	–
[58]	AFM	CB	ZnO[0001] (ϕ 45 nm)	29 ± 8	–
[51]	AFM	TPB	SWCNT rope (ϕ 3–20 nm)	67-1,310[b]	–
[52]	AFM	TPB	MWCNT (ϕ 10–13 nm)	870	–
[41]	AFM	TPB	Si[110] beam (t 255 nm)	170	12-17.5[f]
[21]	AFM	TPB	MoS_2 NT rope (ϕ 10-20 nm)	120	–
[74]	AFM	TPB	Au (ϕ 40-250 nm)	70 ± 11	2.4–5.6 [d]
[75]	AFM	TPB	ZnS[001] NB (t 50–200 nm)	52 ± 7	–
[61]	AFM	TPB	CuO [$\bar{1}$1l](ϕ 80-215 nm)	70–300[b]	–
[70]	AFM	SD	SWCNT rope	–	45 ± 7
[67]	AFM	SD	SWCNT (ϕ 3 nm)	1,200	–
[24]	AFM	NI	Ag twin[110][111] (ϕ 42 nm)	88	0.87[e]
[26]	AFM	NI	ZnS[001] NB (t 50–100 nm)	56 ± 3.5	3.4 ± 0.2[e]
[13]	AFM	MR + CR	Ag (ϕ 40–135 nm)	65–140[b]	–
			Pb (ϕ 30–270 nm)	15-30[b]	–
[25]	OM	MR + PL	Si[110] beam (t 12–300 nm)	53–167[c]	–
[39]	OM	LB	CuO[$\bar{1}$1 1] (ϕ 40–190 nm)	150–270[b]	3–13[b]
[66]	OM	SSB	Pt (ϕ 615–770 nm)	159 ± 38	0.4–0.7 [d]

TEM transmission electron microscopy, *SEM* scanning electron microscopy, *AFM* atomic force microscopy, *OM* optical microscopy, *TV* thermal vibration, *MR* mechanical resonance, *TL* tensile load, *CB* cantilever bend, *BK* buckling, *TPB* three-point bend, *SD* string deflection, *NI* nanoindentation, *CR* contact resonance, *PL* pulsed laser, *LB* large bending, *SSB* small-span bending, *a-C* amorphous carbon, *t* thickness

[a] NWs were laid on a substrate and pinned at one end to the surface
[b] Increases with decreasing diameters
[c] Decreases with decreasing the diameter or thickness
[d] Yield stress, larger than bulk values
[e] Hardness
[f] Increases with decreasing the width

the NW and AFM cantilever easily induce large deflections in the CB method [43]. When the TPB method is applied to a carbon nanotube (CNT) of a few nanometers in diameter, the linear theory of beam bending is ruled out. The nanotube behaves

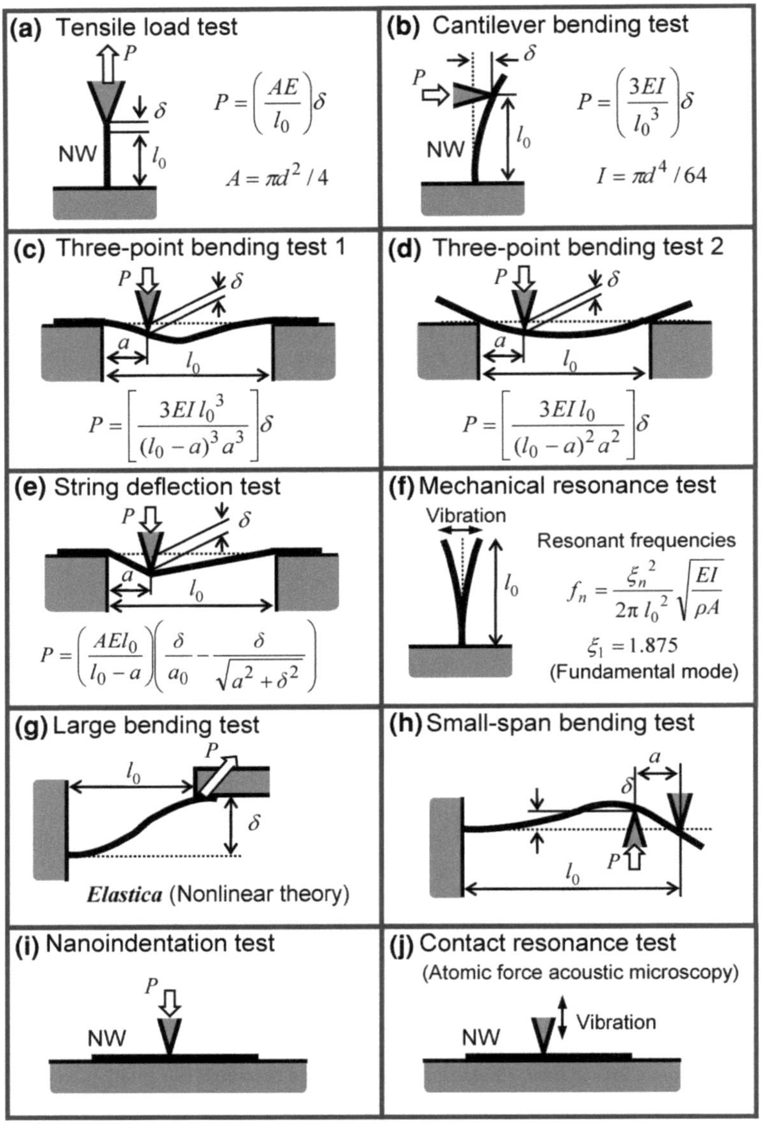

Fig. 1 Schematics of NW tests for mechanical properties, where P is the applied load, δ the deformation, E the Young's modulus, A the sectional area of the NW, I the inertia moment of the cross-section, d the diameter, ρ the density, and $a_0 = a\ (l_{slack}/l_0)$ (l_{slack}: the length of an initially slack NW). The vertical applied load in the test (**c**) can be replaced by a lateral load. Both ends of the NW are not necessarily clamped when P is small

as an elastic string without bending rigidity, where the string deflection obeys simple mechanics under tension, as adopted in the SD method. There still seems to be a lack of modulus and strength measurements for intermediate NWs of tens of

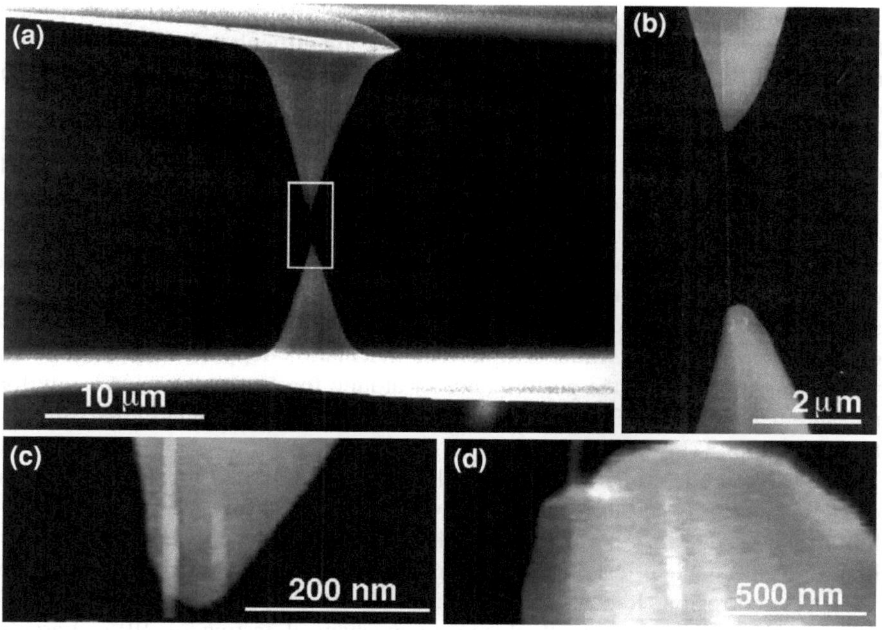

Fig. 2 An individual MWCNT mounted between two opposing AFM tips. **a** An SEM image of two AFM tips holding a MWCNT, which is attached at both ends on the AFM silicon tip surface by electron beam deposition of carbonaceous material. The lower AFM tip in the image is on a soft cantilever, the deflection of which is imaged to determine the applied force on the MWCNT. The top AFM tip is on a rigid cantilever that is driven upward to apply tensile load to the MWCNT. **b** High-magnification SEM image of the indicated region in **a**, showing the MWCNT between the AFM tips. **c** Higher magnification SEM image showing the attachment of the MWCNT on the top AFM tip. There is an apparent thickening of the MWCNT section on the surface. **d** Close-up SEM image showing the attachment of the MWCNT on the lower AFM tip. The MWCNT section is covered by a square-shaped carbonaceous deposit. From [78]. Reprinted with permission from AAAS

nanometers in diameter between the SD and TPB or CB methods. In addition, it is often difficult to establish whether the measured failure is due to the NW itself or the contact between the NW and AFM tip or substrate.

It is also noticed from Table 1 that there is a great lack of evaluating metallic NWs, especially for the yield stress and the post-yielding data, e.g., work hardening modulus. The other mechanical properties such as on fatigue and creep are important, but have not been investigated, as far as we know. In general, plasticity of metals is sensitive to their microstructure. The so-called Hall–Petch equation states that the yield stress increases with decreasing the grain size. However, it holds true for bulk metals with grain sizes ranging from 100 nm to 10 μm [48]. Therefore, polycrystalline metallic solids of micrometer scale are still of concern as well as of nanometer scale.

Mechanical properties obtained by most of the test methods (Fig. 1a–h) are those in the axial direction of NWs. Further, they are difficult to be applied to other types of nano/micromaterials such as nanoballs and nanorods. The nanoindentation (NI) method [24] may be applicable to them, as well as NWs as shown in Fig. 1i. Results obtained by the NI are Young's modulus and hardness in the radial direction of NWs. The NI method measures a relation between the indentation load and depth, and requires an indentation depth of 10 nm at least for reliable data. As a result, the diameters of applicable NWs are limited to larger than 40 nm. Furthermore, it is difficult to extract the yield stress from the measured hardness. With regard to Young's modulus, there is more reliable technique than NI. This is the so-called atomic force acoustic microscopy [46], which detects resonant vibration of a sharp tip in contact with a sample. The indentation depth is 1 nm or less. We will discuss, in Sect. 2, the performance of atomic force acoustic microscopy for measurements of the elastic modulus, together with our enhanced-sensitivity technique [36, 37].

The use of optical microscopy (OM) has not been reported for the observation of NWs, due to its short resolution. However, it is worth investigating the feasibility of OM, due to ease of handling, because this could provide a method for long-term mechanical tests under various environments, including fatigue, creep and stress-corrosion. Recently, Muraoka and Tobe [39] applied the optical diffraction phenomenon of small objects to mechanical testing of NWs. Largely bended elastic NWs are analytically described with some elliptic integrals and functions. The formula of the bent shape, i.e., the so-called *elastica*, was applied to the measurements. In Sect. 3, we will introduce the method based on the large bending (LB) test.

The small-span bending (SSB) test proposed by Tohmyoh et al. [66] is a solution to eliminate effects of the supported ends of samples on the measured strength. They first measured the stress–strain curves of the elastic–plastic region of ultrathin metallic wires, whose diameters ranges from 600 to 800 nm. The final section (Sect. 4) deals with the SSB test.

2 Contact Resonance of Concentrated-Mass Cantilevers

2.1 Short History of Atomic Force Microscopy and Dynamic Modes

The invention of atomic force microscopy (AFM) [5] has raised the hope to extend the atomic resolution capability of scanning tunneling microscopy (STM) [6] to nonconducting surfaces. In AFM, the imaging signal is the force F between a sharp tip and a surface (typically $F = 0.1$ nN) instead of a tunneling current ($I = 0.1$ nA) in STM. The conversion of F into an electrical signal is usually done by measuring the deflection of an elastic micro-cantilever (CL). A topographic image is created by scanning in the x–y plane and keeping F constant by controlling z, where the coordinate z of the Cartesian coordinate system xyz is taken normal to the surface with its origin on a point of the surface. This operation mode is called as contact mode. However, obtaining true atomic resolution by AFM in the contact mode is much more difficult than by STM mainly for the reasons that the conversion of F is more prone to experimental drift and error, and it is difficult to eliminate deformation or damage of the atomic scale due to adhesion of the tip. It was not until the birth of the large-amplitude frequency-modulation (LA-FM) dynamic mode [15] that one achieved true atomic resolution by AFM.

AFM is nothing else than a mechanical system. The most CLs are fabricated from single-crystalline silicon or silicon nitride using photolithographic techniques with integrated sharp tips. Typical lateral-dimensions are on the order of 100 μm, with thickness of order 1 μm. This geometry gives spring constants k_c in the range of 0.01–100 N/m and fundamental resonant frequencies f_1 of 10–500 kHz. These high values of f_1 ensure the immunity of the CLs to external vibrations, such as from the laboratory building. In most of commercially available AFMs, a laser-beam deflection scheme is installed for the detection of CL deflections (see Fig. 3). It is capable of measuring them on the order of 0.01 nm, which provides the force resolution in the range of 0.1 pN–1 nN depending on the value k_c. Light beam from a diode laser is reflected from a mirror-like CL back-surface. The direction of the reflected light beam is sensed with a position-sensitive photodetector (PSD). This scheme is based on the measurement of the CL deflection angle rather than the deflection itself.

While the contact mode AFM, which usually uses soft CLs (typically $k_c = 0.1$ N/m), is difficult to obtain true atomic resolution as mentioned, it is widely used as a nanometer-resolution stylus profilometer. A typical deflection is 1 nm for an exerted force of $F = 0.1$ nN. The contact mode has also triggered

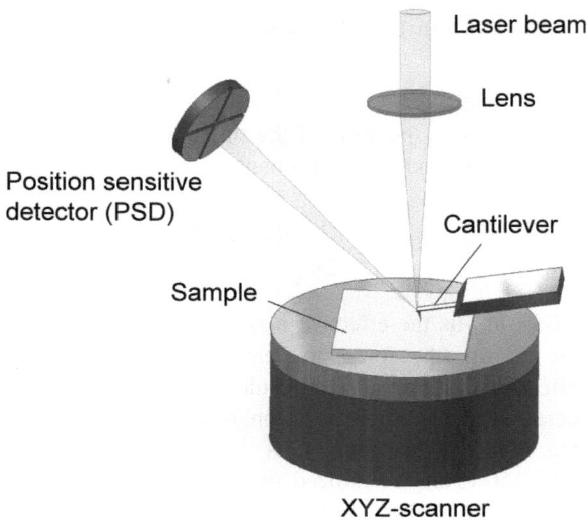

Fig. 3 Schematic diagram of typical AFM. Laser light is focused on a cantilever that reflects it onto a segmented photodiode, i.e., position sensitive detector. The photodiode senses the deflection of the reflected beam and thus the deflection of the cantilever. In operation a feedback loop controls the vertical position of the sample and hence the force exerted on it by the cantilever. This is accomplished by moving the sample up and down as it is scanned

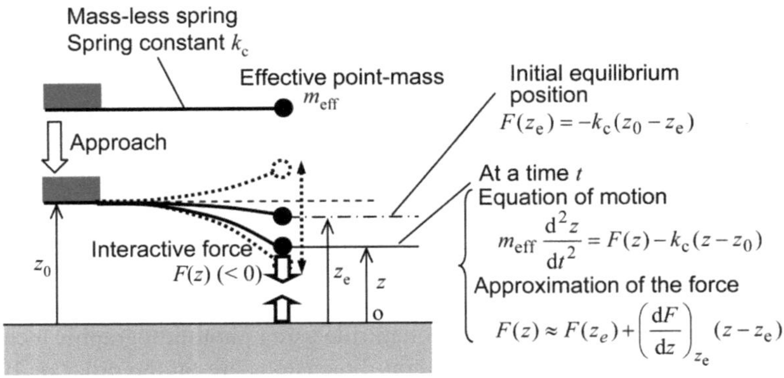

Fig. 4 A point-mass model of free vibrations of AFM cantilevers under the interactive force $F(z)$. Taking the approximation of $F(z)$ for small amplitudes, one sees that the effective spring constant is modified by the force gradient. The natural frequency agrees with the resonant frequency under forced vibrations for cantilevers of high quality-factors

emergence of various modes such as friction-force (or lateral-force) mode [32] and scratch (static plowing) mode [30]. They are all operated under a controlled contact force. By moving the tip away from the sample by 10–100 nm, long range force, such as magnetic, electrostatic and attractive van der Waals forces, become accessible. This type of noncontact (NC) imaging is usually performed using a different method of force detection. Instead of measuring static CL deflections, the CL is driven to vibrate near its resonant frequency by a small piezoelectric element [5]. When the deflection amplitude is small enough for linear vibration, the presence of a force F (negative for attractive force), or more precisely the gradient dF/dz at a tip-sample separation of $z = z_e$, acts to modify the effective spring constant k_{eff} of the CL according to:

$$k_{eff} = k_c - (dF/dz)_{z_e}, \tag{1}$$

where z_e is a z-position of the tip at an initial equilibrium (see Fig. 4). By using a point-mass model, the resonant frequency of the CL is simply given by:

$$f = \frac{1}{2\pi}\sqrt{\frac{k_c - (dF/dz)_{z_e}}{m_{eff}}} = f_1\sqrt{1 - \frac{(dF/dz)_{z_e}}{k_c}}, \tag{2}$$

where m_{eff} is the effective mass of the CL. If the sample exerts an attractive force on the tip (the force gradient is positive, see Fig. 5), the spring constant will effectively soften. As a result, the resonant frequency will decrease, and then detected by measuring the amplitude, phase or frequency change of the vibration. In the NC mode, stiff CLs with high frequencies (typically $k_c = 10$ N/m, $f_1 = 150$ kHz) are utilized to gain the detection sensitivity. It is rare for the NC mode to be applied to topographic imaging, but magnetic-force or electrostatic-force

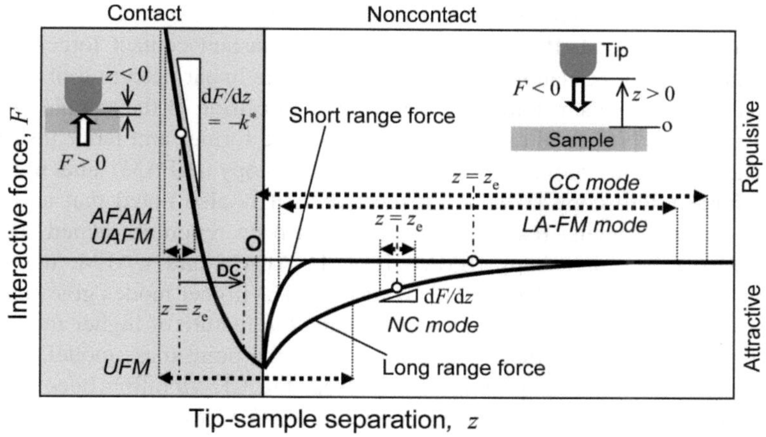

Fig. 5 Schematics of a relation between the interactive force F and the tip-sample separation z (*solid curve*) and of the z-rages (from peak to peak) of the tip vibration (*dotted lines with arrows*) for several dynamic modes based on AFM (noncontact (NC) mode, cyclic-contact (CC) mode, large-amplitude frequency-modulation (LA-FM) mode, atomic force acoustic microscopy (AFAM), ultrasonic atomic force microscopy (UAFM), and ultrasonic force microscopy (UFM)). The *circles* denote the initial equilibrium states. Symbol "DC" means a DC component induced by a nonlinear vibration in UFM (mechanical diode effect)

imaging (e.g., [31, 59]) have been demonstrated with the tips coated with a conducting or magnetic film, respectively.

When increasing the amplitude of the CL, the tip becomes to feel a nonlinear force, i.e., the force gradient varies during a cycle. In this case, Eqs. (1) and (2) are not applicable, but the high-energy vibrations provide two types of stable dynamic modes for nondestructive imaging of surface profiles: the cyclic-contact (CC) mode (or tapping mode) [80] and the LA-FM dynamic mode [15]. The CC mode significantly reduces the frictional force during a scan as compared with the contact mode and is appropriate for delicate soft samples such as biological molecules on a mica surface. In the LA-FM dynamic mode, the tip is precisely in noncontact with the sample, and it periodically approaches close vicinity of the surface atoms to sense the short range force such as covalent bonding force between the sample and tip atoms. It allows one to achieve true atomic resolution in ultrahigh vacuum and liquid conditions.

For all the dynamic modes (NC, CC, and LA-FM modes), the tip is positioned initially (before driving vibration) away from a sample surface ($z_e > 0$). If the tip is initially in contact with the sample ($z_e < 0$) and then driven to vibrate under the contact-mode operation, what will happen? When the amplitude of the tip is small enough for linear vibration, one obtains a counterpart of the NC mode. The resonant frequency is still given by Eq. (2), where the force gradient takes a negative large value. The force gradient significantly increases f. The absolute value of the force gradient is called as contact stiffness, and denoted by $k^* = -(\mathrm{d}F/\mathrm{d}z)_{ze}$. The contact stiffness depends on the elastic modulus of the sample, while k^* is also influenced by the geometry and elastic modulus of the tip, the contact force, etc. as discussed in Sect. 2.4. By measuring f under a constant contact force, one can detect elasticity on the nanometer scale. In 1996, the linear detection of elasticity was first achieved by sinusoidally modulating the position of the sample [46] and of the cantilever [76] at ultrasonic frequencies. The former and latter instrumentations are called as atomic force acoustic microcopy (AFAM) and ultrasonic atomic force microscopy (UAFM), respectively. It is also noted that a scale-up version, i.e., micrometer-resolution version, had been reported, named scanning microdeformation microscopy (SMM) [69]. In AFAM and UAFM, the use of higher-order vibration modes is a key to success. The higher modes give enhanced sensitivity as in a NC mode [19]. To understand the feature of higher modes, one needs the beam vibration theory [63], apart from the point-mass model.

When the excitation amplitude is increased, the CL vibration becomes nonlinear. The nonlinear phenomena in AFAM and UAFM were investigated experimentally and analytically [38, 47, 77]. They suggested that the cyclic jumping of the tips near the contact resonance could be useful to experimentally determine the attractive force. Meanwhile, different detection-techniques of the nonlinear vibrations were developed by using high excitation-frequencies away from the contact resonant-frequencies of the CL: ultrasonic force microscopy (UFM) [22], a nonlinear-version of AFAM [45], and scanning local-acceleration microscopy (SLAM) [10]. The mechanical diode (MD) mode in UFM and the subharmonic (SH) mode in SLAM were proposed to obtain an image reflecting the elasticity. If the excitation frequency is set much lower than the contact resonance, the inertia effect on the CL vibration is reduced and it becomes quasi-static. This quasi-static situation has been applied to image surface elasticity of soft samples [29]. It is the so-called force modulation microscopy (FMM).

In AFAM, the use of higher modes for the enhanced sensitivity requires detection techniques of higher frequencies of cantilever vibrations. Recently, Muraoka [36, 37] proposed an alternative method of the enhancement without high-frequency detection techniques. The novel technique uses a modified cantilever, i.e., concentrated-mass (CM) cantilever. In the rest of Sect. 2, the CM cantilever technique is introduced together with some basics of vibration dynamics.

2.2 Dynamics of Bending

2.2.1 Equations of Bending Motion of Inextensional Thin Rods

Regardless of presence of the nonlinear interactions between a tip and a surface, flexural vibrations of the CLs themselves are treated by means of the linear vibration theory of beams [53, 63]. All of instructive books on AFM start with the linear theory. This section does not take the ordinary way, but introduces general equations of motions taking a geometrical nonlinearity into account, i.e., a large-deflection theory. Then it is reduced into the linear theory. The nonlinear equations will be dealt with again in Sect. 3.2 for a static problem.

Thin rods (including strips such as long rectangular beam) treated here is not extensional but flexible. Consider a thin rod dynamically deforming in the z–x plane, which initially lied on the x axis, as shown in Fig. 6. The rod is supported by dynamic forces and couples (bending moments) at both of the ends, i.e., the outer surface is not subjected to forces or moments along the rod. Let s denote the curvilinear coordinate that is embedded in the rod along the center axis. The tangential angle θ of the curvilinear coordinate s is measured from the x axis, counterclockwise for an observer looking in the direction y. Likewise the relative angle ψ of the coordinate s is taken from the direction of the internal force P that acts at the position s on the infinitesimal segment ds. There is a transverse component Q of P but the shear deformation of the segment is neglected according to the ordinary assumption for thin rods [28]. Then, the segment deforms as an arc, i.e., the cross-sections are kept to be normal to the coordinate s after the deformation. The normal strain ε of a layer parallel to the arc ds at a position η_n, is expressed by the equation $\varepsilon = \eta_n/R = -\eta_n(\mathrm{d}\theta/\mathrm{d}s)$, where η_n is the liner coordinate taken form the neutral surface, i.e., the cylindrical surface containing the arc ds and normal to the z–x plane. The stress σ normal to the cross-section obeys Hooke's law $\sigma = E\varepsilon$, where E is Young's modulus, and produces the couple about

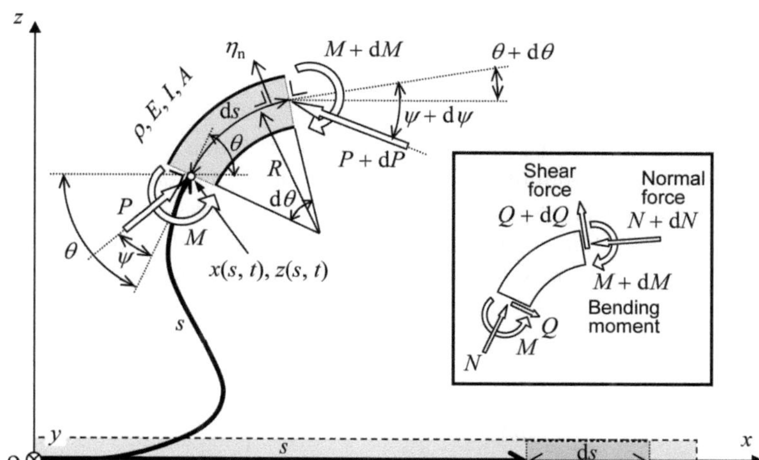

Fig. 6 Positive directions of internal forces and moments that act on a segment of length ds in a dynamically-bending thin rod that initially lied on the x axis. The curvilinear coordinate s is embedded along the central axis of the rod that is not extensional, and it deforms along with the rod in the z–x plane. The tangential angle θ and relative angle ψ of the rod (the coordinate s) are measured from the x axis and from the direction of the internal force P, respectively. These angles are defined positive for those in this figure (see the *left* cross-section). The inset describes the positive direction of the normal and transverse components of P. Note that the normal forces is defined positive for compression

the neutral axis. The condition that the couple is equal to the bending moment M gives Bernoulli–Euler's formula:

$$M = -EI\frac{\partial \theta}{\partial s}, \tag{3}$$

where I is the cross-sectional moment of inertia defined as the surface integral of η_{n}^2 over the cross section. EI is called the flexural rigidity.

The internal force $P = P(s, t)$, moment $M = M(s, t)$, angles $\theta(s, t)$ and $\psi(s, t)$, and coordinates $x = x(s, t)$ and $z = z(s, t)$ of a material point along the central axis of the rod are functions of s and time t. The normal and transverse (shear) forces are expressed by $N = P \cos\psi$ and $Q = P \sin\psi$, respectively. By applying Newton's law of motion to the segment ds, the equations of motion are determined as

$$\left.\begin{aligned} \rho A\frac{\partial^2 x}{\partial t^2} &= -\frac{\partial}{\partial s}[P\cos(\theta - \psi)] \\ \rho A\frac{\partial^2 z}{\partial t^2} &= -\frac{\partial}{\partial s}[P\sin(\theta - \psi)] \\ \rho I\frac{\partial^2 \theta}{\partial t^2} &= -\frac{\partial M}{\partial s} + Q = \frac{\partial}{\partial s}\left(EI\frac{\partial \theta}{\partial s}\right) + P\sin\psi \end{aligned}\right\}, \tag{4}$$

where ρ and A are the density and cross-sectional area, respectively.

2.2.2 Linear Theory of Beam Vibration

When the deflections of the rods (the displacements in the z direction) are small, the normal force N is negligible even in cases of supporting the rods at both of the ends. The condition of no normal forces, i.e., $N = P \cos\psi = 0$, gives $\psi = \pi/2$. Taking small deflection-angles $\theta \ll 1$ ($\sin\theta \approx \theta$ and $\cos\theta \approx 1$) together with $\psi = \pi/2$ and $P = Q$ in Eqs. (4), the equations of motion become

$$\left.\begin{aligned} \rho A\frac{\partial^2 x}{\partial t^2} &= -\frac{\partial (Q\theta)}{\partial s} \\ \rho A\frac{\partial^2 z}{\partial t^2} &= \frac{\partial Q}{\partial s} \\ \rho I\frac{\partial^2 \theta}{\partial t^2} &= \frac{\partial}{\partial s}\left(EI\frac{\partial \theta}{\partial s}\right) + Q \end{aligned}\right\}. \tag{5}$$

Eliminating Q in the second and third equations of Eqs. (5), we obtain

$$\rho A\frac{\partial^2 z}{\partial t^2} + \frac{\partial^2}{\partial s^2}\left(EI\frac{\partial \theta}{\partial s}\right) - \frac{\partial}{\partial s}\left(\rho I\frac{\partial^2 \theta}{\partial t^2}\right) = 0. \tag{6}$$

Further, taking approximations: $s \approx x$, $\theta \approx \partial z/\partial x$, and $\partial\theta/\partial s = (\partial^2 z/\partial x^2)/[1 + (\partial z/\partial x)^2]^{3/2} \approx \partial^2 z/\partial x^2$, the linear equation of motion is written as

$$\rho A\frac{\partial^2 z}{\partial t^2} + \frac{\partial^2}{\partial x^2}\left(EI\frac{\partial^2 z}{\partial x^2}\right) - \frac{\partial}{\partial x}\left[\rho I\frac{\partial^2}{\partial t^2}\left(\frac{\partial z}{\partial x}\right)\right] = 0. \tag{7}$$

The last term of Eq. (7) represents the effect of rotary inertia. The effect becomes important with an increase in the frequency of vibration as well as that of the shear deformation.

Neglecting the effect of rotary inertia and assuming EI is a constant in Eq. (7) yields the equation of motion of beams:

$$\rho A\frac{\partial^2 z}{\partial t^2} + EI\frac{\partial^4 z}{\partial x^4} = 0, \tag{8}$$

whose solution for stationary waves is

$$z(x,t) = Z(x)\cos(\omega_n t + \beta_0), \tag{9}$$

where ω_n is the angular frequency, n is the order of the mode, and β_0 is the phase delay. The function $Z(x)$ defines the shape of the natural mode of vibration and is called a mode function, a principal function, or a normal function. Substitution of Eq. (9) into Eq. (8) results in

$$\frac{\mathrm{d}^4 Z(x)}{\mathrm{d}x^4} - \kappa_n^4 Z(x) = 0, \tag{10}$$

where a parameter κ_n ($n = 1, 2, \ldots$) is defined as

$$\kappa_n^4 = \left(\frac{\rho A}{EI}\right)\omega_n^2. \tag{11}$$

The solution of Eq. (10) has the form

$$\left.\begin{aligned} Z(x) &= A_1\sin\kappa_n x + A_2\cos\kappa_n x + A_3\sinh\kappa_n x + A_4\cosh\kappa_n x \\ &= B_1(\cos\kappa_n x + \cosh\kappa_n x) + B_2(\cos\kappa_n x - \cosh\kappa_n x) \\ &\quad + B_3(\sin\kappa_n x + \sinh\kappa_n x) + B_4(\sin\kappa_n x - \sinh\kappa_n x) \end{aligned}\right\}. \tag{12}$$

The coefficients A_i and B_i ($i = 1$–4) are determined by the boundary conditions. The deflection, slope (deflection angle), moment, and shear force of the deformed beam are proportional to z, $\partial z/\partial x$, $\partial^2 z/\partial x^2$, and $\partial^3 z/\partial x^3$ along with neglect of the rotary inertia), respectively. With regard to the second and third order partial-derivatives, Eq. (3) and the third equation of Eqs. (5) explain the proportionality together with neglect of the rotary inertia and with the approximations $s \approx x$, $\theta \approx \partial z/\partial x$, and $\partial\theta/\partial s \approx \partial^2 z/\partial x^2$. The boundary conditions for a cantilever give $z = \mathrm{d}z/\mathrm{d}x = 0$ at $x = 0$ so that $B_1 = B_3 = 0$. At the free end ($z = l$) of the cantilever of length l, we have $\partial^2 z/\partial x^2 = \partial^3 z/\partial x^3 = 0$, which yields two linear equations on the unknown coefficients B_2 and B_4. The condition of the nonzero values of B_2 and B_4 gives

$$\frac{B_4}{B_2} = -\frac{\cos\xi_n + \cosh\xi_n}{\sin\xi_n + \sinh\xi_n} = \frac{\sin\xi_n - \sinh\xi_n}{\cos\xi_n + \cosh\xi_n}, \tag{13}$$

where $\xi_n = \kappa_n l$. The second equation of Eqs. (13) reduces to the simple equation

$$\cos\xi_n \cosh\xi_n + 1 = 0. \tag{14}$$

Equation (14) is called a frequency equation and only numerically solved. The first five solutions of Eq. (14) and the corresponding values of B_4/B_2, denoted by α_n, are

$$\xi_n = \kappa_n l = 1.875, \quad 4.694, \quad 7.855, \quad 10.996, \quad 14.137 \tag{15}$$

and

$$\alpha_n = B_4/B_2 = -0.7341, \quad -1.0185, \quad -0.9992, \quad -1.0000, \quad -1.0000. \tag{16}$$

For higher values one can use the approximations $\xi_n \approx (2n-1)\pi/2$ and $\alpha_n \approx -1$. The natural frequencies (eigenfrequencies) of vibration are therefore given by

$$f_n = \frac{\omega_n}{2\pi} = \frac{\xi_n^2}{2\pi l^2}\sqrt{\frac{EI}{\rho A}}. \tag{17}$$

The deflection $z(x, t)$ for each mode ($n = 1, 2, 3,\ldots$) can be calculated by using ξ_n and α_n, yielding

$$z(x,t) = A_0[(\cos\xi_n\bar{x} - \cosh\xi_n\bar{x}) + \alpha_n(\sin\xi_n\bar{x} - \sinh\xi_n\bar{x})]\cos(\omega_n t + \beta_0), \tag{18}$$

where $\bar{x} = x/l$, and A_0 and β_0 are determined by the amplitude and phase angle of vibration. Figure 7 shows the first three modes of vibration of a cantilever.

Neglecting the inertia term of Eq. (8), we obtain the equilibrium equation $\mathrm{d}^4z/\mathrm{d}x^4 = 0$, whose solution is a third-order polynomial of x. When the free end of the cantilever is subjected to a shear force $Q = F$, we have the conditions $\mathrm{d}^2z/\mathrm{d}x^2 = 0$

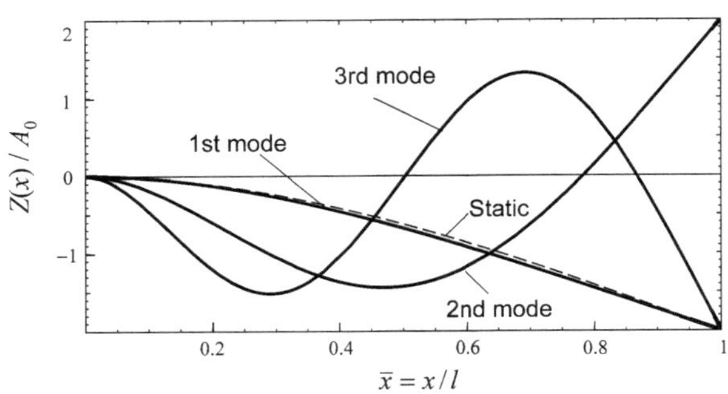

Fig. 7 The first three modes of vibration of a cantilever (*solid curves*) and the static deflection curve (*broken curve*). The vertical axis is $Z(x)/A_0$ for the vibration modes, see Eq. (18), and $-2k_c z(x)/F$ for the static deflection, see Eq. (19)

and $d^3z/dx^3 = -F/(EI)$ (see the third equation of Eqs. (5) along with neglect of the rotary inertia) at $x = l$ together with $z = dz/dx = 0$ at $x = 0$. Using these conditions in the polynomial form yields the deflection curve $z(x)$:

$$z(x) = x^2(3l - x)F/(6EI). \tag{19}$$

Calculating $F/z(l)$ from Eq. (19), we obtain the spring constant k_c:

$$k_c = 3EI/l^3. \tag{20}$$

Cross-section of AFM cantilevers (CLs) is normally a rectangular of height (thickness) h_c and width b_c ($\gg h_c$). For the CLs, the Young's modulus (E) is replaced by the biaxial modulus $E/(1 - \nu)$, where ν is the Poisson's ratio. The cross-sectional moment of inertia is expressed by $I = b_c h_c^3/12$. The frequencies Eq. (17) is rewritten by using Eq. (20):

$$f_n = \frac{1}{2\pi}\sqrt{\frac{k_c}{(3/\xi_n^4)m_{can}}}, \tag{21}$$

where m_{can} is the distributed mass ρAl of the cantilever. The denominator $(3/\xi_n^4)m_{can}$ is the effective mass m_{eff} of the cantilever, giving $m_{eff} \approx 0.24\, m_{can}$ for the fundamental mode ($n = 1$).

2.3 Dynamics of Concentrated-Mass Cantilevers

Atomic force acoustic microscopy (AFAM) provides nanometer resolution images reflecting sample elasticity and a possible technique for measuring elastic modulus of thin films and extremely narrow areas. In AFAM, the resonant frequency of a micro-cantilever equipped with a sensor tip measures the contact stiffness between tip and sample. In the case of stiff samples like metals and ceramics, AFAM exhibits significantly low sensitivity, i.e., the resonant frequency is insensitive to the contact stiffness. Muraoka [36] developed a special cantilever with its mass concentrated as a way of enhancing the sensitivity to contact stiffness without increasing cantilever stiffness. A tungsten particle that served as a concentrated mass was adhesively attached to the free end of a commercially available rectangular cantilever made of silicon. Figure 8 shows an example of the concentrated-mass (CM) cantilever. Here the theoretical base is discussed by means of the beam vibration theory (Sect. 2.2.2).

It is assumed in the analytical model that the CM is a thin rod of mass m_{ad} and length l_m. A tip of length (height) h is attached at a distance l_t from the end of the CM rod, see Fig. 9. The rod is connected to the end of the flexural cantilever of mass m_{can} and length l. The gravitational center of the rod is at a distance l_g from the connected end ($l_g = l_m/2$ for uniform rods). The sample is represented by the vertical stiffness k_v^* and the lateral stiffness k_l^*. The CM cantilever is inclined with

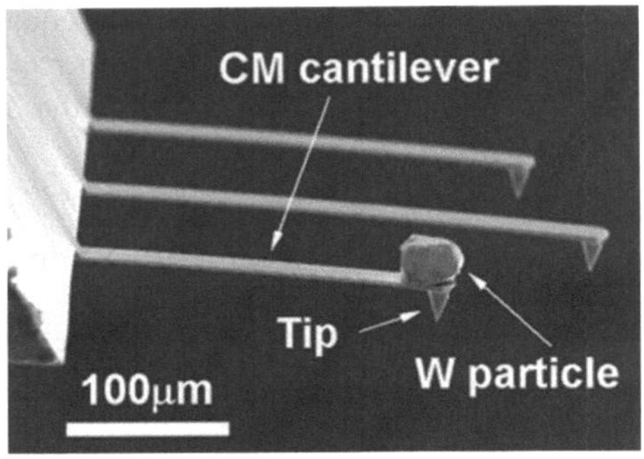

Fig. 8 SEM image of a concentrated-mass (CM) cantilever for sensitivity-enhanced AFAM, where $m_{ad} = 540$ ng and $\alpha = 12$. Reprinted with permission from [37]. Copyright 2005 Institute of Physics Publishing

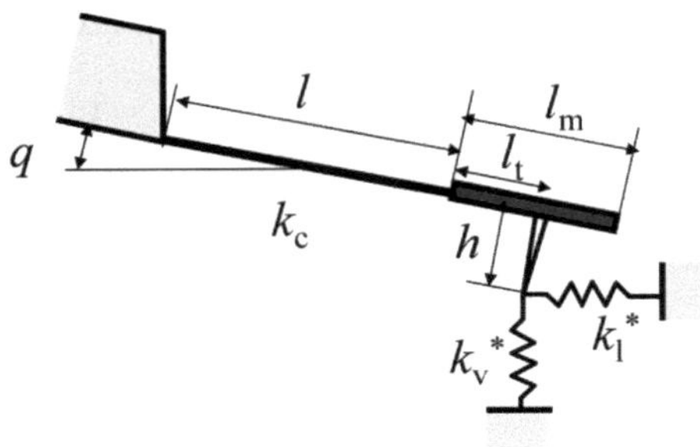

Fig. 9 A theoretical model of vibrations of a CM cantilever whose tip is in contact with a sample. The contact stiffnesses in vertical and lateral directions are expressed by mass-less springs

an angle of q and has a spring constant k_c relating the deflection at the tip site to the force acting in the tip direction.

Natural flexural vibrations of the cantilever are considered under small-amplitude. Using Eq. (20), let a ratio of the flexural rigidity to mass per unit length of the cantilever be described by

$$EI/(\rho A) = l^4 k_{c0}/(3m_{can}), \tag{22}$$

where k_{c0} is the spring constant of the flexural part, related to k_c by $k_{c0} = k_c[1 + 3(l_t/l) + 3(l_t/l)^2]$. The boundary conditions are given by $z = \partial z/\partial x = 0$ at $x = 0$. The conditions for the shear force $F = (k_{c0}l^3/3)\ (\partial^3 z/\partial x^3)$ at $x = l$ and for the bending moment $M = (k_{c0}l^3/3)\ (\partial^2 z/\partial x^2)$ at $x = l$ are derived from the equations of translational motion and rotational motion on the gravitational center of the CM rod:

$$F = \hat{k}_v z_t + \hat{k}_{lv}(\partial z/\partial x)_{x=l} h + m_{ad}(\partial^2 z/\partial t^2)_{x=l+lg}, \tag{23}$$

$$\begin{aligned} M = &-[\hat{k}_v z_t + \hat{k}_{lv}(\partial z/\partial x)_{x=l} h] l_t - [\hat{k}_{lv} z_t + \hat{k}_l(\partial z/\partial x)_{x=l} h] h \\ &- m_{ad}(\partial^2 z/\partial t^2)_{x=l+lg} l_g - J_m \partial^2/\partial t^2 (\partial z/\partial x)_{x=l+lg}, \end{aligned} \tag{24}$$

where z_t is the deflection at the tip site ($x = l + l_t$) and J_m is the moment of inertia on the gravitational center of the CM rod. The effective contact stiffnesses in the conditions are defined by [77]:

$$\left.\begin{aligned} \hat{k}_v &= k_v^* \cos^2 q + k_l^* \sin^2 q \\ \hat{k}_l &= k_l^* \cos^2 q + k_v^* \sin^2 q \\ \hat{k}_{lv} &= (k_l^* - k_v^*) \cos q \sin q \end{aligned}\right\}. \tag{25}$$

Substituting Eq. (9) into Eqs. (23) and (24) and the other boundary conditions, we obtain conditions for the coefficients of the vibration mode $Z(x)$, see Eq. (12), and then determine a frequency equation for $\xi_n = (3\omega_n^2 m_{can}/k_{c0})^{1/4}$.

For simplicity, we deal with the case of $h = q = l_m = 0$, where effects of the tip height, inclination of the cantilever, and size of the CM are ignored. The frequency equation is reduced into:

$$\frac{k_v^*}{k_c} = \left(\frac{\xi_n^3}{3}\right) \frac{\cos \xi_n \cosh \xi_n + 1}{\cos \xi_n \sinh \xi_n - \sin \xi_n \cosh \xi_n} + \frac{\alpha \xi_n^4}{3} \tag{26}$$

where $\alpha = m_{ad}/m_{can}$ is a ratio of mass of the CM to cantilever. In the case of $k_v^* = \alpha = 0$, Eq. (26) coincides with Eq. (14). Figure 10 shows an example of the frequency versus stiffness ratio k_v^*/k_c, predicted by Eq. (26).

Figure 10 is obtained for a normal cantilever and a CM cantilever with a mass ratio (α) of 5. For ceramics and metals, k_v^*/k_c takes a value around (0.5–1) $\times$ 10^3. The broken curves for a normal cantilever (without a concentrated mass) are flat in this region. This implies that normal cantilevers with low spring constants have low sensitivity in detecting the contact stiffness of stiff samples, i.e., the contact resonance frequency does not vary with contact stiffness. By contrast, the CM cantilever (solid curves) has a steep portion in plots of resonant frequency versus k_v^* for each mode. The overall curve consisting of all the steep portions obeys a point-mass model having one degree of freedom. This means that the CM

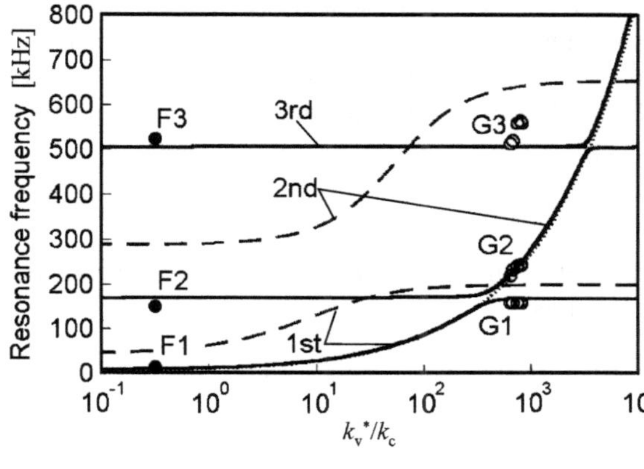

Fig. 10 Relationship between the resonant frequency and the contact stiffness k_v^* normalized by the spring constant k_c of the cantilever. Reprinted with permission from [37]. Copyright 2005 Institute of Physics Publishing

cantilever has the maximum sensitivity for detecting contact stiffness of any sample.

2.4 Evaluation of Elastic Modulus

Flexural vibration of a cantilever having distributed mass exhibits an infinite series of resonance. However, the large inertia of a concentrated mass degenerates all resonant peaks of the deflection at the mass-attached site like a pinned edge, except one particular resonant peak. The remaining resonance corresponds to the translational motion of an effective mass connected with springs having no mass. The resonant frequency f of a CM cantilever increases with the contact stiffness k_v^* in accordance with the point-mass model [36]:

$$f = f_0\sqrt{1 + \frac{k_v^*}{k_c}}, \tag{27}$$

where f_0 is the fundamental resonant frequency of the CM cantilever in the absence of a sample. For stiff samples like metals and ceramics, $k_v^*/k_c \gg 1$, in which case Eq. (27) approximates to:

$$\frac{k_v^*}{k_c} = \left(\frac{f}{f_0}\right)^2. \tag{28}$$

The intermolecular forces between molecules constituting the tip and the sample, and the meniscus forces arising from water films adsorbed on their respective surfaces induce an attractive force between the tip and sample, i.e., an adhesion force. Effects of the adhesion force on the contact stiffness cannot be ignored in conventional AFAM, where profiles of actual tip apexes are not simple but frequently assumed to be spherical. The adhesion force influences the contact area as well as the contact force (F_e). By contrast, a flat tip maintains a constant contact area independent of the adhesion force and the contact force. This ensures that the contact stiffness is also constant. A real flat tip may have a slightly rough and rounded apex surface. However, a constant contact area can be achieved provided the contact force is sufficient, in which case the following simple relation obtained from classical mechanics of linear elasticity [57] becomes applicable to a flat tip:

$$k_v^* = 2a_c E^*, \tag{29}$$

where a_c is the radius of the contact area and E^* is the effective Young's modulus defined as:

$$\frac{1}{E^*} = \frac{1 - \nu_t^2}{E_t} + \frac{1 - \nu_s^2}{E_s}, \tag{30}$$

where E_i and ν_i are the Young's modulus and Poisson's ratio, respectively. The subscripts 't' and 's' indicate the tip and sample, respectively.

Equations (28)–(30) give the relationship between the contact resonant frequency (f) and the effective Young's modulus E_s^* [$=E_s/(1-\nu_s^2)$] of a sample:

$$f = \sqrt{\frac{2A_c E_t^* E_s^*}{E_t^* + E_s^*}} \tag{31a}$$

or

$$E_s^* = \frac{E_t^* f^2}{2A_c E_t^* - f^2}, \tag{31b}$$

where E_t* [$=E_t/(1-\nu_t^2)$] is the effective Young's modulus of the tip. A_c is a factor proportional to the radius of the contact area, defined as:

$$A_c = \frac{a_c f_0^2}{k_c}. \tag{32}$$

Predetermining values of E_t^* and A_c based on measurements of reference samples, we can evaluate the effective Young's modulus of a sample (E_s^*) from Eq. (31b) and measurements of the resonant frequency.

Figure 11a shows spectra obtained for the optical glass slide surface near the second resonance when using a W_2C-coated tip with an apex radius of about 35 nm, i.e., a non-flat geometry. The resonant frequency can be seen to increase with the contact force, and hence, with the contact stiffness. This indicates a contact resonance sensitive to variations in contact stiffness. In addition, all observed resonant frequencies obeyed the theory of CM cantilevers reasonably well, as denoted by circles in Fig. 10, where k_v^* was estimated from a modified Hertzian contact model.

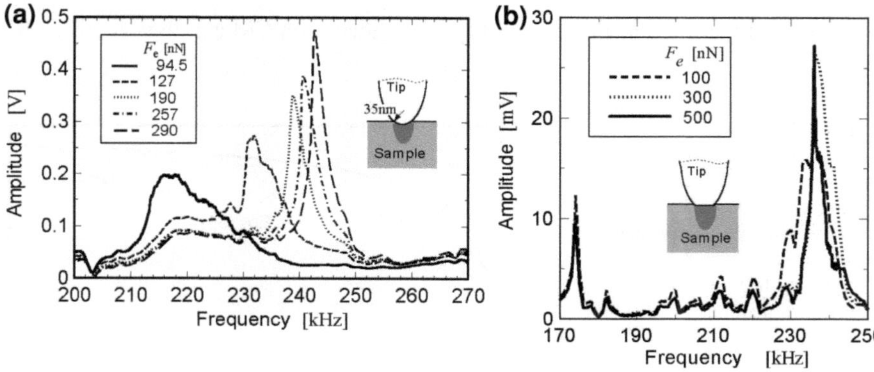

Fig. 11 Contact spectra. **a** A normal tip and **b** a flat tip mounted on CM cantilevers. Reprinted with permission from [37]. Copyright 2005 Institute of Physics Publishing

Figure 11b shows spectra for a sample of etched Si(100) wafer, measured with a Ti/Pt-coated flat tip. The resonant peak sensitive to the contact stiffness can be identified by the fact that the resonant peak is located apart from the trivial resonant frequencies (F2: 171.1 kHz, F3: 530.0 kHz). In contrast to the normal tip (Fig. 11a), the resonant frequency (around 236 kHz) seems to be independent of the contact force for the flat tip. This finding reflects the constant contact area observed in the case of the flat tip.

Elasticity is evaluated for reference samples using the combination of a CM cantilever and a Ti/Pt-coated flat tip. A prerequisite for a reference sample is that the elastic coefficients be well defined for a surface layer about 10 nm in depth, where contact elastic deformation predominates in the case of AFAM. We selected a single-crystalline diamond, and Si(111) and Si(100) wafers as our standard samples. The silicon wafers were etched in saturated KOH solution for tens of hours to remove work-hardening layers or chemically modified layers. In these covalently-bonded materials, interatomic forces act in the extremely short range, and loose, heterogeneous bands that exhibit non-bulk properties are confined to surface layers as little as a few atomic bonds in thickness.

The resonant frequencies for a free end measured $f = f_0 = 9.304$ kHz for the fundamental resonance, and 171.5 and 530.0 kHz for the second and third resonance, respectively, where the latter two were trivial. A contact force of $F_e = 500$ nN resulted in reproducible resonance frequencies $f = f_{(100)} = 236.1$ kHz, $f_{(111)} = 255.4$ kHz and $f_{\mathrm{dia}} = 318.7$ kHz for standard samples of Si(100), Si(111) and diamond, respectively.

Fitting Eq. (31a) to the relationship between the measured resonant frequency and the effective Young's moduli yields the factor A_c $(=a_c f_0^2/k_c)$ and the effective Young's modulus E_t^* $[=E_t/(1-\nu_t^2)]$ of the tip, which are hard to measure or estimate directly. Figure 12 shows the least-squares fit obtained for the reference samples, which yielded $A_c = 0.3815$ m/kg and $E_t^* = 152.3$ GPa. The latter is comparable to the averaged value for bulk platinum (172 GPa) and bulk titanium (129 GPa). Furthermore, using $k_c = 1$ N/m and $f_0 = 9.304$ kHz produced a reasonable contact radius $a_c = 4.4$ nm. The square of the correlation coefficient (=0.9999) of the fit confirms the validity of the theory on CM cantilevers with a flat tip.

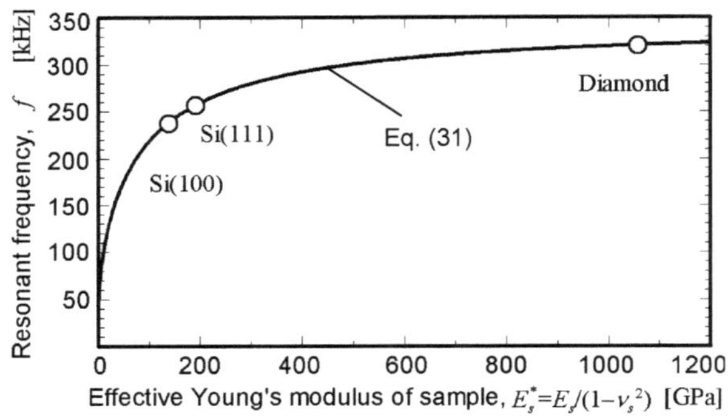

Fig. 12 Relationship between the resonant frequencies and effective Young's modulus of reference samples. Reprinted with permission from [37]. Copyright 2005 Institute of Physics Publishing

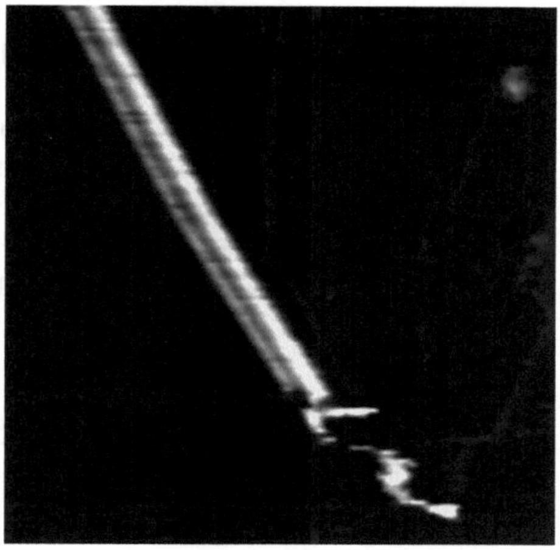

Fig. 13 AFM image (5 × 5 μm scan) of a Cu NW on a polished sapphire wafer

Use of Eq. (31b) with the values of A_c and E_t^* enables evaluation of elastic modulus of unknown samples from measurements of contact resonant frequencies. Figure 13 shows an AFM image of a Cu NW on a sapphire wafer. The NW was fabricated in the same way as Sect. 3.3.1 (stress migration). The diameter of the NW was about 80 nm. The tip of a CM cantilever was positioned on the top surface of the NW and then the contact resonance frequency was measured. The Young's modulus in the direction of the diameter was determined to be about 200 GPa, which was larger than that of bulk value for Cu.

3 Large Bending Test of Nanowires

3.1 Introduction

Here we introduce applications of the *elastica* theory [28] in statically mechanical tests of NWs [39], i.e., the large bending (LB) test. The theory gives analytical formulas of nonlinear relationship between the force and deflection of thin rods under static loads inducing the large deflection or buckling, and makes one free from the restriction on an amount of the deflections measured in the tests, whereas the linear theory that the conventional tests (see Fig. 1b–d) applied requires a small amount of the deflections. Manipulations of NWs for inducing the large deflections and measurements of the force are rather easy as compared with those for the small deflections. In addition OM becomes applicable to observations of the defections and bent shapes. Optics states that incident lights are scattered by small objects in the way called Rayleigh scattering for the objects smaller than one-tenth of the wave length or the other scattering expressed by the Mie solutions to Maxwell's equations, which are applicable for any sizes of the objects [7].

The bent shapes of NWs with diameter more than around 40 nm are visible in OM due to the Mie scattering [39]. The method in OM allows one to perform the mechanical tests in gaseous environments such as in atmospheric air.

The following section (Sect. 3.2) on the *elastica* theory is of an instructive section, which contains fundamental mathematics on Jacobi elliptic functions. In the subsequent section (Sect. 3.3), we demonstrates the test method based on the *elastica* theory to evaluate Young's modulus and fracture strength of copper oxide NWs. Brief discussions on the diameter dependence of Young's modulus are given in Sect. 3.4.

3.2 Theory on Equilibrium Bent-Shape of Nanowires

3.2.1 Preliminary

Let us again start with the equations of motion of large-bending thin rods, i.e., Eqs. (4). Neglecting the inertia terms in the first two equations yields the conditions that $P\cos(\theta - \psi)$ and $P\sin(\theta - \psi)$ are constant on the coordinate s, resulting in that the internal force is constant in the magnitude P and direction $\theta - \psi$ along the central axis of the rod. Let β denote a constant value of the relative angle $\theta - \psi$. The third equation of Eqs. (4) is reduced by neglecting the rotary inertia and assuming the uniformity of flexural rigidity:

$$\frac{\mathrm{d}^2\psi}{\mathrm{d}s^2} + \kappa_P^2 \sin\psi = 0, \tag{33a}$$

where

$$\kappa_P = \sqrt{\frac{P}{EI}}. \tag{33b}$$

Multiplying Eq. (33) by $\mathrm{d}\psi/\mathrm{d}s$ yields

$$\frac{\mathrm{d}}{\mathrm{d}s}\left[\frac{1}{2}\left(\frac{\mathrm{d}\psi}{\mathrm{d}s}\right)^2 - \kappa_P^2 \cos\psi\right] = 0, \tag{34}$$

which results in the constancy:

$$\frac{1}{2}\left(\frac{\mathrm{d}\psi}{\mathrm{d}s}\right)^2 - \kappa_P^2 \cos\psi = C \quad \text{on } s, \tag{35}$$

where C is a constant. This nonlinear differential equation was first solved by Euler in the eighteenth century and is analogous to the kinetics of pendulum [28]. In the kinetic analogue, the flexural rigidity (EI) is the moment of inertia of the pendulum about the axis of suspension, and the center of gravity is at unit distance from this axis. The line drawn from the center of suspension to the center of gravity at the instant s makes an angle ψ with the vertical drawn downwards.

The shape of the curve, called the *elastica*, into which the central-line is bent, is to be determined by means of Eq. (35). The results take different forms according as there are, or are not, inflections [28]. At an inflection $d\psi/ds$ vanishes, and the bending moment vanishes, so that the rod can be held in the form of an *inflectional elastica* by terminal forces alone, without moment. The end points are then inflections, and it is clear that all the inflections lie on the line of action of the terminal force P. The kinetic analogue of an *inflectional elastica* is an oscillating pendulum. Since the interval of time between two instants when the pendulum is momentarily at rest is a constant, equal to half the period of oscillation, the inflection are spaced equally along the central-line of the rod. To hold the rod with its central-line in the form of a *non-inflectional elastica*, terminal moments are required as well as terminal forces. The kinetic analogue is a revolving pendulum. In the particular case where there are no terminal forces, the rod is bent into an arc of a circle ($d\psi/ds = 0$ at $P = C = 0$). The kinetic analogue in this case is a rigid body revolving about a horizontal axis which passes through its center of gravity.

Assuming small angles ψ reduces Eq. (33) into the linear equation of ψ, resulting in the solution of sine waves. This coincides with the linear theory of buckling [62]. Therefore the solutions of Eq. (35) are expressed by functions that become trigonometric as the limit of small angles. They are elliptic functions, see Sects. 3.2.2 and 3.2.3.

In an *inflectional elastica*, let s be measured from an inflection, and let α be the value of ψ at the inflection $s = 0$, see Fig. 14. We rewrite Eq. (35) in the form:

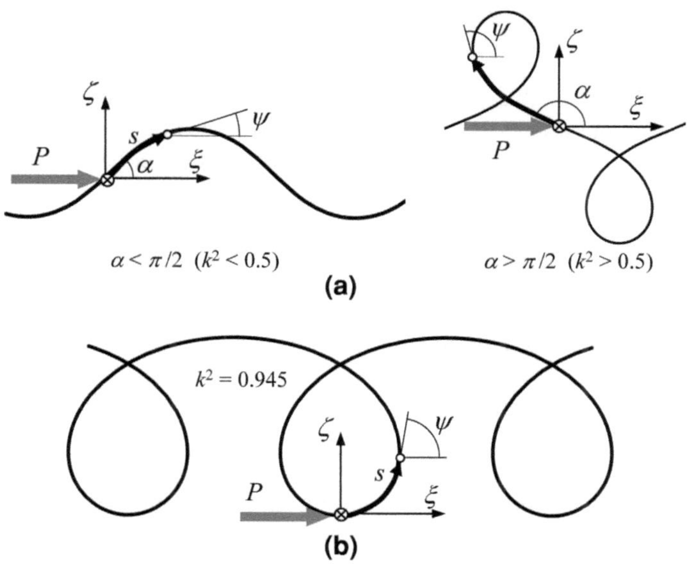

Fig. 14 The two forms of Elastica. **a** Inflectional elastica, **b** Non-inflectional elasitca

$$\left(\frac{\mathrm{d}\psi}{\mathrm{d}s}\right)^2 = 2\kappa_P^2(\cos\psi - \cos\alpha). \tag{36}$$

Integrating Eq. (36) together with using $\cos\psi = 1 - 2\sin^2(\psi/2)$ yields

$$s = \frac{1}{2\kappa_P \sin(\alpha/2)} \int_\alpha^\psi \frac{\mathrm{d}\psi}{\sqrt{1 - \sin^2(\psi/2)\sin^{-2}(\alpha/2)}}. \tag{37}$$

Introducing $y = \sin(\psi/2)\sin^{-1}(\alpha/2)$, we rewrite Eq. (37) into

$$\kappa_P s = \int_1^y \frac{\mathrm{d}y}{\sqrt{1-y^2}\sqrt{1-\sin^2(\alpha/2)y^2}}. \tag{38}$$

We see that the integral of Eq. (38) is the elliptic integral of the first kind with taking $\sin(\alpha/2)$ as the modulus k whose absolute is less than unit. It is also noticed that when taking the limit of $k \to 0$, the integral becomes the inverse function of sine, i.e., $\sin^{-1}y$.

In a *non-inflectional elastica*, let s be measured from a point where the absolute value of the curvature $\mathrm{d}\psi/\mathrm{d}s$ takes the maximum, see Fig. 14. At the maximum ($s = 0$), we see $\mathrm{d}^2\psi/\mathrm{d}s^2 = 0$ and then $\mathrm{d}M/\mathrm{d}s = 0$, resulting in $Q = 0$. The zero shear force means $\psi = 0$ at $s = 0$, i.e., the internal force acts tangent to the central line of the rod. Let $\kappa_{\max}$ be the maxima of the absolute of $\mathrm{d}\psi/\mathrm{d}s$. We rewrite Eq. (35) into the form:

$$\left(\frac{\mathrm{d}\psi}{\mathrm{d}s}\right)^2 = 2\kappa_P^2\cos\psi + (\kappa_{\max}^2 - 2\kappa_P^2). \tag{39}$$

From Eq. (39) we see that $(\mathrm{d}\psi/\mathrm{d}s)^2$ takes the non-zero minimum $\kappa_{\min}^2 = \kappa_{\max}^2 - 4\kappa_P^2$ at $\psi = \pi$, and $\kappa_{\max}^2(1 - 4\kappa_P^2/\kappa_{\max}^2)$ is positive, resulting in that $(2\kappa_P/\kappa_{\max})$ is less than unit. Integrating Eq. (39) together with using $\cos\psi = 1 - 2\sin^2(\psi/2)$ yields

$$s = \frac{1}{\kappa_{\max}} \int_0^\psi \frac{\mathrm{d}\psi}{\sqrt{1 - (2\kappa_P/\kappa_{\max})^2 \sin^2(\psi/2)}}. \tag{40}$$

Introducing $y = \sin(\psi/2)$, we rewrite Eq. (40) into

$$\frac{\kappa_{\max}s}{2} = \int_0^y \frac{\mathrm{d}y}{\sqrt{1-y^2}\sqrt{1-(2\kappa_P/\kappa_{\max})^2 y^2}}. \tag{41}$$

We see again that the integral is the elliptic integral of the first kind with taking $2\kappa_P/\kappa_{\max}$ as the modulus k whose absolute is less than unit.

We have seen above that the variable y related to the angle ψ is expressed by the inverse function of the elliptic integral of the first kind with the variable proportional to s and the modulus k. What we want finally to obtain is the form of the *elastica* in the Cartesian coordinates. The elliptic functions, which are introduced in the following section, are significantly helpful to work out the formulation in the Cartesian coordinates.

3.2.2 Elliptic Functions

Let $f\left(x, \sqrt{p(x)}\right)$ be a rational function. If $p(x)$ is a linear expression or quadratic polynomial in x, indefinite integrals of f can be expressed in terms of rational functions and their logarithms. Otherwise, i.e., a polynomial $p(x)$ of higher degree, it is generally impossible to express the integrals in terms of elementary functions. In physical problems we often encounter the integrals where $p(x)$ is a cubic or quartic polynomial, which are known as elliptic integrals. Any elliptic integral can be expressed in terms of the three standard kinds of Legendre–Jacobi elliptic integrals [8].

The elliptic integral of the first kind, defined as

$$u = \int_0^y \frac{dy}{\sqrt{(1-y^2)(1-k^2y^2)}} = \mathrm{sn}^{-1}y, \tag{42}$$

is especially important because it defines a Jacobi elliptic function. The coefficient k is called the modulus. By analogy with the sine function, the inverse function of Eq. (42) is denoted by

$$y = \mathrm{sn}\,u = \mathrm{sn}(u,k). \tag{43}$$

By using sn$u$, other Jacobi elliptic functions cn$u$ and dn$u$ are defined as

$$\left.\begin{aligned} \mathrm{cn}^2u &= 1-\mathrm{sn}^2u \\ \mathrm{dn}^2u &= 1-k^2\mathrm{sn}^2u \end{aligned}\right\}. \tag{44}$$

Equation (42) defines u in the limited range from $-K$ to K, where $K = K(k)$ is the complete elliptic integral of the first kind, i.e., the value of u at $y = 1$. However, the range of u in the definition of the Jacobi elliptic functions is periodically extended to from minus to plus infinity so that K is a quarter period of sn$u$ and cn$u$, and a half period of dn$u$. The complete integral K becomes $\pi/2$ as $k \to 0$. Figure 15 shows a comparison of the elliptic function sn$u$ with the trigonometric function sin$u$.

These elliptic functions take elementary functions in the limiting cases $k \to 0$ and $k \to 1$:

$$\left.\begin{array}{llll} \mathrm{sn}u \to \sin u, & \mathrm{cn}u \to \cos u, & \mathrm{dn}u \to 1 & \text{as } k \to 0 \\ \mathrm{sn}u \to \tanh u, & \mathrm{cn}u \to \mathrm{dn}u \to \mathrm{sech}u & & \text{as } k \to 1 \end{array}\right\}. \tag{45}$$

The derivatives are

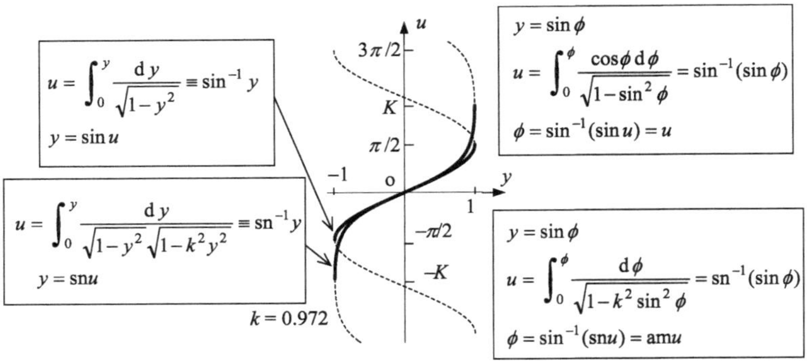

Fig. 15 Comparison of the elliptic function sn with the trigonometric function sin

$$\frac{\mathrm{d\,sn}u}{\mathrm{d}u} = \mathrm{cn}u\ \mathrm{dn}u, \quad \frac{\mathrm{d\,cn}u}{\mathrm{d}u} = -\mathrm{sn}u\ \mathrm{dn}u, \quad \frac{\mathrm{d\,dn}u}{\mathrm{d}u} = -k^2\mathrm{sn}u\ \mathrm{cn}u. \tag{46}$$

Substituting $y = \sin\phi$ into Eq. (42) yields another expression of the elliptic integral of the first kind. Then the inverse function is defined as

$$\left.\begin{array}{l}\phi = \sin^{-1}y = \sin^{-1}(\mathrm{sn}u) = \mathrm{am}\,u = \mathrm{am}(u,k) \\ \mathrm{sn}\,u = \sin(\mathrm{am}\,u), \quad \mathrm{cn}\,u = \cos(\mathrm{am}\,u)\end{array}\right\}. \tag{47}$$

The function am is called the amplitude function. Jacobi elliptic and amplitude functions are now used as if they are elementary functions in commercial software such as *Mathematica* [72].

3.2.3 Formula of the *Elastica*

Let introduce Cartesian coordinate system (ξ, ζ) whose origin coincide with that of the coordinate s. The direction of the axis ξ is set equal to that of P, and the axis ζ is taken so that $\mathrm{d}\zeta/\mathrm{d}s$ is equal to $+\sin\psi$ (see Fig. 14).

For an *inflectional elastica*, referring to Eq. (38), we introduce Jacobi elliptic functions of the argument u with a modulus k which are given by the equations:

$$u = \kappa_P s, \quad k = \sin(\alpha/2). \tag{48}$$

Then Eq. (38) is rewritten into

$$\sin(\psi/2) = k\,\mathrm{sn}(u + K), \tag{49}$$

where the phase of the sn function is shifted by K to let Eq. (49) coincide with the boundary conditions $\psi = \alpha$ at $u = 0$ and $\psi = 0$ at $u = K$. Using Eqs. (44), (46) and (49) gives

$$\frac{\mathrm{d}\psi}{\mathrm{d}u} = 2k\,\mathrm{cn}(u + K). \tag{50}$$

To determine the coordinate ζ, we calculate $\mathrm{d}\zeta/\mathrm{d}s$ as follows:

$$\left.\begin{aligned}\frac{\mathrm{d}\zeta}{\mathrm{d}s} &= \sin\psi = 2\sin\frac{\psi}{2}\cos\frac{\psi}{2} = 2k\,\mathrm{sn}(u+K)\sqrt{1-k^2\mathrm{sn}^2(u+K)}\\ &= 2k\,\mathrm{sn}(u+K)\mathrm{dn}(u+K)\\ \kappa_P\mathrm{d}\zeta &= 2k\,\mathrm{sn}(u+K)\mathrm{dn}(u+K)\mathrm{d}u = -2k\frac{\mathrm{d}}{\mathrm{d}u}[\mathrm{cn}(u+K)]\mathrm{d}u\\ &= -2k\,\mathrm{d}[\mathrm{cn}(u+K)]\end{aligned}\right\}. \tag{51}$$

Further we calculate $\mathrm{d}\xi/\mathrm{d}s$ as follows:

$$\left.\begin{aligned}\left(\frac{\mathrm{d}\xi}{\mathrm{d}s}\right)^2 &= 1-\left(\frac{\mathrm{d}\zeta}{\mathrm{d}s}\right)^2 = 1-(2k)^2\mathrm{sn}^2(u+K)\mathrm{dn}^2(u+K)\\ &= 1-(2k)^2\mathrm{sn}^2(u+K)[1-k^2\mathrm{sn}^2(u+K)] = [1-2k^2\mathrm{sn}^2(u+K)]^2\\ \kappa_P\mathrm{d}\xi &= [1-2k^2\mathrm{sn}^2(u+K)]\mathrm{d}u = [-1+2\mathrm{dn}^2(u+K)]\mathrm{d}u\end{aligned}\right\}. \tag{52}$$

Integrating Eqs. (51) and (52) yields the form of the *inflectional elastica*:

$$\left.\begin{aligned}\xi &= \frac{1}{\kappa_P}[-u+2E\mathrm{am}(u+K)-2E\mathrm{am}\,K]\\ \zeta &= -\frac{2k}{\kappa_P}\mathrm{cn}(u+K)\end{aligned}\right\}, \tag{53}$$

where $E\mathrm{am}u$ denotes the elliptic integral of the second kind expressed by the formula

$$E\mathrm{am}\,u = \int_0^u \mathrm{dn}^2 u\,\mathrm{d}u. \tag{54}$$

The inflections are given by $\psi = \alpha$ or $\mathrm{sn}(u+K) = 1$, and therefore the arc between two consecutive inflections is $2K/\kappa_P$, and the inflections are spaced equally along the axis of ξ at the intervals $2(2E\mathrm{am}\ K - K)/\kappa_P$.

For a *non-inflectional elastica*, referring to Eq. (41), we introduce Jacobi elliptic functions of the argument u with a modulus k which are given by the equations:

$$u = \kappa_P s/k, \quad k = 2\kappa_P/\kappa_{\max}. \tag{55}$$

Then Eq. (41) is rewritten into

$$\sin(\psi/2) = \mathrm{sn}\,u. \tag{56}$$

Using Eqs. (44),(46) and (56) gives

$$\frac{\mathrm{d}\psi}{\mathrm{d}u} = 2\mathrm{dn}\,u. \tag{57}$$

To determine the coordinate ζ, we calculate dζ/ds as follows:

$$\left.\begin{aligned}\frac{\mathrm{d}\zeta}{\mathrm{d}s} &= \sin\psi = 2\sin\frac{\psi}{2}\cos\frac{\psi}{2} = 2\mathrm{sn}\,u\;\mathrm{cn}\,u\\ (\kappa_P/k)\mathrm{d}\zeta &= 2\mathrm{sn}\,u\;\mathrm{cn}\,u\,\mathrm{d}u = -\frac{2}{k^2}\frac{\mathrm{d}(\mathrm{dn}u)}{\mathrm{d}u}\mathrm{d}u = -\frac{2}{k^2}\mathrm{d}(\mathrm{dn}u)\end{aligned}\right\}. \tag{58}$$

Further we calculate dξ/ds as follows:

$$\left.\begin{aligned}\left(\frac{\mathrm{d}\xi}{\mathrm{d}s}\right)^2 &= 1-\left(\frac{\mathrm{d}\zeta}{\mathrm{d}s}\right)^2 = 1-4\mathrm{sn}^2u\;\mathrm{cn}^2u = (1-2\mathrm{sn}^2u)^2\\ (\kappa_P/k)\mathrm{d}\xi &= (1-2\mathrm{sn}^2u)\mathrm{d}u = \left[\left(1-\frac{2}{k^2}\right)+\frac{2}{k^2}(1-k^2\mathrm{sn}^2u)\right]\mathrm{d}u\\ &= \left[\left(1-\frac{2}{k^2}\right)+\frac{2}{k^2}\mathrm{dn}^2u\right]\mathrm{d}u\end{aligned}\right\}. \tag{59}$$

Integrating Eqs. (58) and (59) yields the form of the *non-inflectional elastica*:

$$\left.\begin{aligned}\xi &= \frac{k}{\kappa_P}\left[\left(1-\frac{2}{k^2}\right)u+\frac{2}{k^2}E\mathrm{am}\,u\right]\\ \zeta &= \frac{2}{\kappa_P k}(1-\mathrm{dn}u)\end{aligned}\right\}. \tag{60}$$

The curve consists of a series of loops lying altogether on one side of the axis ζ.

3.3 Evaluation of Elastic Modulus and Strength

3.3.1 Procedure

The LB test in OM made it possible to evaluate thin cupric-oxide (CuO) NW of tens of nanometers in diameter. Furthermore, it was shown that the bending strength of CuO NWs could be measured by applying intense bend-like buckling to localize the bending strain on the suspended part of the NW and observing the radius of curvature of the bent NW at fracture [39].

Copper forms a stable oxide surface layer by heating in air. Interestingly, the heating temperature range from 673 to 973 K induces the growth of CuO NWs on the oxide surface layers [20, 79]. The NWs are of 30–200 nm in diameter and up to tens of micron in length. Each NW is a single crystal of monoclinic system, where the wire axis is parallel to the [$\bar{1}$11] direction of CuO crystal [61]. The growth mechanism is unclear, but the vapor–solid (VS) mechanism has been proposed. In this experiment, CuO NWs were prepared by heating a 0.1 mm diameter Cu wire in air at 673 K for 4 h. The NWs were up to 20 μm in length, most of which were uniform in diameter, although some NWs were tapered.

The load sensor was composed of an AFM cantilever made of silicon and a laser interferometer. The cantilever was attached to the arm head of a three-dimensional hydraulic micromanipulator, which was able to position the cantilever. The cantilever was set vertically to the Cu wire so that the cantilever was parallel to the CuO NWs, as shown in Fig. 16. A laser interferometer was equipped on a manual *XYZ*-stage, so as to be positioned independent of the NWs and the cantilever. The laser focal spot was set on the aluminum-coated back surface of the cantilever for the detection of cantilever deflection. The Cu wire was precisely displaced by means of a piezo-actuated *XY*-stage for the loading tests. These components were integrated in an optical microscope.

The low resolution of the optical microscope (ca. 200 nm) did not allow clear observation of the NWs. However, transmitted illumination was able to show images of the NWs as diffraction images when the focal plane was set to the *Z*-position containing the cross-sectional center of the Cu wire, and NWs were observed on both side edges of the Cu wire.

For the measurement of Young's modulus, the Cu wire was displaced in the axial direction with the cantilever in a fixed position, see Fig. 16. The spring constant of the cantilever was $k_c = 0.082$ N/m. When the tip of an NW approached the aluminum-coated back surface of the cantilever, the NW displayed a jump-in-contact with the back surface, due to attractive interactions such as van der Waals and electrostatic forces. The NW was deflected in the same way as a both-end-clamped beam with end-offset, where the deflection angle was fixed at both ends, but the end adhered to the cantilever was allowed to slide on the cantilever back surface. After the jump-in-contact, the specimen was periodically displaced by applying a saw-tooth voltage to the piezo-actuated *XY*-stage. The rate of displacement was approximately 1 μm/s. The bent shape of the NW was recorded using a digital video camera. The amount of the end offset (the deflection of the NW) and the deflection angle at the center were measured from the video frames. The output voltage of the laser interferometer, i.e., the deflection of the AFM cantilever, was acquired using an A/D converter. The cantilever deflection multiplied by the value of k_c provided the transverse component of the reaction force.

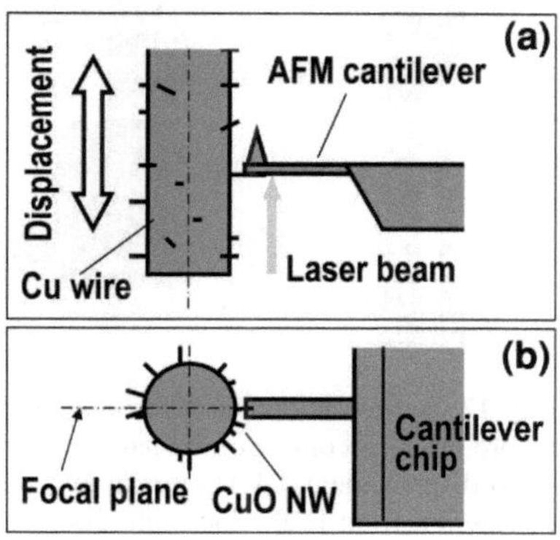

Fig. 16 Schematics of experimental setup for the large bending (LB) test. **a** Top view, **b** side view. The deflection of the AFM cantilever is measured by means of a laser interferometer. The Cu wire with CuO NWs on it is precisely displaced by a piezo-driven XY stage. Reprinted with permission from [39]. Copyright 2009 American Scientific Publishers

In the fracture test, with the Cu wire in a fixed position, a stiff AFM cantilever was manually moved by means of the three-dimensional micromanipulator. The micromanipulator enabled application of diversiform bending to the NWs. To provide an intense bending strain, the NW was displaced transversely and then pushed toward its base, as in buckling. The process was recorded on video until failure. The failure strain was determined from the bending radius of the NW at the failure site, based on the bent shape NW image just before failure.

Optical diffraction images of NWs observed through the transmitted illumination do not provide the actual diameters. Nevertheless, the luminance profile across a NW depends on the actual diameter. Therefore, in order to obtain a relation between the profile and the actual diameter, some of CuO NWs were observed with both field-emission-SEM and the optical microscope. The NWs were scraped from the surface of the Cu wire with an AFM cantilever edge in the same way as for the fracture test. Most of the broken NWs were adhered to the back-surface of the cantilever and protruded from the cantilever edge. The protruding part of the NWs was observed with both microscopes.

3.3.2 Experimental Results and Evaluation

Figure 17 shows a series of optical images of a CuO NW during the LB test. When the NW approaches the back surface of the cantilever with a distance of a few micrometers (Fig. 17a), the NW exhibits an unstable jump-in-contact with the

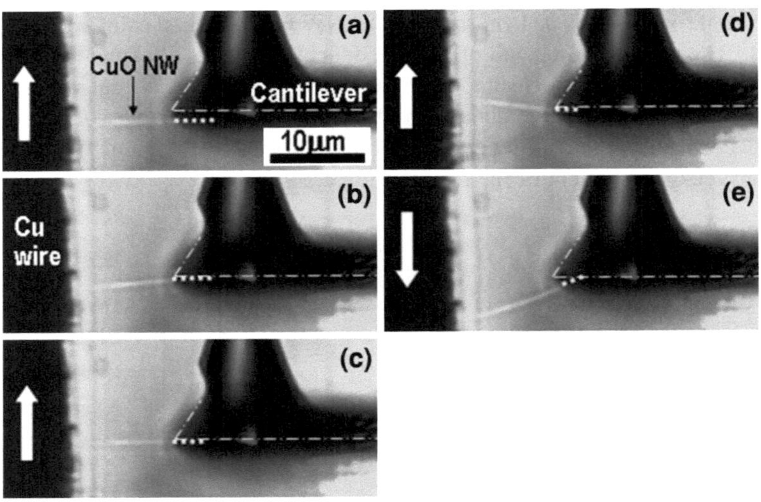

Fig. 17 Series of optical micrographs under transmitted illumination for the LB test. **a** Just before jump-in-contact of the nanowire. **b** Just after the jump-in-contact. **c** The reference position (zero displacement). **d** The maximum displacement. **e** Peeling of the nanowire. Reprinted with permission from [39]. Copyright 2009 American Scientific Publishers

back surface (Fig. 17b). This instability is similar to that observed in force curve measurements using AFM, where an AFM cantilever tip jumps in contact with a sample surface [71]. This is because the force gradient between the NW and the back surface exceeds the spring constant of the NW. Figure 17c shows the reference position of the NW displacement, where the base of the NW is on the extrapolated line on the back surface of the AFM cantilever. Since the distance between the base and the cantilever edge is smaller than the suspended length of the NW just after jump-in-contact (Fig. 17b), the NW is slightly buckled at the reference position. After further displacement of the Cu wire, the NW becomes subjected to tension (Fig. 17d). The NW slides on the back surface of the AFM cantilever during the process, shown from Fig. 17c, d, but the following cyclic displacement within the maximum displacement (Fig. 17d) does not induce sliding, because the tangential force does not exceed the critical (static) frictional force required for sliding. When the Cu wire is displaced backward from the reference position, the NW is easily peeled from the back surface of the cantilever, as shown in Fig. 17e.

Figure 18a shows an example of the synchronous records for the displacement δ of a CuO NW and the deflection of the AFM cantilever when the zero displacement corresponds to the reference position. During the first displacement cycle, there seems to be a drift of the cantilever, but the following cycles appear relatively stable. The drift data are excluded from the data analyses. Around the maximum displacement of the CuO NW, from 2 to 3 μm, the AFM cantilever deflection is detected, which measures the shear force P_z applied to the NW at the fixed end. The maximum deflection is approximately 60 nm, which corresponds to $P_z = 5$ nN. Figure 18b shows P_z and the deflection angle α_c at the center of the NW as a function of δ. In the P_z-detectable region, the NW forms a straight line except near both fixed ends, as schematically shown in Fig. 19. This means that tension dominates the NW, but the bending moment is also significant near the fixed ends. In the P_z-undetectable region around the reference position, the NW is relaxed and slightly buckled.

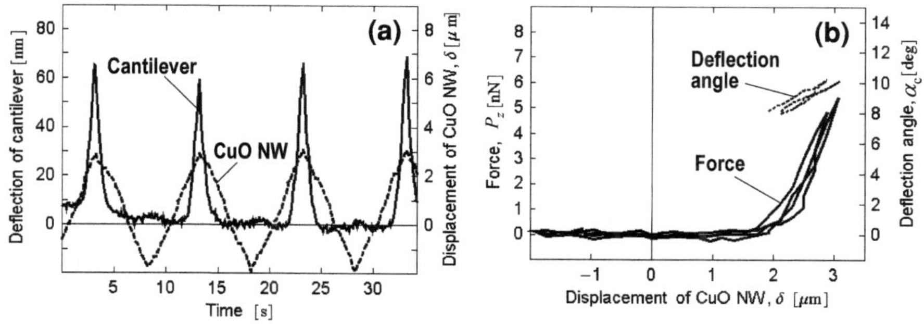

Fig. 18 Experimental data of the LB test. **a** Time series of the cantilever deflection and nanowire displacement. **b** The force and deflection angle as a function of the displacement. Reprinted with permission from [39]. Copyright 2009 American Scientific Publishers

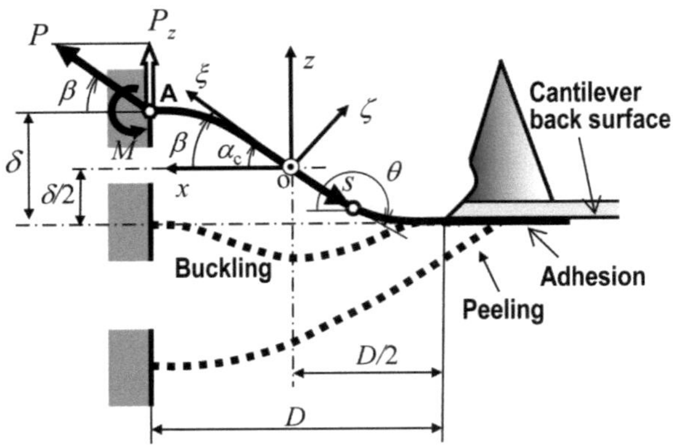

Fig. 19 Geometry of the nanowire in the LB test. Reprinted with permission from [39]. Copyright 2009 American Scientific Publishers

The P_z versus δ relation shown in Fig. 18b depends on the flexural rigidity EI of the NW. In the P_z-detectable region, the displacement is so large that the linear mechanics theory is inapplicable. The bent shape of NWs in the test has an inflection at the center, as shown in Fig. 19 (point o). The Cartesian coordinate systems (x, z) and (ξ, ζ) are set so that the origin coincides with the inflection. The directions of x and z are fixed. The axis ξ lies in the direction of the reaction force (P) at end A, with an inclination of angle β. When the tension dominates the NW, except near both ends, the angle of the reaction force (β) becomes close to the deflection angle α_c at the center of the NW. In the same way as in Sects. 2.2 and 3.2, set a curvilinear coordinate s along the NW originating at the inflection and let θ be the tangential angle of s measured from the axis x counterclockwise for an observer looking in the y direction. We are ready to apply the formulas of the *inflectional elastica*. It is noted that in the test the value of α is more than $\pi/2$ and equals to $\pi - (\beta - \alpha_c)$. The angle ψ equals to $\theta - \beta$. For convenience, we express the coordinates (ξ, ζ) of a point on the NW with the angle θ as a parameter instead of u in Eqs. (53):

$$\left.\begin{aligned} \xi &= \frac{1}{\kappa_P}[K(k) - F(\phi,k) + 2E(\phi,k) - 2E(k)] \\ \zeta &= -\frac{2k}{\kappa_P}\mathrm{cn}[F(\phi,k)] \end{aligned}\right\}, \tag{61a}$$

where

$$\phi = \sin^{-1}\left(\frac{1}{k}\sin\frac{\theta-\beta}{2}\right), \quad k = \sin\frac{\alpha}{2} = \cos\frac{\beta-\alpha_c}{2}. \tag{61b}$$

The functions $F(\phi, k)$ and $E(\phi, k)$ denote elliptic integrals of the first and second kind with a modulus k, respectively. They are explicitly expressed by substituting $y = \sin\phi$ in Eq. (53) and then by using Eqs. (44) and (54):

$$F(\phi,k) = \int_0^{\phi} \frac{\mathrm{d}\phi}{\sqrt{1-k^2\sin^2\phi}}, \quad E(\phi,k) = \int_0^{\phi} \sqrt{1-k^2\sin^2\phi}\,\mathrm{d}\phi. \tag{62}$$

$K(k) = F(\pi/2, k)$ and $E(k) = E(\pi/2, k)$ are the complete elliptic integrals of the first and second kind, respectively. When the modulus (k) approaches 1, the bent shape becomes straight.

Both the Cartesian coordinate systems are related by

$$\left.\begin{aligned} x^*(k,\beta,\theta) &= \kappa_P x = \kappa_P(\xi\cos\beta - \zeta\sin\beta) \\ z^*(k,\beta,\theta) &= \kappa_P z = \kappa_P(\xi\sin\beta + \zeta\cos\beta) \end{aligned}\right\} \tag{63}$$

Some of the boundary conditions at $\theta = \pi$ (end A) and $\theta = \alpha_c + \pi$ (the center) impose

$$\left.\begin{aligned} &\frac{\delta}{D} = \frac{z^*(k,\beta,\pi)}{x^*(k,\beta,\pi)} \\ &x^*(k,\beta,\alpha_c+\pi) = 0 \\ &\delta^2 P_z = 4EI[z^*(k,\beta,\pi)]^2\sin\beta \end{aligned}\right\}, \tag{64}$$

where D is the distance between the Cu wire surface and the cantilever edge, as shown in Fig. 19.

Values of δ/D, α_c and $\delta^2 P_z$ are known at each sampling time during the test. Equations (64) can be used to determine the solutions k, β and EI. Since k and β should vary with the deformation, corresponding values k_i and β_i were determined at each sampling time t_i by substituting the experimental data δ_i/D and α_{ci} into the first and second equations of Eqs. (64). The root-finding analyses were carried out using the software *Mathematica* [72], with initial values $k \approx 1$ and $\beta \approx \alpha_c$. To obtain flexural rigidity (EI), the error Err was minimized:

$$Err = \sqrt{\frac{1}{m}\sum_{i=1}^{m}\left\{\delta_i^2 P_{zi} - 4EI[z^*(k_i,\beta_i,\pi)]^2\sin\beta_i\right\}^2}, \tag{65}$$

where m is the sampling number. Minimization was performed using *Mathematica*. Figure 20 shows an example of the experimental $\delta^2 P_z$ vs. δ/D relationship,

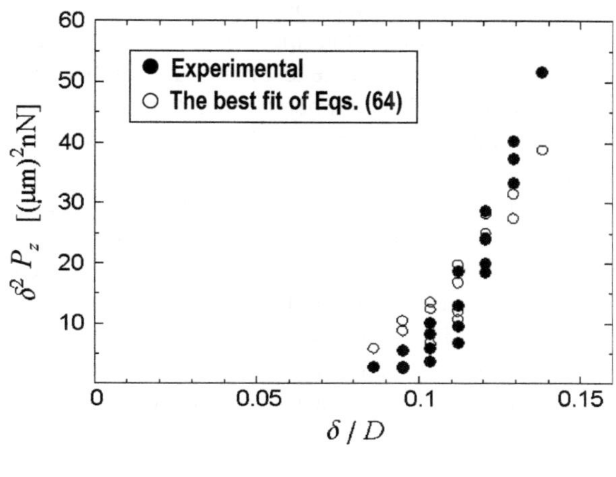

Fig. 20 An example of the best fit using the inverse analysis to determine the Young's modulus. Reprinted with permission from [39]. Copyright 2009 American Scientific Publishers

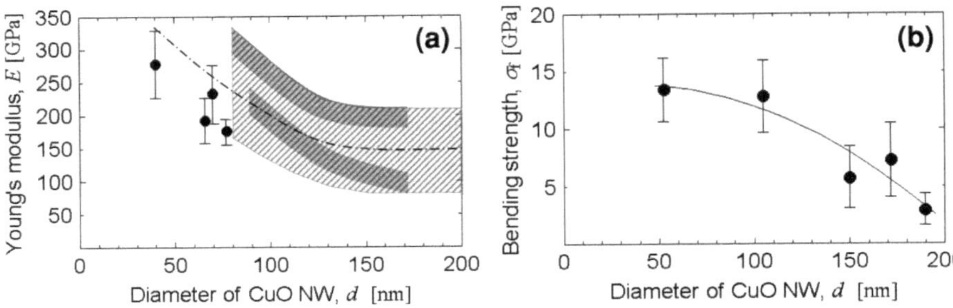

Fig. 21 Results of the mechanical properties of CuO nanowires. **a** The Young's modulus, where the hatched region represents the reported data scattering obtained by the three point bend test [61] and the *upper* and *lower grey bands* represent uniform NWs with smooth and rough surfaces, respectively. **b** The bending strength. Reprinted with permission from [39]. Copyright 2009 American Scientific Publishers

after Fig. 18b, and the best fit of Eqs. (64). The fitting resulted in $EI = 179$ (μm)^{2}nN with $Err = 6.5$ (μm)^{2}nN. The standard deviation of EI was estimated to be approximately 30 (μm)^{2}nN. The diameter of the NW was determined to be $d = 66$ nm from the luminance of the diffraction image, see Figs. 7 and 8 in the report by [39]. Using $I = \pi d^4/64$ provided the Young's modulus $E = 192$ GPa.

Figure 21a shows the results of the LB test as a plot of Young's modulus against the diameter of the NW. The error bars are standard deviation, based on the error propagation of EI. The hatched area indicates reported results [61], evaluated from TPB tests for uniform and tapered CuO NWs. The present results (circles) reasonably demonstrate the extrapolation (dot-dashed curve representing the average of the scattered band) to NWs thinner than approximately 80 nm in diameter. The Young's modulus still displays diameter dependence, i.e., the modulus value increases as the diameter decreases.

The CuO NWs tested had uniform diameters along the longitudinal axis. The results in Fig. 21a seem lower than the extrapolation obtained for the uniform NWs by the TPB test (the upper and lower gray bands in Fig. 21a). This could be due to the difference in the test configuration between the present test (the LB test) and the TPB test. The diameter dependence due to the surface stress effect depends on the extent of surface area increase induced during the deformation, see Sect. 3.4. Another reason could be the error in the diameter estimation. The estimation using the luminance profile may have an error in the diameter (d) of $\pm 5\%$. Because I is proportional to d^4, the error propagation to the Young's modulus would be $\pm 20\%$. Therefore, the difference in the Young's modulus between the LB and the TPB test could be due to the error in d.

Figure 22 shows optical images of a fracture test example. The fracture of the CuO NW is brittle and occurs at the maximum-curvature site in the bent shape. The radius of curvature at fracture, measured for six NWs of different thickness, ranges from 0.55 to 4.6 μm. On the assumption that the NWs behave as a linear

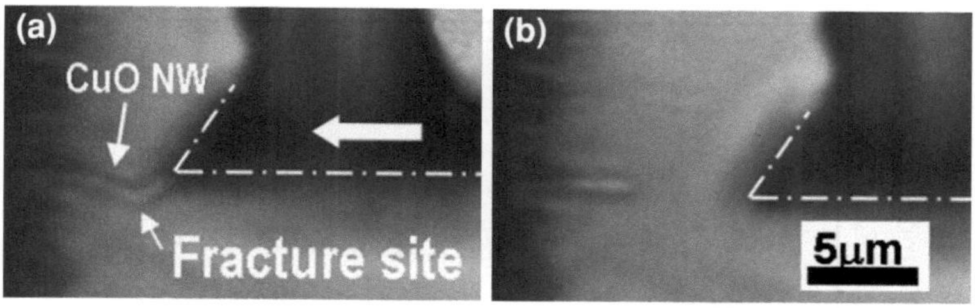

Fig. 22 Optical micrographs of the NW in the fracture test under transmitted illumination, **a** just before fracture, and **b** after fracture. Reprinted with permission from [39]. Copyright 2009 American Scientific Publishers

elastic beam until fracture, the relation between the bending strength σ_f and the radius R_f of curvature at fracture is expressed by

$$\sigma_f = E\varepsilon_f = Ed/(2R_f), \tag{66}$$

where ε_f [$=d/(2R_f)$] is the fracture strain. Equation (66) comes from the Bernoulli–Euler's theory, see Sect. 2.2.1. The fracture strain ranged from approximately 2–7%, depending on the diameter. The Young's modulus corresponding to the diameter was obtained from the dot-dashed curve in Fig. 21a. Figure 21b shows the results of the fracture test. The bending strength of the CuO NWs increases as the diameter decreases and a constant value is displayed for diameters smaller than approximately 100 nm. The constant value (ca. 13 GPa) is only a factor of 2 lower than the theoretical cohesive strength, $E/10$ [16]. The diameter dependence of strength for NWs larger than 100 nm in diameter could be due to the presence of fewer defects in the smaller NWs. For the NWs with diameters smaller than 100 nm, surface or volume defects seem to play only a minor role in fracture initiation. Effects of the surface stress and nonlinear stress–strain relation on the bending strength were discussed, and they were estimated to be errors less than 10% in the strength measured [39].

3.4 Diameter Dependency of Elastic Modulus

Experimentally, Young's modulus of CuO, ZnO, Ag and Pb NWs is found to increase drastically with decreasing diameters [12, 13, 39, 61], see Fig. 23. However, investigations on GaN NWs, Cr and Si nanocantilevers show an opposite tendency [25, 40, 42]: their moduli decrease sharply with decreasing diameters or thickness. The investigations on SiC and Au NWs suggest that their Young's moduli do not depend on diameters essentially [73, 74].

Theoretical studies are through atomistic simulations [9, 23, 27, 34, 54, 56, 60, 81] or corrections to continuum theory [13, 55]. The elastic response of

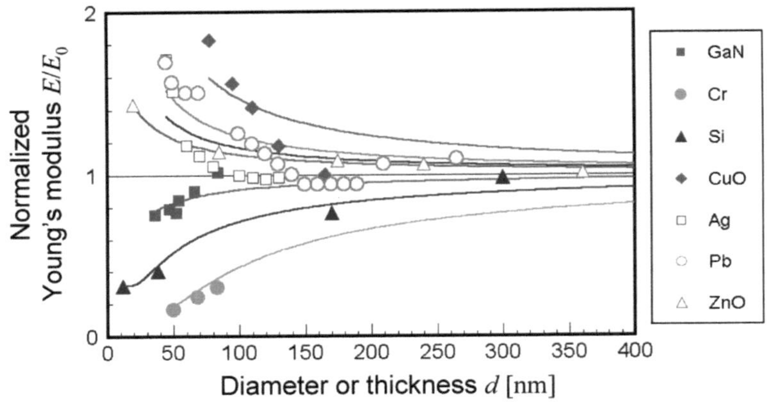

Fig. 23 Dependence of Young's modulus on the diameter or thickness. GaN [120] NWs [40], Cr thin beams [42], Si[110] thin beams [25], CuO[-111] NWs [61], Ag NWs [13], Pb NWs [13], and ZnO NWs [12]. E_0 is bulk values of the Young's modulus E. *Curves* are nonlinear-least-squares fits of Eq. (68)

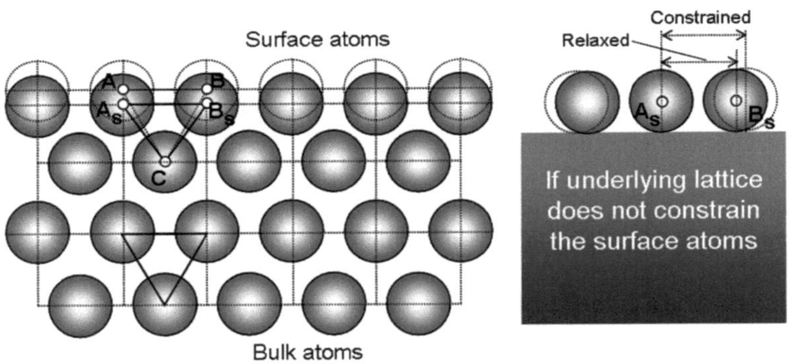

Fig. 24 Schematic explanation for origins of surface elasticity and surface stress

nanostructures has been explained by nonlinear effects, i.e., non-Hookean elasticity [27], surface stress [23, 55, 56], surface elasticity [9, 34, 60, 81].

The surface elasticity, surface stress and nonlinear effects are encompassed by effects of imperfect coordination of surface atoms. Sun et al. [60] has presented the following understandable and consistent insight: The termination of the lattice periodicity in the surface normal has two consequences. First, the imperfection of the coordination numbers of a surface atom causes the lengths of the remaining bonds of this lower-coordinated surface atoms to relax. As the relaxation (both contraction and expansion) is a spontaneous process, the binding energy of the relaxed bond will reduce (rise in absolute value) to minimize the system energy. The relaxed bond is also stronger. The contraction case is schematically illustrated in Fig. 24, where the bond length A–C contracts into A_s–C. Such a bond-order-bond-length-bond-strength (bond-OLS) correlation has been observed in various materials. The Young's modulus is very sensitive to the relaxed bond length. According to the bond-OLS analysis of [60], the correlation between surface modulus and the bond contraction can be described as

$$(E_i - E_0)/E_0 = b_i^{-m_b} - 3b_i + 2, \tag{67}$$

where E_i and E_0 are the Young's moduli of the ith atomic layer and bulk materials, respectively; b_i is defined as d_i/d_0, with d_i and d_0 the bond length in the ith layer and the bulk value, m_b is a parameter used to describe the change of the binding energy, for compound and alloys $m_b \approx 4$.

An NW with modified surface layers can be treated as a composite with a core–shell structure [12], composed of a cylinder core having modulus of bulk material E_0 and a surface shell coaxial with the core but having a surface modulus E_s which is correlated to the surface bond length contraction, Eq. (67). For simplicity the relaxed surface layers are approximated as a uniform shell processing an average bond contraction through the overall thickness r_s. The size-related elastic properties can be analyzed in terms of the approximated core–shell composite NW model. The model predicts:

$$\frac{E}{E_0} = 1 + 8\left(\frac{E_s}{E_0} - 1\right)\left[a_1\left(\frac{r_s}{d}\right) + a_2\left(\frac{r_s}{d}\right)^2 + a_3\left(\frac{r_s}{d}\right)^3 + a_4\left(\frac{r_s}{d}\right)^4\right], \tag{68}$$

where $a_1 = 1$, $a_2 = -3$, $a_3 = 4$, and $a_4 = -2$. For nano-cantilevers, a corresponding sandwich-panel model is applicable with $a_1 = 3/4$, $a_2 = -3/2$, $a_3 = 1$, and $a_4 = 0$. The curves shown in Fig. 23 are results of fitting of Eq. (68). It can be seen that the model of surface elasticity fits reasonably well with the experimental results except for CuO, Ag, and Pb.

Surface stress f_s is a reversible work required to elastically stretch unit surface area. For isotropic surfaces suffering from elastic strain ε_s, it is defined as [11]:

$$f_s = \frac{\partial \gamma_L}{\partial \varepsilon_s} = \gamma + \frac{\partial \gamma}{\partial \varepsilon_s}, \tag{69}$$

where γ is surface free energy per unit area, and $\gamma_L = (S/S_L)\gamma$ is that defined in the Lagrangian coordinate system. S_L is the surface area measured with respect to a standard state of strain, and remains unchanged during elastic deformation. A surface element with area S_L in the reference state is stretched into area $S = (1 + \varepsilon_s)S_L$, but the number of lower-coordinated surface-atoms is unchanged. Therefore the surface density of atoms decreases after stretching, and thus the value γ is reduced. The value γ_L excludes this purely geometrical effect.

Surface atoms would have an equilibrium length different from that of the interior atoms as mentioned above. This is not the case for the bond length in the lateral direction, see the length A_s–B_s in Fig. 24. The underling lattice constrains the contraction of the lateral bond. As a result, the interior of the solid can be considered as exerting a stress on the surface, which is the origin of surface stress. Stretching a surface is accompanied by the work done by the surface stress.

Tensile tests, which result in an increase of total surface area of NWs, may be influenced by the surface stress. Further, nanostructures would be internally stressed by the surface stress. The internal stress increases with decreasing the thickness, and may induce a modification of the elastic modulus measured due to a nonlinearity of the stress–strain relation.

4 Small-Span Bending Test

4.1 Importance of Bending Test

The tensile test may be the simplest one and the mechanical properties of materials can be directly determined from the stress–strain relationship. However, the tensile test for small-scaled materials is not yet easy task because of many difficulties in testing. For example, usually the elongation of such materials in testing, especially for metallic materials, is very small, and therefore, strain is very difficult to be accurately measured. Moreover, in case of thin wire specimens, they are likely to be broken at the grips due to unacceptable scratch and the stress concentration at these regions.

The very-thin wires prepared by vapor–liquid–solid [33], spuattering [35], or atomic diffusion [50] techniques are usually on the substrate and one end of the wires is rigidly fixed. Moreover, one-end-fixed beam structure of thin wire can be obtained by advanced welding technology for small-scaled materials, e.g., electron-beam-induced deposition of metal [74], Joule heat welding [66], etc. Therefore, the bending tests are one of the best choices to determine the mechanical properties of small-scaled objects. In this section, small-span bending (SSB) test is introduced, which was developed to characterize the mechanical properties of local region of thin wires.

4.2 Testing Scheme

Let us consider the linear hardening elastic–plastic material. The relationship between stress, σ, and strain, ε, for the material can be expressed as

$$\varepsilon = \frac{\sigma}{E}, \quad \text{for} \quad \sigma \leq \sigma_Y, \tag{70}$$

and

$$\varepsilon = \frac{\sigma_Y}{E} + \frac{\sigma - \sigma_Y}{E'}, \quad \text{for} \quad \sigma > \sigma_Y, \tag{71}$$

where E, σ_Y and E' are Young's modulus, yield stress and hardening modulus, respectively, and these three parameters should be decided for describing the elastic–plastic properties of the material.

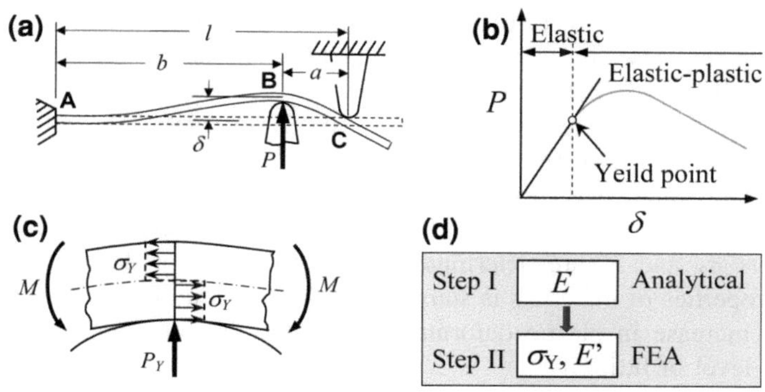

Fig. 25 Conception for testing methodology. **a** Schematic of local small-span bending configuration. **b** Elastic and elastic–plastic regions on a typical force–displacement curve. **c** Assumed distribution of yield stress along diameter of the wire at yield point. **d** Flow of determining elastic–plastic properties of materials

The configuration of the SSB test is illustrated in Fig. 25a. A one-end-fixed sample is considered to be bended at its free end by two opposite probes. As the loading is considered close to a support C and far from the fixed end A, higher stress field due to large bending occurs locally across the loading probe B and this eliminates the problem of failure at the root of the sample.

Figure 25b illustrates a typical relationship between the load applied by the probe at B (P) and the displacement at the loading point in the loading direction (δ) obtained by the bending test, and the relationship contains the elastic and elastic–plastic regions. The property E can be obtained from the load–displacement relation in the elastic zone as

$$E = \frac{Pb^3}{3I\delta} - \frac{Pb^4(2l+a)^2}{12l^3I\delta}, \tag{72}$$

where b the distance between the fixed end A and the loading point B, a the distance between the loading point B and the simply-supported end C, and l the total span length. The symbol I is the moment of inertia of area of the sample and is given by $\pi d^4/64$ for wire, where d is the diameter of the wire.

If we consider the whole cross-sectional area of the wire yielded as shown in Fig. 25c, the moment at the yielding point corresponding to yield load, P_Y, can be described as,

$$M = \frac{4}{3}\sigma_Y\left(\frac{d}{2}\right)^3 = \frac{P_Y ab^2}{2l^3}(2l+a). \tag{73}$$

From Eq. (73), yield stress σ_Y is given by

$$\sigma_Y = \frac{3P_Y ab^2}{l^3d^3}(2l+a). \tag{74}$$

Remained parameter, E' is difficult to be determined analytically because it is non-linear problem. Also, the distribution of yield stress as considered (Fig. 25c) is valid for fully plastic behavior of materials.

Therefore, for accurate estimation of σ_Y and E' we employ elastic–plastic finite element analysis (FEA). In this case, the value for σ_Y obtained from Eq. (74) is taken as the initial guess for FEA. By searching the best fitting of the load–displacement relationship obtained by FEA with the experimental one [65], the parameters σ_Y and E' are determined. The procedure for determining elastic–plastic properties of materials is summarized in Fig. 25d. The smaller a gives an effective increase in plastic deformation, and this is important to improve the accuracy level of fitting.

4.3 Example of Small-Span Bending Test

4.3.1 Sample

Extremely thin Pt wires with the nominal diameter of 625 nm were examined. The ultrathin wires were Ag-coated with the overall diameter of about 75 μm. This type of wire is known as Wollaston wire [49]. The Wollaston wires were cut into 10 mm lengths. The Ag coatings around the tips of the cut pieces were removed by HNO_3 to expose Pt.

To obtain the one-end-fixed beam structure on a suitable substrate for Pt wire as a testing sample, we adopted the welding and cutting techniques utilizing Joule heating [64]. In this experiment, Indium (In) wire of diameter 500 μm was taken as substrate material. The thin Pt wire was welded onto the In wire substrate by Joule heat welding technology. Here the current was locally supplied to the Pt–In system to avoid the structural changes in testing section of Pt wire. The details of cutting and welding thin wires are described in chapter titled Modification of

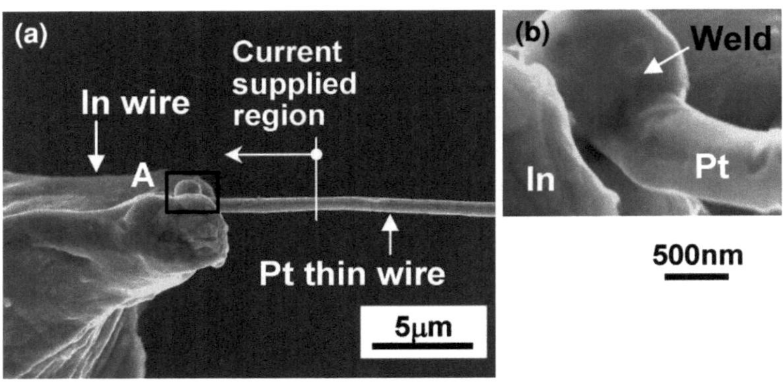

Fig. 26 SEM micrographs for Pt-In joint. **a** Whole view of the fabricated sample. **b** Magnified view of the joint

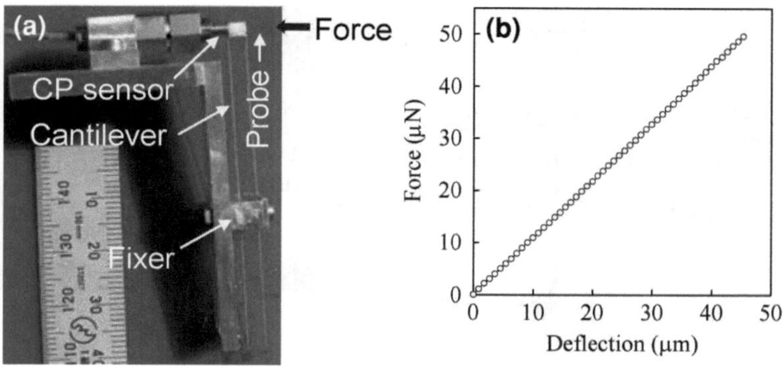

Fig. 27 **a** Compact force sensor. **b** Calibration curve for force sensor

Nano/Micromaterials. Figure 26 shows an SEM micrograph of a typical weld joint between the Pt wire and the In substrate. The weld joint was found to be rigid. In the bending experiment, the welded end of the Pt wire is, therefore, considered as the fixed end of the testing sample.

4.3.2 Experimental Setup

Figure 27a shows the force sensor developed to use with the combination of various types of microscopes for mechanical testing of small-scaled objects [1]. This compact, integrated force sensor is composed of a cantilever and a capacitive (CP) sensor. The force is determined from the deflection of the cantilever measured by the CP sensor (Fig. 27b). The range for CP sensor used was up to 25 μm, and the maximum load as sensor capacity was 27.5 μN. It was experimentally confirmed that the measurement uncertainty level of this type of force sensor was found within 0.4% of maximum load. The resolution of the present force sensor was about 10 nN. More details about the compact, integrated force sensor are described in the literature [1].

Figure 28a shows the test setup. A force sensor, a piezo stage and two manipulators (one carrying the sample and the other carrying the piezo stage), were set on a single platform. The sample was taken on a carrier and was placed just touching the fixed probe by using manual manipulation. The loading probe of the force sensor was placed against the sample as shown in Fig. 28b. The deformation and load were recorded in a computer directly. The deformation pattern was also monitored by a high-resolution digital microscope.

4.3.3 Results

Figure 29a shows an example of the experimental force–displacement (P–δ) relationship obtained under the condition of $l = 60$ μm and $a = 10$ μm. Also the

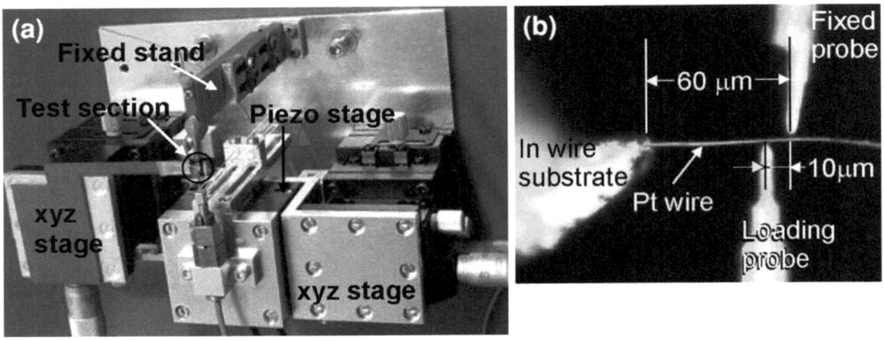

Fig. 28 **a** Whole view of the testing setup. **b** Digital microscope image of loading configuration

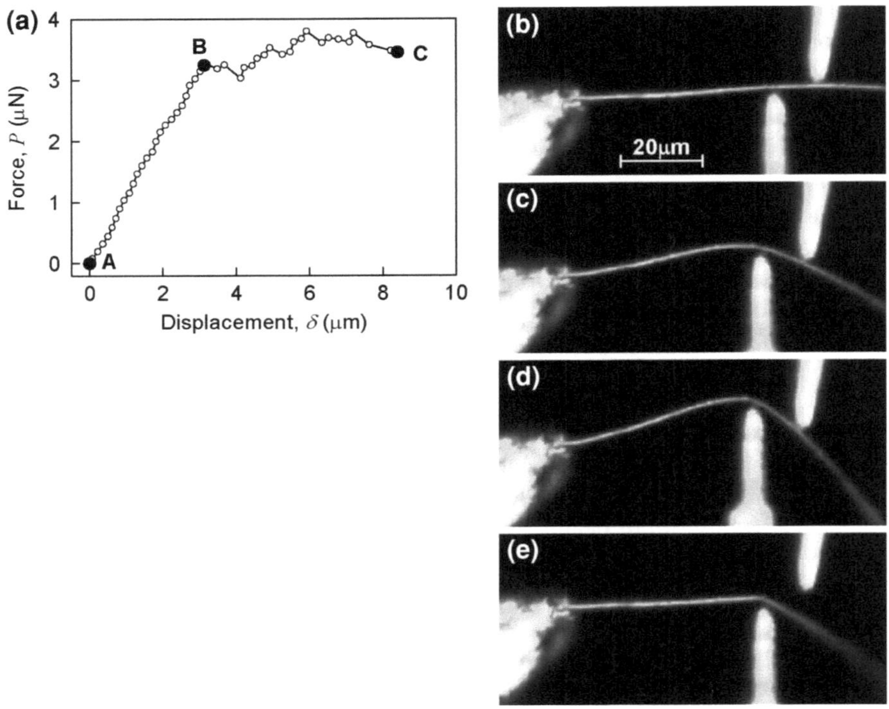

Fig. 29 **a** An example of the load–displacement relationship (l = 60 μm, a = 10 μm). The deformation images of the wire at the states A to C in **a** are, respectively, shown in **b–d**, and that after unloading is shown in **e**. Note that the wire was yielded under the condition of the point B. Reprinted from Tohmyoh et al. [66] with kind permission of Springer Science and Business Media

deformation patterns are displayed in Figs. 29b–d for the loading states A, B and C in Fig. 29a, respectively. The pattern after unloading is shown in Fig. 29e. From the bending patterns, it was clear that less bending occurred at the root of the wire and the maximum bending took place across the tip of loading probe. After unloading, the root region of the wire returned to its original shape whereas a permanent bending remained at the loading point, see Fig. 29e. This confirmed that large plastic deformation in the wire was successfully achieved across the closely-coupled loading probes without affecting the root of the wire. In P–δ

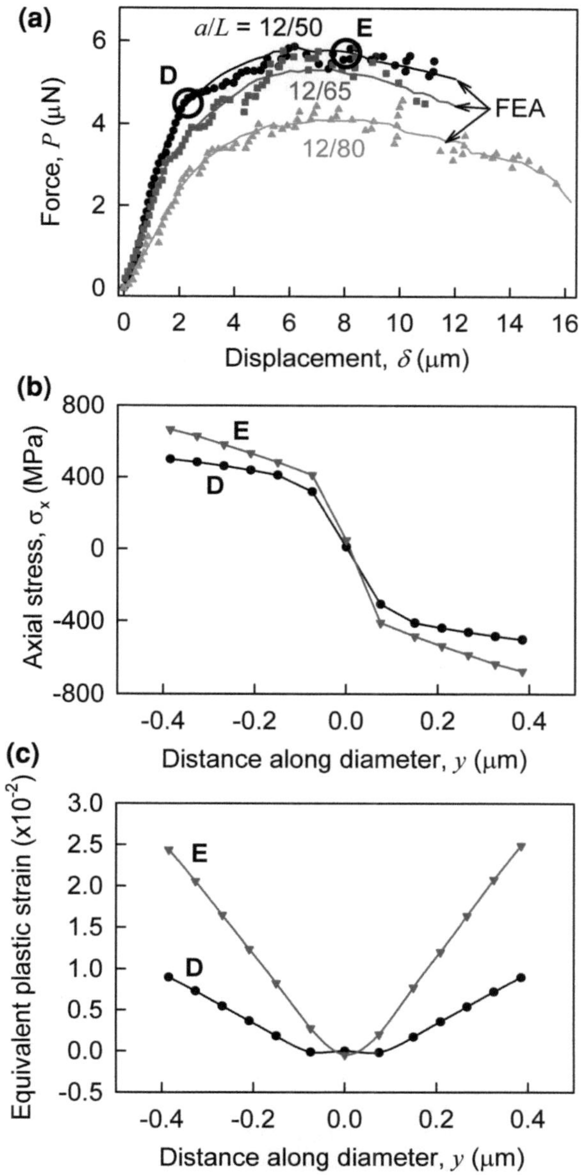

Fig. 30 **a** The P–δ relations for various combination of l and a together with the results of FEA after non-linear fitting. The distributions of **b** stress and **c** plastic strain of the wire along diameter at the location of bending are obtained by FEA. Reprinted from Tohmyoh et al. [66] with kind permission of Springer Science and Business Media

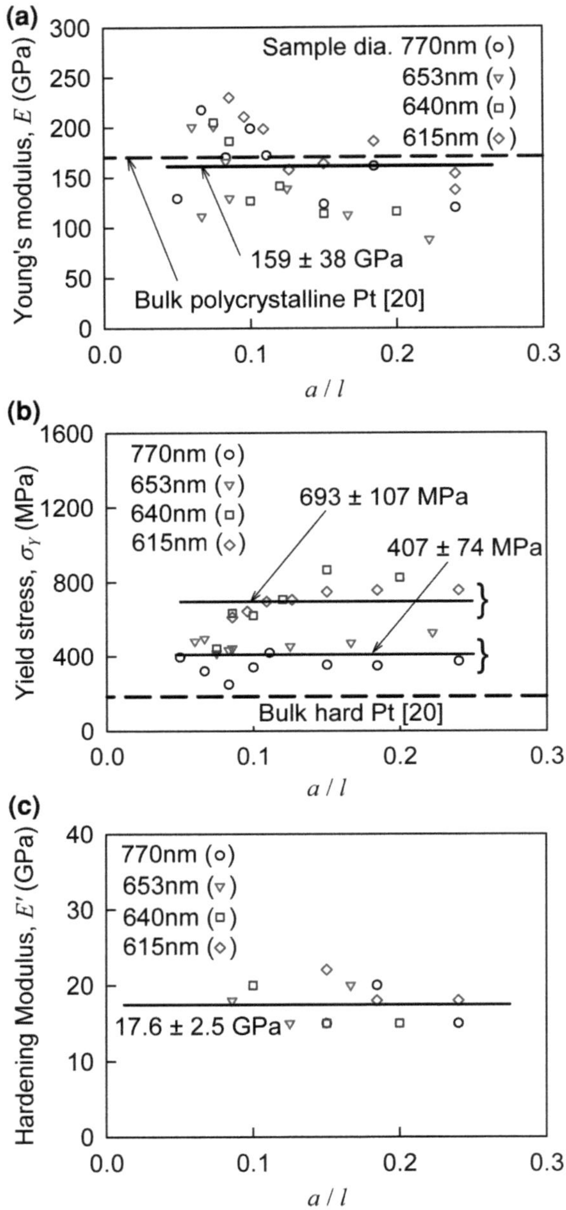

Fig. 31 **a**E versus a/l. **b**σ_Y versus a/l. **c**E' versus a/l. The available values for bulk Pt are shown in the figures. Reprinted from Tohmyoh et al. [66] with kind permission of Springer Science and Business Media

relationships for all conditions, the proportional limit was clearly observed, e.g., point B in Fig. 29a.

Figure 30a shows the P–δ relations for various combinations of l and a together with the results of FEA after non-linear fitting. In the FEA the values of σ_Y and E'

are searched for which the curve gives the best fit to the experimental data. In FEA the probe-sample friction coefficient was assumed to be 0.2. The results of FEA are in good agreement with their corresponding experimental measurements, and this fact indicated that the non-linear fittings up to a large deformation range were successfully achieved. In other words, Eqs. (70) and (71) were valid for describing the mechanical behavior of the examined Pt wires. In P–δ relation as shown in Fig. 30a, the point D indicates the onset of nonlinearity, which is considered here as the yield point and the point E indicates a state of large plastic loading. The distributions of axial stress along the diameter parallel to the direction of loading for the two states D and E were determined by FEA and they are presented in Fig. 30b. The stress distribution in the wire at the yield point has slight difference with that considered for determining σ_Y analytically; see Fig. 25c. Therefore, without using FEA, the value of σ_Y determined from Eq. (74) will be rough and it is found 14% higher than the FEA estimation. Figure 30c shows the distributions of the equivalent plastic strain of these loading states D and E. The distribution for state D indicates that the full cross-section except a narrow part at the center of the wire suffers plastic strain. From the figure, it is also clear that the plastic strain at the loading point increases with increasing probe displacement. The nonlinear relationships for P and δ after yielding are mainly found to be due to the plastic deformation of the wire.

Figures 31a–c display the determined values of E, σ_Y and E' as a function of a/l, respectively. In the figures, the values reported for bulk Pt are also shown. The value of E is determined analytically from Eq. (72), and the values of σ_Y and E' are determined by fitting the experimental results with the corresponding FEA results. The average value of E was 159 ± 38 GPa, and the value is close to that of the reported bulk Pt [3]. On the other hand, the values of σ_Y vary and the values were divided into two groups. The average value of σ_Y for group 1 was 407 ± 74 MPa and that for group 2 was 693 ± 107 MPa. Both the average values of σ_Y were quite higher than the reported yield stress of bulk Pt. The values of E' did not vary and the average value was found as 17.6 ± 2.5 GPa. In the present experiments, the values of E and E' were independent of the samples. From the experimental fact, it was concluded that the difference of σ_Y for each group might be brought by the difference in the crystalline structure. Although the bulk Pt usually shows perfectly-plastic behavior, the examined ultrathin Pt wire showed work hardening effect. This may be due to the special crystalline structure of the Pt wire, i.e., the bamboo structure. Few grain boundaries were existed in the testing section on the wire and this attributed higher resistance for movement of atoms.

Acknowledgments M.Muraoka. acknowledges partial support from the Japan Society for the Promotion of Science (JSPS), through the Grant-in-Aid for Scientific Research (B) Grant No. 20360049 and Ms. Y. Ishigami and K. Kanazawa for their help in preparing the manuscript. H.Tohmyoh. acknowledges partial support from the Ministry of Education, Culture, Sports, Science and Technology (MEXT), Japan under Grant-in-Aid for Young Scientists (A) Grant No. 21686012 and Dr. M.A.S. Akanda for his thankful discussion in preparing the manuscript.

References

1. Akanda, M.A.S., Tohmyoh, H., Saka, M.: An integrated compact unit for wide range micro-newton force measurement. J. Solid Mech. Mater. Eng. **4**, 545–556 (2010)
2. Akita, S., Nishijima, H., Kishida, T., Nakayama, Y.: Influence of force acting on side face of carbon nanotube in atomic force microscopy. Jpn. J. Appl. Phys. **39**, 3724–3727 (2000)
3. ASM International: Metals Handbook, 10th edn, vol. 2, Properties and Selection: Nonferrous Alloys and Special-Purpose Materials. Materials Park, OH (1990)
4. Bai, X.D., Gao, P.X., Wang, Z.L.: Dual-mode mechanical resonance of individual ZnO nanobelts. Appl. Phys. Lett. **82**, 4806–4808 (2003)
5. Binnig, G., Quate, C.F., Gerber, C.: Atomic force microscope. Phys. Rev. Lett. **56**, 930–933 (1986)
6. Binnig, G., Rohrer, H.: Scanning tunneling microscopy. Helv. Phys. Acta. **55**, 726–735 (1982)
7. Bohren, C.F., Huffman, D.R.: Absorption and Scattering of Light by Small Particles. Wiley, New York (1983)
8. Bowman, F.: Introduction to Elliptic Functions with Applications. Dover, New York (1961)
9. Broughton, J.Q., Meli, C.A., Vashishta, P., Kalia, R.K.: Direct atomistic simulation of quartz crystal oscillators: bulk properties and nanoscale devices. Phys. Rev. B **56**, 611–618 (1997)
10. Burnham, N.A., Gremaud, G., Kulik, A.J., Gallo, P.-J., Oulevey, F.: Materials' properties measurements: choosing the optimal scanning probe microscope configuration. J. Vac. Sci. Technol. B **14**, 1308–1312 (1996)
11. Cammarata, R.C.: Surface and interface stress effects in thin films. Prog. Surf. Sci. **46**, 1–38 (1994)
12. Chen, C.Q., Shi, Y., Zhang, Y.S., Zhu, J., Yan, Y.J.: Size dependence of Young's modulus in ZnO nanowires. Phys. Rev. Lett. **96**, 075505(1–4) (2006)
13. Cuenot, S., Frétigny, C., Demoustier-Champagne, S., Nysten, B.: Surface tension effect on the mechanical properties of nanomaterials measured by atomic force microscopy. Phys. Rev. B **69**, 165410(1–5) (2004)
14. Gao, R., Wang, Z.L., Bai, Z., de Heer, W.A., Dai, L., Gao, M.: Nanomechanics of individual carbon nanotubes from pyrolytically grown arrays. Phys. Rev. Lett. **85**, 622–625 (2000)
15. Giessibl, F.J.: Atomic resolution of the silicon (111)-(7 $\times$ 7) surface by atomic force microscopy. Science **267**, 68–71 (1995)
16. Gilman, J.J.: Cleavage, ductile, and tenacity in crystals. In: Averbach, B.L., Felbeck, D.K., Hahn, G.T., Thomas, D.A. (eds.) Fracture. Technology Press of MIT, Cambridge (1959)
17. Hoffmann, S., Utke, I., Moser, B., Michler, J., Christiansen, S.H., Schmidt, V., Senz, S., Werner, P., Gösele, U., Ballif, C.: Measurement of the bending strength of vapor-liquid-solid grown silicon nanowires. Nano Lett. **6**, 622–625 (2006)
18. Hoffmann, S., Östlund, F., Michler, J., Fan, H.J., Zacharias, M., Christiansen, S.H., Ballif, C.: Fracture strength and Young's modulus of ZnO nanowires. Nanotechnology **18**, 205503–205507 (2007)
19. Hoummady, M., Farnault, E.: Enhanced sensitivity to force gradients by using higher flexural modes of the atomic force microscope cantilever. Appl. Phys. A **66**, S361–S364 (1998)
20. Jiang, X., Herricks, T., Xia, Y.: CuO nanowires can be synthesized by heating copper substrates in air. Nano Lett. **2**, 1333–1338 (2002)
21. Kis, A., Mihailovic, D., Remskar, M., Mrzel, A., Jesih, A., Piwonski, I., Kulik, A.J., Benoit, W., Forró, L.: Shear and Young's moduli of MoS_2 nanotube ropes. Adv. Mater. **15**, 733–736 (2003)
22. Kolosov, O., Yamanaka, K.: Nonlinear detection of ultrasonic vibrations in an atomic force microscope. Jpn. J. Appl. Phys. **32**, L1095–L1098 (1993)
23. Kulkarni, A.J., Zhou, M., Ke, F.J.: Orientation and size dependence of the elastic properties of zinc oxide nanobelts. Nanotechnology **16**, 2749–2756 (2005)

24. Li, X.D., Gao, H.S., Murphy, C.J., Caswell, K.K.: Nanoindentation of silver nanowires. Nano Lett. **3**, 1495–1498 (2003)
25. Li, X., Ono, T., Wang, Y., Esashi, M.: Ultrathin single-crystalline cantilever resonators: fabrication technology and significant specimen size effect on Young's modulus. Appl. Phys. Lett. **83**, 3081–3083 (2003)
26. Li, X., Wang, X., Xiong, Q., Eklund, P.C.: Mechanical properties of ZnS nanobelts. Nano Lett. **5**, 1982–1986 (2005)
27. Liang, H., Upmanyu, M., Huang, H.: Size-dependent elasticity of nanowires: nonlinear effects. Phys. Rev. B **71**, 241403(1–4) (2005)
28. Love, A.E.H.: A Treatise on the Mathematical Theory of Elasticity, 2nd edn. Cambridge University, Cambridge (1906)
29. Maivald, P., Butt, H.J., Gould, S.A.C., Prater, C.B., Drake, B., Gurley, J.A., Elings, V.B., Hansma, P.K.: Using force modulation to image surface elasticities with the atomic force microscope. Nanotechnology **2**, 103–106 (1991)
30. Marrian, C.R.K.: Technology of proximal probe lithography. SPIE Optical Engineering, Bellingham (1993)
31. Martin, Y., Wickramasinghe, H.: Magnetic imaging by 'force microscopy' with 1000 Å resolution. Appl. Phys. Lett. **50**, 1455–1457 (1987)
32. Mate, C.M., McClelland, G.M., Erlandsson, R., Chiang, S.: Atomic-scale friction of a tungsten tip on a graphite surface. Phys. Rev. Lett. **59**, 1942–1945 (1987)
33. Miao, W.G., Wu, Y., Zhou, H.P.: Morphologies and growth mechanisms of aluminium nitride whiskers. J. Mater. Sci. **32**, 1969–1975 (1997)
34. Miller, R.M., Shenoy, V.B.: Size-dependent elastic properties of nanosized structural elements. Nanotechnology **11**, 139–147 (2000)
35. Motoyama, M., Fukunaka, Y., Sakka, T., Ogata, Y.H., Kikuchi, S.: Electrochemical processing of Cu and Ni nanowire arrays. J. Electroanal. Chem. **584**, 84–91 (2005)
36. Muraoka, M.: Sensitive detection of local elasticity by oscillating an AFM cantilever with its mass concentrated. JSME Int. J. A **45**, 567–572 (2002)
37. Muraoka, M.: Sensitivity-enhanced atomic force acoustic microscopy with concentrated-mass cantilevers. Nanotechnology **16**, 542–550 (2005)
38. Muraoka, M., Arnold, W.: A method of evaluating local elasticity and adhesion energy from the nonlinear response of AFM cantilever vibrations. JSME Int. J. A **44**, 396–405 (2001)
39. Muraoka, M., Tobe, R.: Mechanical characterization of nanowires based on optical diffraction images of the bent shape. J. Nanosci. Nanotechnol. **9**, 4566–4574 (2009)
40. Nam, C.-Y., Jaroenapibal, P., Tham, D., Luzzi, D.E., Evoy, S., Fischer, J.E.: Diameter-dependent electromechanical properties of GaN nanowires. Nano Lett. **6**, 153–158 (2006)
41. Namazu, T., Isono, Y., Tanaka, T.: Evaluation of size effect on mechanical properties of single crystal silicon by nanoscale bending test using AFM. J. Microelectromech. Syst. **9**, 450–459 (2000)
42. Nilsson, S.G., Borrisé, X., Montelius, L.: Size effect on Young's modulus of thin chromium cantilevers. Appl. Phys. Lett. **85**, 3555–3557 (2004)
43. Okada, S., Mukawa, T., Kobayashi, R., Ishida, M., Ochiai, Y., Kaito, T., Matsui, S., Fujita, J.: Comparison of Young's modulus dependency on beam accelerating voltage between electron-beam- and focused ion-beam-induced chemical vapor deposition pillars. Jpn. J. Appl. Phys. **45**, 5556–5559 (2006)
44. Poncharal, P., Wang, Z.L., Ugarte, D., de Heer, W.A.: Electrostatic deflections and electromechanical resonances of carbon nanotubes. Science **283**, 1513–1516 (1999)
45. Rabe, U., Arnold, W.: Acoustic microscopy by atomic force microscopy. Appl. Phys. Lett. **64**, 1493–1495 (1994)
46. Rabe, U., Janser, K., Arnold, W.: Vibrations of free and surface-coupled atomic force microscope cantilevers: theory and experiment. Rev. Sci. Instrum. **67**, 3281–3293 (1996)
47. Rabe, U., Kester, E., Arnold, W.: Probing linear and non-linear tip-sample interaction forces by atomic force acoustic microscopy. Surf. Interface Anal. **27**, 386–391 (1999)

48. Rogers, B., Pennathur, S., Adams, J.: Nanotechnology: Understanding Small Systems. Taylor & Francis, Boca Raton (2008)
49. Sacharoff, A.C., Westervelt, R.M.: Physical properties of ultrathin drawn Pt wires. Phys. Rev. B **29**, 6411–6418 (1984)
50. Saka, M., Yamaya, F., Tohmyoh, H.: Rapid and mass growth of stress-induced nanowhiskers on the surfaces of evaporated polycrystalline Cu films. Scr. Mater. **56**, 1031–1034 (2007)
51. Salvetat, J.-P., Briggs, G.A.D., Bonard, J.-M., Bacsa, R.R., Kulik, A.J., Stöckli, T., Burnham, N.A., Forró, L.: Elastic and shear moduli of single-walled carbon nanotube ropes. Phys. Rev. Lett. **82**, 944–947 (1999)
52. Salvetat, J.-P., Kulik, A.J., Bonard, J.-M., Briggs, G.A.D., Stöckli, T., Méténier, K., Bonnamy, S., Béguin, F., Burnham, N.A., Forró, L.: Elastic modulus of ordered and disordered multiwalled carbon nanotubes. Adv. Mater. **11**, 161–165 (1999)
53. Sarid, D.: Scanning Force Microscopy with Applications to Electric, Magnetic and Atomic Forces. Oxford University, New York (1994)
54. Segall, D.E., Ismail-Beigi, S., Arias, T.A.: Elasticity of nanometer-sized objects. Phys. Rev. B **65**, 214109(1–10) (2002)
55. Sharma, P., Ganti, S., Bhate, N.: Effect of surfaces on the size-dependent elastic state of nano-inhomogeneities. Appl. Phys. Lett. **82**, 535–537 (2003)
56. Shenoy, V.B.: Atomistic calculations of elastic properties of metallic fcc crystal surfaces. Phys. Rev. B **71**, 094104(1–11) (2005)
57. Sneddon, I.N.: The relation between load and penetration in the axisymmetric Boussinesq problem for a punch of arbitrary profile. Int. J. Eng. Sci. **3**, 47–57 (1965)
58. Song, J., Wang, X., Riedo, E., Wang, Z.L.: Elastic property of vertically aligned nanowires. Nano Lett. **5**, 1954–1958 (2005)
59. Stern, J.E., Terris, B.D., Mamin, H.J., Rugar, D.: Deposition and imaging of localized charge on insulator surfaces using a force microscope. Appl. Phys. Lett. **53**, 2717–2719 (1988)
60. Sun, C.Q., Tay, B.K., Zeng, X.T., Li, S., Chen, T.P., Zhou, J., Bai, H.L., Jiang, E.Y.: Bond-order-length-bond-strength (bond-OLS) correlation mechanism for the shape-and-size dependent of nanosolid. J. Phys. Condens. Mater. **14**, 7781–7795 (2002)
61. Tan, E.P.S., Zhu, Y., Yu, T., Dai, L., Sow, C.H., Tan, V.B.C., Lim, C.T.: Crystallinity and surface effects on Young's modulus of CuO nanowires. Appl. Phys. Lett. **90**, 163112(1–3) (2007)
62. Timoshenko, P.S., Young, D.H.: Elements of Strength of Materials, 5th edn. Van Nostrand, Tokyo (1981)
63. Timoshenko, P.S., Young, D.H., Weaver, W. Jr.: Vibration Problems in Engineering, 4th edn. Wiley, New York (1974)
64. Tohmyoh, H., Imaizumi, T., Hayashi, H., Saka, M.: Welding of Pt nanowires by Joule heating. Scr. Mater. **57**, 953–956 (2007)
65. Tohmyoh, H., Yamanobe, K., Saka, M., Utsunomiya, J., Nakamura, T., Nakano, Y.: Analysis of solderless press-fit interconnections during the assembly process. ASME J. Electron. Packag. **130**, 031007(1–6) (2008)
66. Tohmyoh, H., Akanda, M.A.S., Saka, M.: Small-span bending test for determination of elastic–plastic properties of ultrathin Pt wires. Appl. Phys. A (in press)
67. Tombler, T.W., Zhou, C., Alexseyev, L., Kong, J., Dai, H., Liu, L., Jayanthi, C.S., Tang, M., Wu, S.Y.: Reversible electromechanical characteristics of carbon nanotubes under local-probe manipulation. Nature **405**, 769–772 (2000)
68. Treacy, M.M., Ebbesen, T.W., Gibson, J.M.: Exceptionally high Young's modulus observed for individual carbon nanotubes. Nature **381**, 678–680 (1996)
69. Vairac, P., Cretin, B.: Scanning microdeformation microscopy in reflection mode. Appl. Phys. Lett. **68**, 461–463 (1996)
70. Walters, D.A., Ericson, L.M., Casavant, M.J., Liu, J., Colbert, D.T., Smith, K.A., Smalley, R.E.: Elastic strain of freely suspended single-wall carbon nanotube ropes. Appl. Phys. Lett. **74**, 3803–3805 (1999)

71. Wiesendanger, R.: Scanning Probe Microscopy and Spectroscopy. Cambridge University, Cambridge (1994).
72. Wolfram, S.: The Mathematica Book, 3rd edn. Wolfram Media and Cambridge University, Cambridge (1996)
73. Wong, E.W., Sheehan, P.E., Lieber, C.M.: Nanobeam mechanics: elasticity, strength, and toughness of nanorods and nanotubes. Science **277**, 1971–1975 (1997)
74. Wu, B., Heidelberg, A., Boland, J.J.: Mechanical properties of ultrahigh-strength gold nanowires. Nat. Mater. **4**, 525–529 (2005)
75. Xiong, Q., Duarte, N., Tadigadapa, S., Eklund, P.C.: Force-deflection spectroscopy: a new method to determine the Young's modulus of nanofilaments. Nano Lett. **6**, 1904–1909 (2006)
76. Yamanaka, K., Nakano, S.: Ultrasonic atomic force microscope with overtone excitation of cantilever. Jpn. J. Appl. Phys. **35**, 3787–3792 (1996)
77. Yamanaka, K., Noguchi, A., Tsuji, T., Koike, T., Goto, T.: Quantitative material characterization by ultrasonic AFM. Surf. Interface Anal. **27**, 600–606 (1999)
78. Yu, M.-F., Lourie, O., Dyer, M.J., Moloni, K., Kelly, T.F., Ruoff, R.S.: Strength and breaking mechanism of multiwalled carbon nanotubes under tensile load. Science **287**, 637–640 (2000)
79. Yu, T., Zhao, X., Shen, Z.X., Wu, Y.H., Su, W.H.: Investigation of individual CuO nanorods by polarized micro-Raman scattering. J. Cryst. Growth **268**, 590–595 (2004)
80. Zhong, Q., Inniss, D., Kjoller, K., Elings, V.B.: Fractured polymer/silica fiber surface studied by tapping mode atomic force microscopy. Surf. Sci. Lett. **290**, L688–L692 (1993)
81. Zhou, L.G., Huang, H.: Are surfaces elastically softer or stiffer? Appl. Phys. Lett. **84**,1940–1942 (2004)

Evaluation of Electrical Properties

Yang Ju

Abstract Electrical property is one of the important intrinsic properties of materials, which strongly affects the functionality of materials, especially for metallic nanomaterials. The evaluation of electrical properties plays a significant role in distinguishing the electrical factors of metallic nanomaterials. In this chapter, the methods for evaluating the electrical properties of metallic nanomaterials are described. First, the recent researches on the measurement of electrical properties of metallic nanomaterials are summarized. Then, four-point probe (FPP) method, which is the most common and effective method to measure the resistivity of metallic nanowires, is introduced. Next, a four-point AFM probe method which combines the conventional FPP method with the atomic force microscope thereby providing a capability to characterize the local resistivity of metallic nanomaterials is represented. Finally, a next-generation technique, microwave AFM method which can measure the topography and distribution of electrical properties of nanomaterials simultaneously, is explained.

1 Introduction

In general, nanomaterials refer to low-dimensional materials whose crystal structures arrange into zero-dimensional dots, one-dimensional chains or two-dimensional planes, such as nanoparticle, nanowire, nanowhisker, nanorod, nanobelt, nanotube and nanofilm. Among which, the one-dimensional metallic nanostructures display a multitude of electrical, optical, chemical and magnetic properties with diverse applications. For example, the important role of metallic

Y. Ju (✉)
Department of Mechanical Science and Engineering, Nagoya University, Furo-cho, Chikusa-ku, Nagoya 464-8603, Japan
e-mail: ju@mech.nagoya-u.ac.jp

M. Saka (ed.), *Metallic Micro and Nano Materials*, Engineering Materials, DOI: 10.1007/978-3-642-15411-9_5, nanowires in semiconductor industry or integrated circuits (IC) may be one of the most attractive topics for scientists and engineers. In this chapter, we focus on the evaluation of the electrical conductivity (resistivity) of one-dimensional metallic nanomaterials, especially metallic nanowires.

Recently, the scale of the interconnects used in IC is continually shrinking towards dimensions comparable with the electronic mean free path (tens of nanometers) or even smaller. Electrical properties such as the conductivity of metallic nanowires will become size-dependent and quite different from that of the bulk materials even at room temperature. Although a lot of worthful works have been done by many researchers, there are still many questions in evaluating conductivity of metallic nanowires on the aspects of either theoretical analysis or experimental measurement. Some typical works and literatures reported by many researchers are summarized as follows.

It is well known that the electrical resistivity of metallic nanomaterials increases once the dimensions decrease below the bulk electronic mean free path. Initial works on thin metallic films by Fuchs and Wills [11] and Sondheimer [52] attributed this effect to diffuse scattering at the film boundaries (surface scattering), which essentially imposes a restriction on the mean free path (FS theory). Their works were then extended to the case of nanowires of different shapes of cross sections including square [36], circular [8], trapezoidal [19] and even arbitrary ones [5]. In this theoretical framework, the Boltzmann transport equation subjects to the condition that a proportion of electron distribution function is independent of direction (diffuse scattering) at the surface. However, the empirical specularity parameter p (the probability of electrons specularly scattering from the surface) in the FS theory has never been interpreted clearly with values varying over a wide range. Moreover, the size-dependent conductivity of nanomaterials seems not to be dominated by surface scattering only.

Towards the end of the 1960s, significant departures from the FS theory were observed [41]. A new mechanism, namely the grain boundary scattering of electrons, was formulated by Mayadas et al. [43] and Mayadas and Shatzkes [42] as MS model. Electron grain boundary scattering occurs for electrons scattering at grain boundary interfaces of electron conductors, such as polycrystalline metallic films and wires. As thinner films or wires are concerned, the mean grain size decreases, leading to the presence of more grain boundaries and hence an increase in the resistivity [10], which makes the conductivity become size-dependent. The MS model introduces another empirical parameter r, which is the reflectivity or the mean probability for an electron to be reflected by a grain boundary. Grain boundary scattering and surface scattering are recognized to be the two most important mechanisms for size-dependent conductivity of metallic nano-conductors at room temperature. However, when both of these two mechanisms are present, the weights of their respective influences are difficult to determine, which calls further systematic experimental investigations.

Other factors that impact electron transport in fine metal wires include contributions by impurities and defects, surface roughness, on-chip contamination, and materials in the cross sections of the wire. The corresponding works were reviewed by Josell et al. [22]. Electron–electron interactions, localization, and electron–phonon interactions known as intrinsic contributions can also influence the conductivity of metallic nanomaterials, especially at extremely low temperature. Quantized effects in metallic nanowires and thin films are significant only for dimensions approaching the Fermi wavelength of materials (a few nanometers), and they are not presently of practical concern [22].

1.1 Conductivity Measurement of Metallic Nanowires

To study the temperature- and size-dependent resistivity of metallic nanowires, many experimental measurements have been carried out since the beginning of 1980s. The general methods of these experiments are introduced as follows.

Evaluating the size-dependent resistivity of metallic nanowires and quantifying the causes of it require that experiments, materials, and device structures are amenable to accurate interpretation and evaluation. As it is known to us, the conductivity is an intrinsic property of material and is difficult to be measured directly. In general, the resistivity of a metallic nanowires is given by

$$\rho = R\frac{A}{l}, \tag{1}$$

where R is the measured resistance. While A and l denote the cross section area and length of the nanowire respectively. As a result, the values which should be measured directly include the geometry, dimensions of the nanowires, as well as the resistance of a single nanowire or several parallel and non-contact nanowires. Sometimes, the grain size should also be measured for the purpose of quantifying grain boundary scattering.

1.2 Geometry (Cross-section Shape) of Nanowires

The nanowires used in early experiments were fabricated by various kinds of lithographic techniques [3, 6, 13, 14, 60]. These wires were completely laid on substrates instead of free-standing. The cross-section shapes which were always with non-smooth angles such as square [3] and triangle [13, 14], can be observed by a scanning electron microscope (SEM). In the experiment reported by Williams and Giordano [61], nanowires were synthesized by electroplating Au into etched nuclear tracks in mica, which were quite different from the lithographic ones. The cross-sections were observed to be circular due to the hollow shapes of the porous mica template. Normally, the metallic nanowires used for resistivity measurements which have been fabricated by various methods can be classified into four kinds. The first kind following the work of Williams is depositing nanowires into pores of porous membrane templates including polycarbonate membranes [1, 7, 35, 58] and anodic alumina membranes [66]. Nanowires fabricated by this method can be free-standing with circular cross-sections which are preferred for their regularity. The second kind of method is depositing metallic nanowires in pre-manufactured trenches on the substrates, which has been frequently utilized by a corporate research group in Infineon [51, 53–56] and others [21, 34, 38, 39, 62, 65]. These trenches can be made by lithography or so-called spacer technique [56]. The cross-sections of nanowires fabicated by this kind of method are always trapezoidal or rectangular. In the third kind of method, the nanowires are directly deposited onto substrates by evaporation [19] or sputtering [18]. The cross-sections are always trapezoidal and difficult to control. Finally, the fourth one is the DNA-templated metallic nanowires [4, 32, 49, 50]. Although the cross-sections of these nanowires constructed by metallic particle clusters attached to the DNA templates are always irregular, they were mostly assumed to be circular for convenience. Also, Yan et al. [63] performed an experiment on more regular nanowires synthesized using more complicated DNA nanostructure templates.

The characterization of cross-sections can be performed by various types of equipments including the transmission electron microscope (TEM), the scanning tunneling microscope (STM) and the atomic force microscope (AFM), as well as the SEM.

1.3 Measuring the Dimensions (Length and Cross-Section Area) of Nanowires

The measurement of the cross-section sizes becomes meaningful when the cross-section shapes have been determined. In general, the dimensions are determined by microscopy observation. However, there are still some tricks in the measurements which should be noted.

For circular nanowires fabricated by porous template assisted synthesis, of which the cross-sections are always regular and nearly constant along the length direction, diameter a the only lateral dimension to be determined. The diameters of circular nanowires can be easily read from the top views perpendicular to the length direction. However, for rectangular and trapezoidal nanowires, microscopic pictures of their cross-sections should be examined carefully to determine the width, height and especially the trapezoid angle. This is the reason why circular nanowires are preferred in experiment [7, 58].

When the cross-sections cannot be kept constant due to fabrication process, the average cross-section areas also can be estimated by calibration measurements. In the works of Hinode et al. [18], Steinhögl et al. [54] and Maîtrejean et al. [38], different resistances of the same nanowire at different temperatures were tested. Then the average cross-section area was estimated as

$$A = \left(\frac{\mathrm{d}\rho}{\mathrm{d}T}\right) \Big/ \left(\frac{\mathrm{d}R}{\mathrm{d}T}\right) L, \tag{2}$$

where A and L denote the cross-section area and wire length, respectively. The term (dR/dT) can be determined by tested resistances at different temperatures. The calibration measurements were always performed around room temperature, at which the term ($d\rho/dT$) can be determined by well-known material constants [54]. This semi-empirical method should be examined carefully by microscopic observation when it is utilized.

On the other hand, when the nanowires are tested in the porous template without being extracted, the pore size should be measured before depositing nanowires into it to pre-determine the diameter of nanowires. This was done by measuring the resistance of a porous template filled with an electrolyte of known resistivity and estimating the number of pores in the template [7]. This method is based on the assumption that all the pores are the same size and should also be carefully verified by microscopic observation.

The length of nanowires under testing ranges from several micrometers to hundreds of micrometers, which can be steadily measured by microscopes.

1.4 Measuring the Resistance of Nanowires

The resistance measurement is the most important step for resistivity evaluation, in which the preparation of nanowires with electrodes connected to a circuit is limited by manipulation difficulties in nano-scale. Most literatures did not give detailed descriptions of the very process. Here, we summarize and present some examples in this section.

For nanowires tested on the same substrate or trench where they are synthesized, or nanowires tested in the porous template without being extracted, the process may be straightforward. The probes can be connected to the nanowires directly or to electrodes deposited on the nanowires. A micrograph of a silver nanowire under four-point probe (FPP) measurement in the experiment of Huang et al. [19] is shown in Fig. 1a. In the experiment performed by Vazquez-Mena et al. [59], the electrodes were deposited together with the nanowire to a substrate through a stencil membrane containing a nanoslit in between two side micrometric apertures, as shown in Fig. 1b.

This process will become somewhat complex for free-standing nanowires. In the experiment performed by Toimil-Molares et al. [58], wires were contacted using two different configurations: wires "*below*" and wires "*above*" the contacts. In the first case, a few drops of solution containing the wires were first placed on a SiO_2 substrate. After evaporation of the solvent, an array of contacting structures, each consisting of four finger-like electrodes was created by means of optical lithography and metal evaporation on top of the wires. In the second case, the electrodes were deposited first, and the wires were subsequently placed on top. In both cases, the number of wires on the substrate was controlled by varying the concentration of wires in the solution and/or the number of drops deposited on the substrate. Measurements can then be performed by use of a straightforward FPP

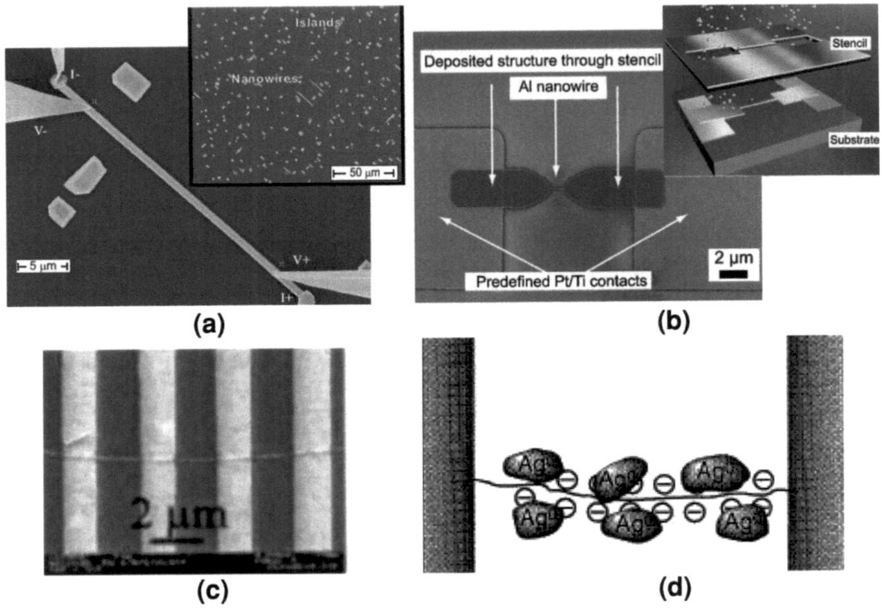

Fig. 1 The micrographs of **a** a nanowire under a four-point probe measurement, [1]**b** the microslit stencil-assisted nanowire and the electrodes, [2]**c** the nanowire connected to four electrodes, [3] and **d** the scheme of assembling a DNA-templated metallic nanowire with electrodes[4]

method with the four electrodes shown in Fig. 1c. For DNA-templated nanowires, the DNA molecules can be stretched and connected to electrodes by molecular combing before the metallic nanowires were deposited [4], as shown in Fig. 1d.

When the nanowire and electrodes are well connected, resistance measurement can be carried out by the FPP method which is able to cancel the influences of contact resistances. The details of the FPP method will be interpreted in Sect. 2. Some of the experiments presented in the literatures were performed by two-point

---

1 Reprinted with permission from Huang, Q.J., Lilley, C.M., Bode, M.: Surface scattering effect on the electrical resistivity of single crystalline silver nanowires self-assembled on vicinal Si (001). Appl. Phy. Lett. **95**, 103112-1. ©2009, American Institute of Physics.

2 Reprinted with permission from Vazquez-Mena, O., Villanueva, G., Savu, V., Sidler, K., van den Boogaart, M.A.F., Brugger, J.: Metallic nanowires by full wafer stencil lithography. Nano Lett. **8**, 3676. ©2008, American Chemical Society.

3 Reprinted with permission from Toimil-Molares, M.E., Höhberger, E.M., Schaeflein, C, Blick, R.H., Neumann, R., Trautmann, C.: Electrical characterization of electrochemically grown single copper nanowires. Appl. Phys. Lett. **82**, 2139. ©2003, American Institute of Physics.

4 Reprinted by permission from Braun, E., Eichen, Y., Sivan, U., Ben-Yoseph, G.: DNA-templated assembly and electrode attachment of a conducting silver wire. Nature **39**, 775. ©1998, Macmillan Publishers Ltd.

probe measuring with contact resistance correction by linear fits [59] or calibrations [1]. In the measurements, the current should be limited to below several microamperes to avoid Joule heating.

1.5 Determining the Grain Size of Nanowires

In order to quantify the grain boundary scattering, the grain structure of the nanowires should also be determined, which can be done by the various microscopies and X-ray diffraction method [7, 10, 35, 39, 65, 66].

2 The FPP Method

2.1 The Advantage of the FPP Method

Two-point probe system is utilized in the conventional resistance measurement. A sketch of a typical two-point probe system is shown in Fig. 2a, where "DUT" stands for device under test. On the micro-scale or the nano-scale, the contact resistance between the DUT and the probes always becomes nonnegligible due to the extremely small contact area. In addition, for DUTs made of conductors, the resistance of the leading wires may also be of comparable magnitude. Therefore, the two-point probe system should be treated as an equivalent circuit in Fig. 2b when the resistance of the DUT is comparable with the contact resistance, or even to the resistance of the leading wire. The current flowing through the DUT is nearly the same as the total current supplied by the current source due to the extremely large internal resistance of the voltmeter (which is on the order of 1MΩ). However, the voltage across the DUT seems to be significantly smaller than that indicated by of the voltmeter due to the voltage drop across the equivalent

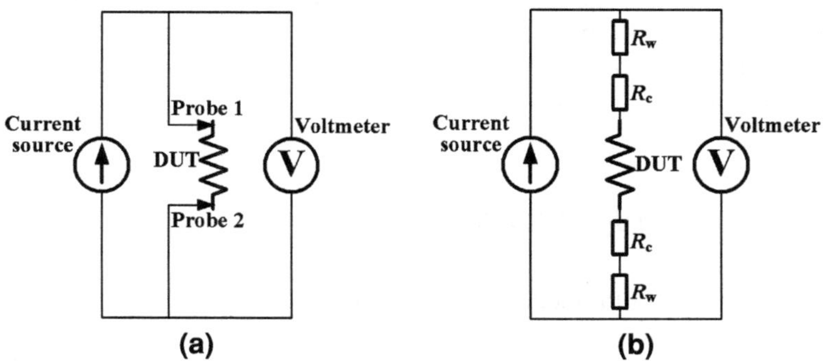

Fig. 2 **a** Sketch of the two-point probe system and **b** its equivalent circuit

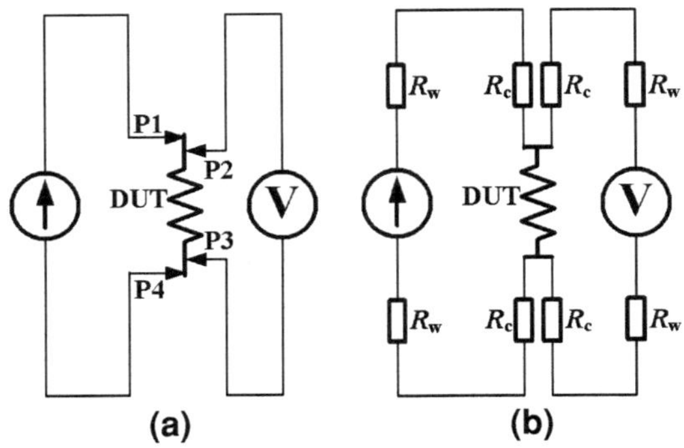

Fig. 3 **a** Sketch of the four-point probe system and **b** its equivalent circuit

resistors R_c and R_w which indicate the contact and wire resistances, respectively. As a result, the measured resistance R_m will be

$$R_m = V_m/I = 2R_c + 2R_w + R_{DUT}, \tag{3}$$

where V_m denotes the indication of the voltmeter, and R_{DUT} is the resistance of the DUT. Thus, it is difficult to extract the exact value of R_{DUT} from the interference factors.

The FPP method has become very popular since it was introduced by William Thomson (Lord Kelvin), who invented the Kelvin bridge in 1861 to measure very small resistances. It is also known as four-terminal sensing or Kelvin sensing.

The probes for the current supply and the voltage measurement are separated in the FPP system, as shown in Fig. 3a. As can be seen from the equivalent circuit shown in Fig. 3b, the voltmeter together with four equivalent resistors are connected in parallel with the DUT. The current mainly flows through the DUT branch and the voltage drop across the equivalent resistors in the other branch should be extremely small due to the large internal resistance of the voltmeter. Therefore, the voltage across the DUT is nearly the same as the indication of the voltmeter. Moreover, the current flowing through the DUT remains the same as the indication of the current source, since a constant current source is commonly used. As a result, the DUT resistance R_{DUT} can be derived directly by dividing the voltmeter indication V_m by the current indication I in a single measurement. In most experimental works referred to in Sect. 1, the resistances of nanowires were measured by the FPP method.

2.2 The Modified FPP Method

Recently, a modified FPP method for measuring the resistances of nanowires was proposed by Gu et al. [15].

In this method, the four metal pads numbered as pads 1, 2, 3, and 4 are connected, by the four leading wires, to the nanowire to be tested, as show in

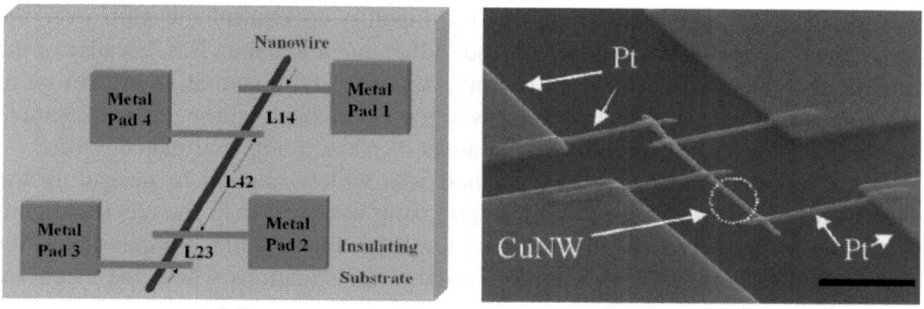

(a) (b)

Fig. 4 **a** Sketch and **b** micrograph of the modified four-point probe system[5]

Fig. 4. Let the contact resistance for pads 1, 2, 3, and 4 be R_{C1}, R_{C2}, R_{C3}, and R_{C4}, which can also include the probing resistance and the wire resistance. The resistance of each section of nanowire between every two pads is R_{L12}, R_{L13}, R_{L14}, R_{L23}, R_{L24} and R_{L34}, respectively.

The typical current–voltage (*I–V*) measurements should be performed between each and every pair of pads. Such measurements are repeated at each applied voltage with the polarities of the electrodes in a given pair switched for all probing paths. Then the exact resistance of the nanowire can be derived from the measured results [15].

The tested resistance cannot be read directly from a single measurement, which may be the main shortcoming of this modified method. However, it is mentioned to be able to cancel the influences of the contact resistance completely no matter they are ohmic or not, thereby being suitable for more accurate measurements.

3 The Four-Point AFM Probe Method

Metallic micro and nano wires have been the subject of intensive studies because of their fundamental properties and potential applications for fabricating nanocircuits and nanodevices, which have provided a revolution in high-density data storage, thin-film displays, sensors and medical diagnostics. So far, however, the development of applications such as nanocircuits and nanodevices has been limited by several critical issues. One of the most important issues is the ambiguous electrical conductivity inside nanowires, with the consequence that nanocircuits and nanodevices do not, usually, function as expected. The inhomogeneity in the

[5] Reprinted with permission from Gu, W., Choi, H., Kim, K.: Universal approach to accurate resistivity measurement for a single nanowire: theory and application. Appl. Phys. Lett. **89**, 253102-1. ©2006, American Institute of Physics.

conductivity of such wires is due to the inability to control the gain over the movement of the electrons within the individual metallic nanowires, because of the lack of chemical and physical uniformity. Accordingly, quantitative evaluation of the nanowires' electrical conductivity is urgently needed to allow the development of nanotechnologies, especially in the areas of nanocircuits and nanodevices.

As mentioned above, the FPP method was widely applied in measuring the electrical properties of metallic micro and nano wires, in that, the electrical conductivity is measured as the current-to-voltage ratio while compensating for geometrical effects. This technique is suitable for situations where the electrical conductivity is homogeneous in a volume that is large compared with the electrode spacing, but is rendered ineffective in the presence of homogeneity over a smaller spacing. In addition, to know the geometrical information of nanowires, a dimension measurement such as SEM observation is required. One way to resolve the intractable problems of the FPP method is to reduce the electrode spacing below the dimensions of the inhomogeneous. Recently, various FPPs have been fabricated by means of photolithography in order to obtain a higher spatial resolution [45, 46]. The minimum spacing between the electrodes achievable in commercial probes is roughly 5.0 μm [16], while micro-FPPs with electrode spacing of several hundreds of nanometers are also reported [2, 44]. However, these probes are difficult to be used for scanning because of the difficulty in manipulating the force between the probe and the sample surface due to their structure and rigidity. Also, at present, in situ dimension measurement at the nanolevel is still a complicated problem.

In this section, we demonstrate a four-point AFM probe technique [24, 29], which is a combination of the principles of the FPP method and standard AFM. The probe not only retains the ability of surface profile imaging but is also capable of characterizing the local conductivity simultaneously. Therefore, the in situ measurement of the electrical properties of metallic micro and nano wires can be realized. The probes can have an electrode spacing as small as several hundreds nanometers and work at AFM contact mode. The average force applied to the surface is smaller than 10 pN, which provides a nondestructive measurement and still ensures good electrical conduction between probe and sample. At the same time, by applying the four-point AFM measurement, the influence of the contact resistance between the electrodes and the sample can be eliminated.

3.1 Fabrication of the Four-Point AFM Probe

Figure 5 shows a micrograph of the four-point AFM probe [29]. The device was fabricated based on modification of a conventional silicon nitride bio-AFM probe with dimension 100 μm × 30 μm × 0.18 μm. A V-shaped tip with a two-dimensional sliced structure is located at the tip of the cantilever. The tip height is 7.0 μm and the apex angle is less than 90°. This unique structure facilitates the

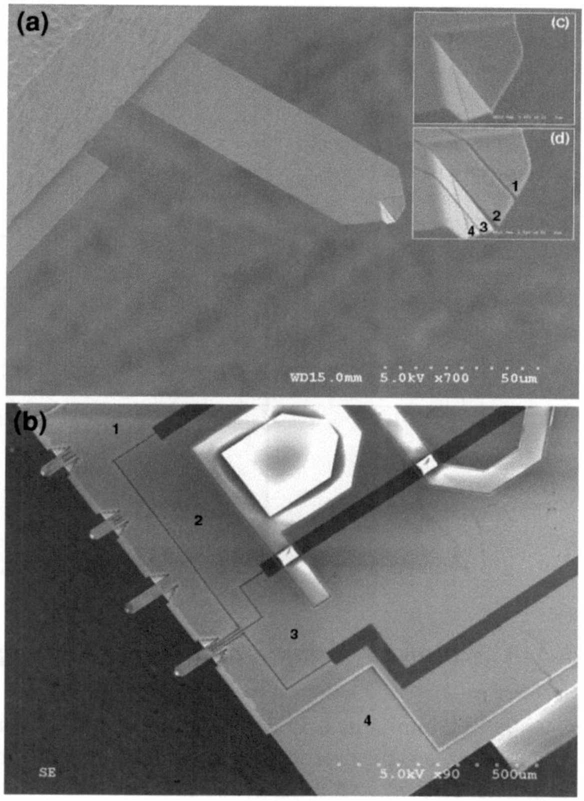

Fig. 5 Scanning electron microscope (SEM) images of the four-point AFM probe, which was fabricated from the V-shaped tip of a conventional AFM probe.[6] **a** The finally patterned four electrodes and their individual shapes. The inset shows the original tip geometry, **c** before and **d** after FIB fabrication. **b** The well-defined circuitry on the probe surface and substrate for the purpose of applied current and electrical potential drop measurements

microfabrication of the four electrodes. The probe is coated with a 30 nm gold film and its cantilever has a spring constant around 6 pN/nm. As shown in Fig. 5a, the electrodes were introduced by fabricating three slits at the tip of the cantilever utilizing a focused ion beam (FIB) system. The spacing of the inner electrode pair (electrode 2, 3) is approximately 300 nm and those of the outer pairs (electrodes 1, 2 and 3, 4) is approximately 1.0 μm. In Fig. 5a, the gold film on the cantilever was etched by FIB, thereby four independent conducting parts were introduced. Figure 5b shows the fabricated substrate of the probe in detail. The gold film was etched to form four disconnected conducting paths. The device allows simultaneous current transmission and detection of the electrical potential drop.

[6] Reprinted with permission from Ju, B.-F., Ju, Y., Saka, M.: Quantitative measurement of submicrometer electrical conductivity. J. Phys. D Appl. Phys. **40**, 7467. ©2007, IOP Publishing Ltd.

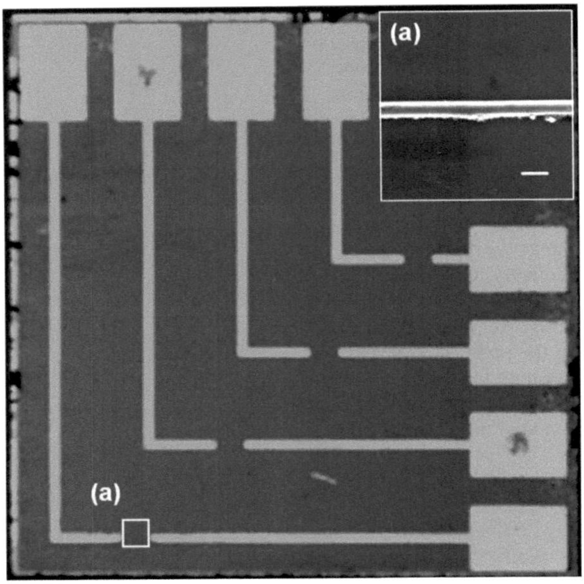

Fig. 6 Optical image showing the 99.999% Al wire that was prepatterned using sputering and ion etching techniques (see footnote 6). The *inset* is an SEM image of the wire that gives more details of its profile and dimensions (*scale* bar 500 nm)

3.2 A Nanowire Sample and Its AFM Image

A 99.999% aluminum wire with a width of 400 nm and a thickness of 200 nm was patterned by a typical procedure of sputtering and ion etching, as shown in Fig. 6 [29]. The wire was on a TiN (200 nm)/SiO_2 (200 nm)/Si (525 μm) substrate.

We carried out simultaneous measurements of both surface topography and electrical conductivity of the Al wire within only one scanning procedure. The image obtained is virtually identical to a standard AFM one, as shown in Fig. 7 [29], which demonstrates the capability of the four-point AFM probe for imaging surface topography. The scanning area is 4.0 × 4.0 μm, and the scanning rate is 0.25 Hz. The measurements were reproducible, but a few fuzzy scratches were observed in Fig. 7. Splitting the tip of the probe into four electrodes may generate potential wobble induced by the force inhomogeneity between the neighboring electrodes and the consequent offset of the reflected AFM laser signal. It is noted that during the shape and dimensions are different for each electrode, they may work at different forces, which may larger than the average value

3.3 Measuring the Conductivity of the Nanowire

The current–voltage relationship acquired during the surface topography scanning of the Al wire is shown in Fig. 8 [29]. In fact, the significant voltage drop clearly indicates simultaneous contact of all four electrodes. Since the

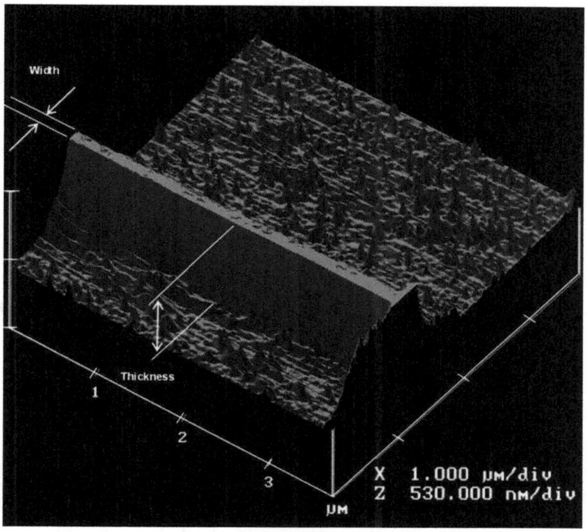

Fig. 7 AFM topography images of the Al wire, which were obtained by the four point AFM probe with a scanning rate of 0.25 HZ, under constant force contact mode (see footnote 6), working environment temperature at 24.4°C, and the relative humidity 46.7%. The area scanned is 4.0 × 4.0 μm

mechanical contact force is small, the spacings between the electrodes are assumed to be constant. During scanning, the two outer electrodes act as the current source and drain, while the inner ones measure the electrical potential drop using a digital voltmeter. A constant current signal is applied in our experiments in order to remove the influence of the contact resistance between the electrode tips and the nanowire surface. During scanning, the current was increasing linearly and the voltage was recorded at the same time. The curve can be divided into five interrelated stages that also distinctly describe the AFM probe progressive scanning procedure: (a) along path *S–A*, the four-point AFM probe scanned the surface of the TiN substrate, where the four electrodes move together parallel to the wire and the row of the electrodes is perpendicular to the wire; the applied current and potential drop measurements were done simultaneously, consequently, the slope of *S–A* is corresponding to the conductivity of the TiN; (b) along path *A–B*, the electrodes were moving across the edge of the nanowire, therefore, the slope of *A–B* indicates the average of the conductivities of the TiN substrate and of the Al nanowire; the electrical potential drop was clearly observed to swing, which is believed to result the instabilities caused by the height difference between the TiN substrate and the Al nanowire; (c) in the case of path *B–C*, the two inner electrodes were located on the Al nanowire and the two outer electrodes were located on the surface of the TiN substrate; since a constant current signal was applied, the slope of *B–C* is corresponding to the conductivity of the Al nanowire; (d) along path *C–D*, the electrodes went down from the nanowire in sequence, a situation similar to that of case (b); (e) in the case of path *D–E*, the electrodes scanned the TiN substrate again, similar to case (a).

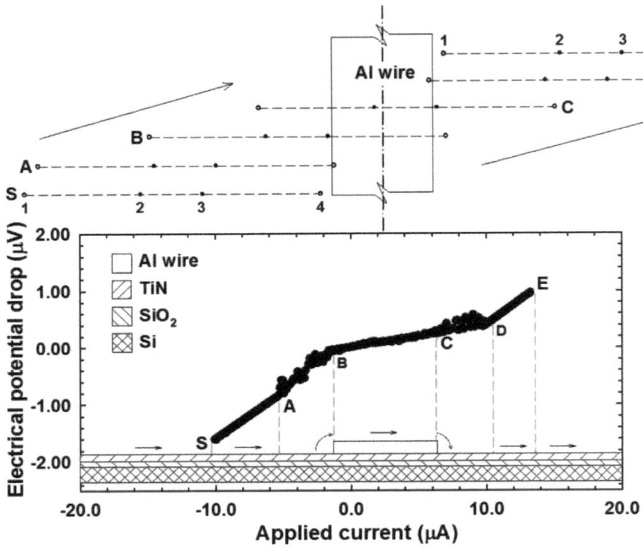

Fig. 8 Typical current–voltage relationship of an Al wire obtained by the four-point AFM probe technique (see footnote 6), which is also capable of revealing the relative position of the four electrodes at each instant during the four-point AFM probe scanning over the wire. The AFM scanning direction was parallel to the Al wire's length. At point *S*, the four electrodes were located on the surface of the TiN substrate, until they reached point *A*, the measured electrical potential drop was only the one of TiN. From instant *A*, electrode *4* began to cross the wire, the current flow through both the TiN substrate and the Al wire. Since electrical potential electrodes *2* and *3* were still on the surface of the TiN substrate, it caused the current–voltage curve to be very close to that of the *S–A* stage. After point *B*, the two potential electrodes began to move across the Al wire until they arrived at point *C*. Starting from point *D*, the four-point AFM probe was leaving the Al wire, and at point *E*, it was located on the surface of the TiN substrate again. The current–voltage relationship of the *D–E* stage remains linear and its slope is the same as that obtained in *S–A* stage

It is worth noting that the current–voltage slopes of *S–A* and *D–E* are almost identical, but the slopes of *A–B* and *C–D* are different from one another. We can conjecture that: at stage (b), one of the outer electrodes of the four-point AFM probe has touched the Al nanowire, and the inner potential electrodes were still remaining on the surface of the TiN substrate, which gave a potential signal very similar to the one of the substrate; at stage (d), the outside electrode has left the Al nanowire, but the potential electrodes were still remaining on the nanowire, so the measured potential was very close to the one of the nanowire. This also clearly demonstrates that the inner two potential electrodes mainly dominate the final measurement of the conductivity.

A series of scanning has also been conducted by applying this four-point AFM probe to other Al microwires with the same thickness but different widths (1.2, 5.0, 12.0 and 25.0 μm). The calculated conductivity versus width of Al wires is plotted in Fig. 9 [29], where the calculation was carried out using the

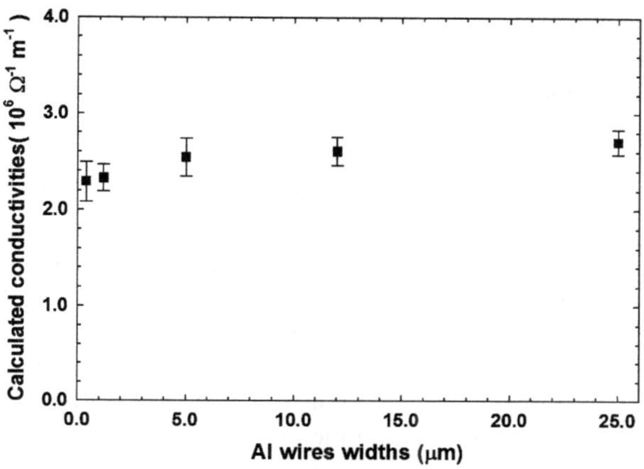

Fig. 9 The calculated conductivities of the Al wires with the same thickness but different widths of 400 nm, 1.2 μm, 5.0 μm, 12.0 μm and 25.0 μm (see footnote 6). The current–voltage measurement for each wire was carried out ten times, the error bars correspond to the mean

semi-infinite model[7] and considering the geometrical effects of the wires. Since our AFM image can provide simultaneously the dimensions of the wire at the nano-level, this ensures the precision of the calculation. As shown in the figure, the calculated conductivities are comparable with each other.

Our four-point AFM probe technique opens the way to in situ quantitative evaluation of the electrical conductivity of metallic micro and nano wires. The technique can self-define the location of the nanowire and facilitates nondestructive measurements. This work provides a basis for fast in situ characterization and fault-finding of submicron interconnects in nanocircuits and nanodevices.

4 The microwave AFM Method

As an evaluation apparatus which has the spatial resolution is on the nanometer scale, the scanning probe microscope (SPM) has become increasingly important. After STM, AFM and near-field scanning optical microscope (NSOM) were invented, various kinds of SPMs based on these microscopes have been developed. The improved SPMs made it possible to measure not only the topography of materials but also the thickness of the oxidized membrane, the profile of the two-dimensional dopant [33], the distribution of the electrical potential and the magnetic field on the material surface [40], the distribution of the hardness and stiffness on the material surface [47] and so on in nanometer order.

On the other hand, development of a technology which is able to measure the electrical properties, such as conductivity, permittivity and permeability, on

[7] $\sigma = \frac{I}{2\pi V}\left[\frac{1}{S_1} + \frac{1}{S_3} - \frac{1}{S_1+S_2} - \frac{1}{S_2+S_3}\right]$. Here, σ is the conductivity of undetermined samples; I and V represent the applied current and corresponding electrical potential drop; S_1, S_2 and S_3 are the spacings between the four electrodes in sequence.

nanometer scale is far in the rear comparing the development of the SPMs noted above. Electrical property is one of the most important basic properties, which has an influence on the functionality of materials. Especially, the electrical properties of materials in a nanoregion are affected by not only the structure and composition of materials but also the mechanical factors of stress and strain related to the lattice oscillation unlike macroscopic characteristics. The measurement of the electrical properties in nanoregion is expected to be used for various fields of creation of nanomaterials, development and evaluation of nanodevice, mechanism elucidation of living tissues and so on.

Recently, microwave microscopies for the measurement of electrical properties and the detection of defects in the microscopic region have been studied. Duewer et al. [9] developed a scanning evanescent microwave microscope (SEMM). They succeeded in measuring the resistivity of Cr, Zr and Mn using the characteristic of microwaves that the resonant frequency changes depending on the capacitance between the probe tip and the material surface. Tabib-Azar and Akiwande [57] were successful in detecting and imaging depletion regions in solar cell p-n junctions in real time with the evanescent microwave probe. Ju et al. [27] were successful in the detection of delamination in integrated circuit packages by applying the properties of microwaves signals that change depending on the electrical properties of the materials.

To evaluate the electrical properties of materials using microwaves, it is necessary to keep the standoff distance between the microwave probe and the sample constant because microwave signals in the near-field are extremely sensitive to the standoff distance. Otherwise, it will be difficult to distinguish that the change of microwave signals is due to a difference of the properties of a measured object or the changing of the standoff distance. Especially, to evaluate the electrical properties of materials in nanometer scale and with high resolution, it is indispensable to control the standoff distance precisely in nanometer order.

To solve these problems, Ju et al. [25, 26, 30, 31] proposed a microwave atomic force microscope (M-AFM). The M-AFM has the characteristics that to keep the standoff distance to be constant by an AFM and that to evaluate the electrical properties of materials quantitatively by a microwave microscope. Combined with these characteristics, M-AFM is able to realize the evaluation of electrical

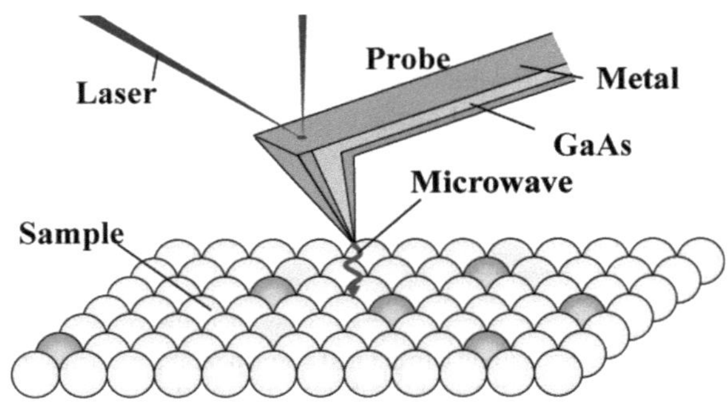

Fig. 10 Schematic diagram of an M-AFM probe used for measuring the electrical properties of materials

properties as well as the measurement of topography of materials simultaneously in nanometer order. In the development of M-AFM, the most important is to fabricate an M-AFM probe constructed with a parallel plate waveguide, in which microwave signals can propagate well. Figure 10 shows a schematic diagram of the M-AFM probe.

4.1 Fabricating the Tip of the M-AFM Probe

To restrain the attenuation of microwaves in the M-AFM probe, GaAs was used as the substrate of the probe. On the other hand, to obtain the desired structure, wet etching was used to fabricate the tip of the probe [25]. Different with dry etching, side-etching will occur under the etching mask. Utilizing this property, a micro tip can be fabricated by etching a wafer, of which a small mask was introduced on the surface in advance. In the case of single crystalline wafer, such as Si and GaAs, the chemical activities are different for different crystalline planes, thereby, the etch rates are also different. Therefore, the side plane obtained on the side of the mask pattern is the most inactive plane (that is the plane having the lowest etching speed) which is parallel to the side of the mask pattern. Consequently, the result of the etching is strong affected by the direction of the mask pattern. On the other hand, GaAs has a sphalerite structure that is more complex than that of Si, which has a similar structure as diamond. Therefore, prediction of the etch effects is very difficult [17, 20, 37].

In the experiment, it was found that only square resist pattern can form a sharp tip. In the case of a hexagonal pattern, the reason that tip was not formed well may be due to the side of the etching mask to be too short. The reason for triangular pattern may due to that there is no crystalline plane parallel to the side of the etching mask. In addition, it was found that one side of the square mask being 45° to the $\langle 011 \rangle$ direction can form a tip with a higher aspect ratio comparing with the case of one side of the resist pattern being parallel to $\langle 011 \rangle$ direction [25].

Figure 11a–c shows the SEM photographs of the tips formed from square etching masks with different side lengths of 13, 14 and 15 μm, respectively. A sharp tip having high aspect ratio (1.4) was fabricated in the case of 14 μm etching mask. An enlarged photograph of the tip is shown in Fig. 11d. The tip is 7 μm high and has a radius of curvature of 50 nm. In the case of 13 μm etching mask, the aspect ratio is 0.85 that is smaller than that of the 14 μm etching mask. In the case of 15 μm etching mask, the expected tip was not obtained and a column shaped structure was formed [25].

4.2 Fabrication of the M-AFM Probe

The process of probe fabrication is shown in Fig. 12 in details [26]: (a) patterning the etching mask for tip generation, (b) forming the tips by wet etching, (c)

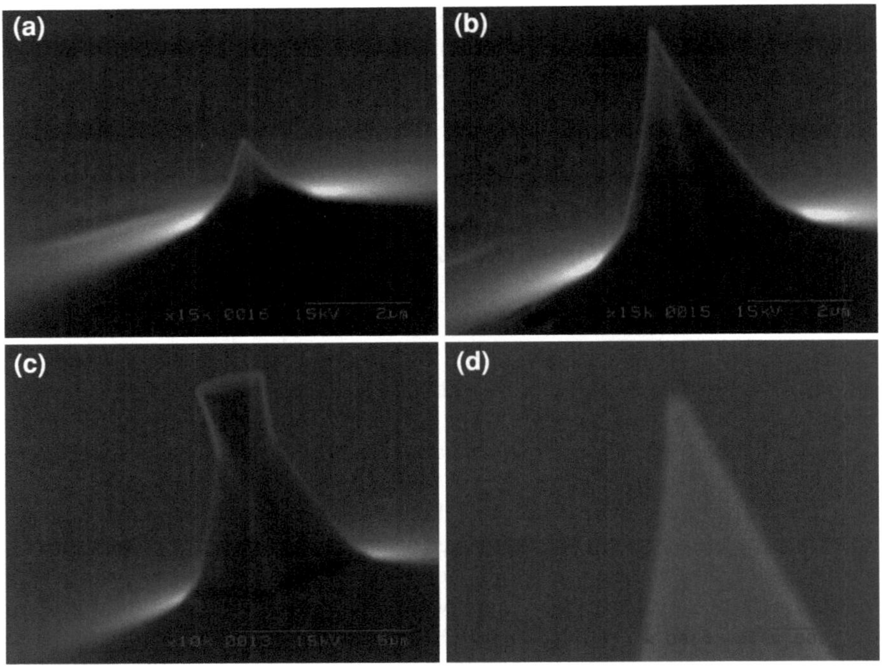

Fig. 11 SEM photograph of the tips fabricated by etching masks with different dimensions:[8] **a** by the13 μm etching mask, **b** by the 14 μm etching mask, **c** by the 15 μm etching mask and **d** enlarged part of the tip fabricated from 14 μm etching mask

patterning the stencil mask for the waveguide and evaporating the metal film, (d) removing the resist and the film, (e) patterning the etching mask for the beam of the cantilever, (f) forming the beam of cantilever by wet etching, (g) patterning the etching mask on the back side for the fabrication of the holder, (h) forming the holder, (i) evaporation of a metal film on the back side and (j) introducing a slit aperture at the tip of the probe.

In the experiment, semi-insulated GaAs wafer having (100) oriented surface and 350 μm thickness was used. At first, the tips were formed by etching the wafer for 100 s to reach the etching depth of 7.7 μm. After that, Au film used to construct the waveguide was evaporated on the substrate. The film thickness was about 50 nm (Fig. 12c). After the deposition, the pattern of the waveguide was formed by the lift-off process, where the film on the resist mask corresponding to the area without the waveguide pattern was removed (Fig. 12d).

[8] Reprinted with permission from Ju, Y., Sato, H., Soyama, H.: Fabrication of the tip of GaAs microwave probe by wet etching. Proceedings of IPACK2005. Page 73140-1. ©2005, American Society of Mechanical Engineers.

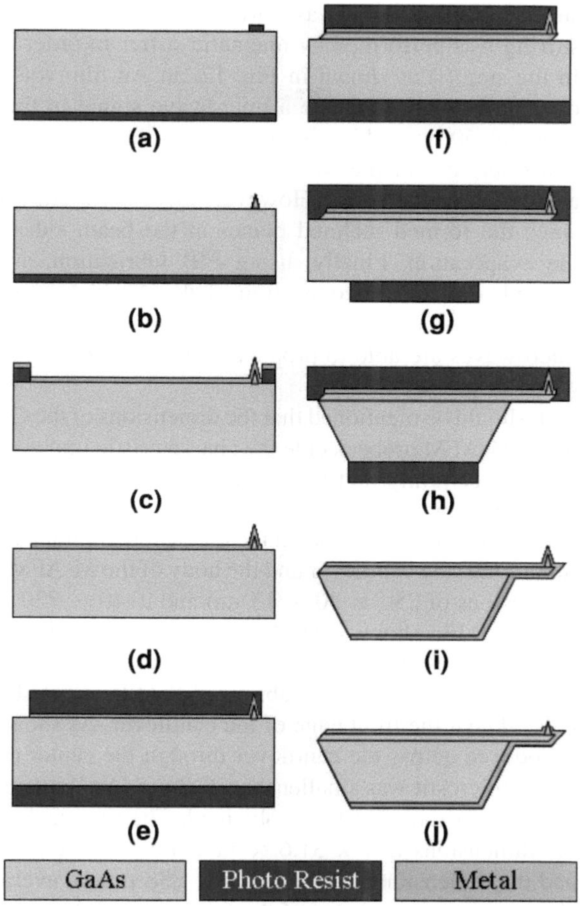

Fig. 12 Fabrication processes of the M-AFM probe.[9] **a** Patterning the etching mask for the generation of tip. **b** Forming the tip by wet etching. **c** Patterning the resist mask for the waveguide. **d** Evaporating the metal film. **e** Removing the resist and metal film. **f** Patterning the etching mask for the beam of cantilever. **g** Forming the beam of cantilever by wet etching. **h** Patterning the etching mask on back side for fabrication of the holder. **i** Forming the holder. **j** Evaporation of metal film on the back side. **k** Introducing the micro slit at the tip of probe

Then, in order to form the beam of the cantilever, the beam etching mask was patterned. Here, by considering the chemical activities at different crystalline planes, the length direction of the etching mask was patterned along the ⟨011⟩ direction. In consequence, the side-etching occurred under the resist mask, and mesa type planes appeared on both sides of the beam (45° inclined plane). On the other hand, inverse-mesa type plane was formed at the end of the beam (60°–75° inclined plane). Etching depth of the beam was about 20 μm [26].

In the same conditions as the beam fabrication process, holder was formed by back side etching (Fig. 12f). Here, the etching mask was patterned on the bottom

---

9 Reprinted with permission from Ju, Y., Kobayashi, T., Soyama, H.: Fabrication of a GaAs microwave probe used for atomic force microscope. Proceedings of IPACK2007. Page 33613-1. ©2007, American Society of Mechanical Engineers.

surface, and etching was carried out until the substrate was penetrated. The stirring was performed by magnetic stirrer in order to etch the sample uniformly. In the step (i) as shown in Fig. 12, an Au film was deposited on bottom surface of the probe to propagate a microwave signal in the probe. The thickness of the film was 50 nm, which is the same as that on the top surface of the probe. Both plane surfaces of the waveguide which were evaporated Au films are connected at the end of the beam. However, there is no Au film on the sides of the beam, since the formed inclined planes at the beam sides do not face the direction of the evaporation. Finally, using FIB fabrication, a slit at the tip of probe was formed to open the connection of the Au film on the two surfaces of the probe [26]. Consequently, a homogeneous parallel plate waveguide was formed and microwaves are able to propagate along the probe and emit at the tip apex of the M-AFM probe.

It should be mentioned that the dimensions of the GaAs substrate and the Au films of the M-AFM probe decide the characteristic impedance of the waveguide, in order to make certainly that microwave signals can propagate properly in the M-AFM probe for maximum sensitivity, the waveguide should have a characteristic impedance of 50Ω (to match the characteristic impedance of a coaxial transmission line). Thus, the cantilever and the body of the M-AFM probe were designed with the dimensions of $250 \times 30 \times 15$ μm and $2740 \times 720 \times 340$ μm, respectively.

Figure 13a shows a SEM photograph of the fabricated M-AFM probes. There 44 probes were fabricated in one process for the one substrate. Figure 13b shows a SEM photograph of the fabricated cantilever of the M-AFM probe. The tip is located near the front edge of the cantilever. As shown in Fig. 13c, a microslit was introduced across the cantilever through the center of the tip by a FIB. The width of the microslit was smaller than 100 nm. As shown in Fig. 13d, a sharp tip with a high aspect ratio (2.0) was obtained. The tip was 8 μm high. The cantilevers had the dimensions $254 \times 31.6 \times 11.1$ μm on average and the bodies of the probes had the dimensions $2{,}743 \times 721 \times 338$ μm on average. The average value of the characteristic impedance of the fabricated M-AFM probes was 35.7 Ω, which was smaller than the desired value.

4.3 Measuring Topography by M-AFM Probe

In order to confirm the spatial resolution of the fabricated M-AFM probes, the AFM topography of two grating samples having 2,000 line/mm and 17.9 nm step height were measured by the M-AFM probes and a commercial Si AFM probe, respectively [30]. Figure 14a–c show the non-contact mode AFM topographies of the grating sample having 2,000 lines/mm obtained by the M-AFM probes. The scan area was 3×3 μm and the white spots in these figures are due to micro-dust on the sample surface. Higher resolution topography was obtained by probe C, which has the highest Q values as compared with probes A and E. Figure 14d shows the non-contact mode AFM topography obtained using the commercial Si

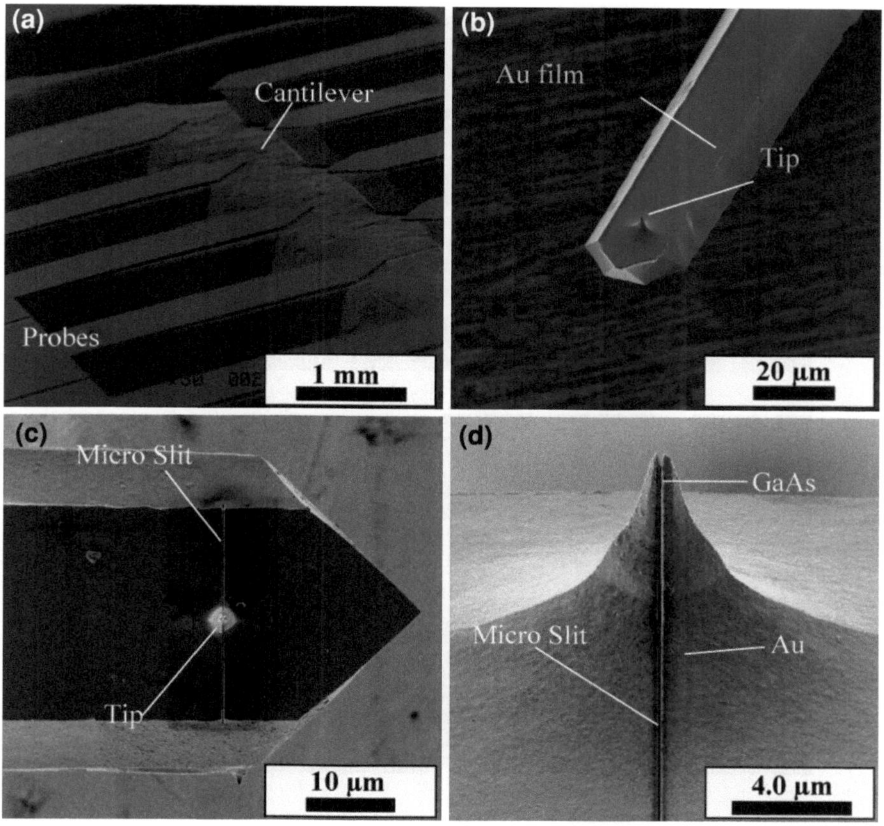

Fig. 13 SEM photograph of the fabricated M-AFM probes: **a** the M-AFM probes on a quarter of GaAs wafer (see footnote 9), **b** a cantilever of the M-AFM probe, **c** a micro slit introduced across the cantilever through the center of the tip with a width no more than 100 nm and **d** high-magnification image of the tip

cantilever. Even though the Q value is lower than that of M-AFM probe C, the commercial Si probe still can obtain higher resolution topography due to the higher aspect ratio of the tip. These results illustrate that the M-AFM probe has a similar capability for sensing the surface topography of materials as that of commercial AFM probes.

In order to evaluate the accuracy of the height measurement, a grating sample having 17.9 ± 1 nm step height was measured using the M-AFM probe C and the Si probe, respectively. As shown in Fig. 15a, by analyzing the scanning profile of the sample, M-AFM probe C reported that a step height of 18.60 nm. In contrast, as shown in Fig. 15b, measurement results obtained by the commercial Si probe indicate that the step height of 18.62 nm. These results suggest that the M-AFM probe has a similar height evaluation capability as that of commercial AFM probes.

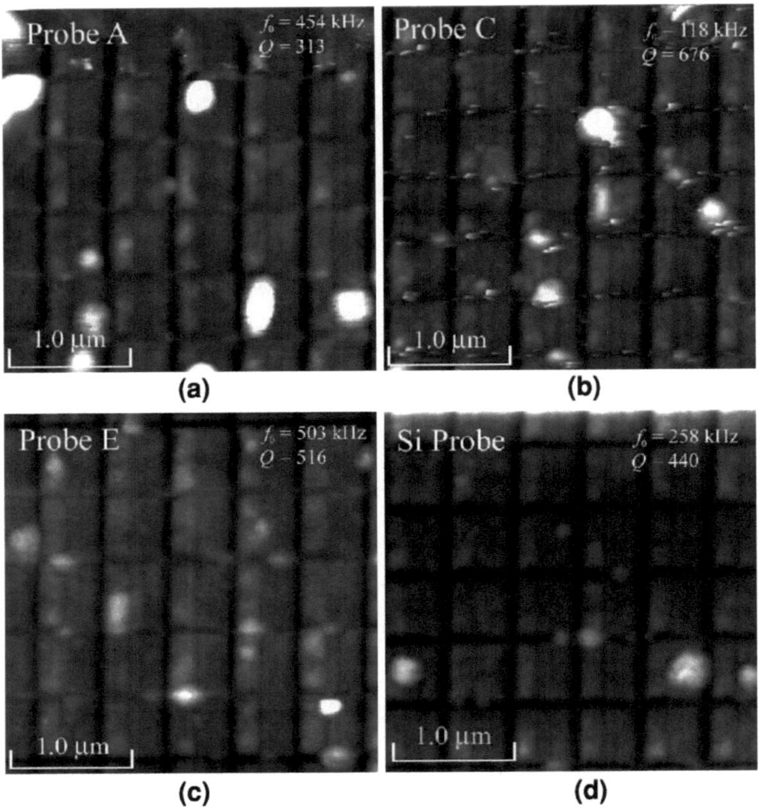

Fig. 14 Surface topography of the grating sample measured by different probes:[10] **a** M-AFM probe A, **b** M-AFM probe C, **c** M-AFM probe E and **d** commercial Si probe

4.4 Microwave Image Obtained by the M-AFM

For the microwave measurement, a compact microwave instrument working at a fixed frequency [64] or a network analyzer working at different sweeping frequencies [12, 31] can be used. In the experiment, a coaxial line with an inner diameter of 1 mm was fixed on the AFM probe holder, in order to connect the M-AFM probe to the microwave measurement instrument. Therefore, the microwave transmission line changed from a coaxial line to a parallel plate waveguide (in the M-AFM probe) [31].

[10] With kind permission from Ju, Y., Kobayashi, T., Soyama, H.: Development of a nanostructural microwave probe based on GaAs. Microsyst. Technol. **14**, 1021 (2008), Figs. 2, 3, 5, 6. ©2008, Springer Publishing Ltd., Springer Science and Business Media.

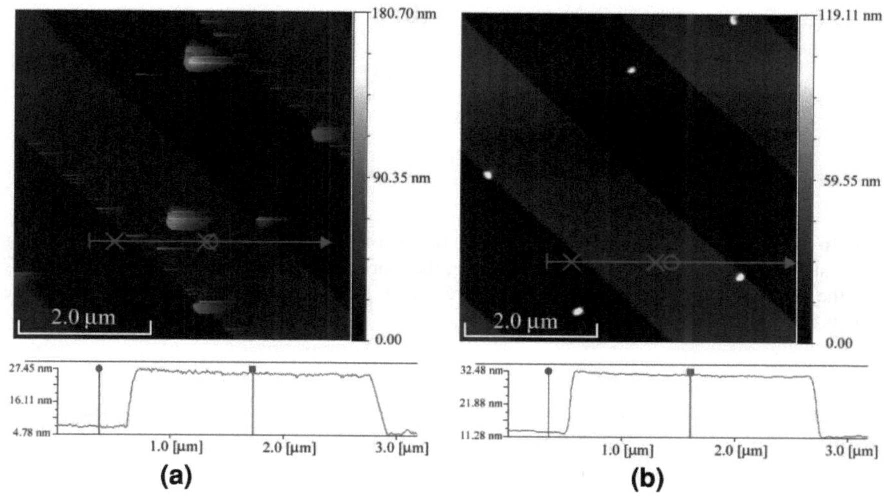

Fig. 15 Surface topography and scanning profile of the grating sample measured by different probes:[11] **a** M-AFM probe C and **b** commercial Si probe

Figure 16 shows the scanning results for a sample having Au and glass step structure obtained by a M-AFM, which combines a compact microwave instrument with a commercial AFM [64]. Since the microwave instrument was synchronized well with the AFM scanner, the microwave AFM image was created simultaneous with the AFM topography. The M-AFM worked on the noncontact mode where the scanning speed is 1.5 μm/sec, and the scanning area 3 μm × 3 μm.

Figure 16a shows the three-dimensional surface topography of the sample measured by the M-AFM at the step area between the Au film and the glass wafer substrate. In this figure, the left side is Au film and the right side is the glass substrate. By analyzing the scanning profile obtained from Fig. 16a, the thickness of the Au film was evaluated to be 150 nm on average. On the other hand, Fig. 16b shows the microwave image of the sample obtained by converting the measured microwave signals into a voltage, acquired by the M-AFM probe at the corresponding position of Fig. 16a. The experimental result demonstrates that the microwave image has two spatial phases. Since the standoff distance between the tip of the M-AFM probe and the surfaces of the Au film and glass substrate is certainly constant controlled by the atomic force, it is only due to the reason that the response of the microwave signals changed with the different electrical characteristics of the measured materials. In Fig. 16b, the outputting voltage over

[11] With kind permission from Ju, Y., Kobayashi, T., Soyama, H.: Development of a nanostructural microwave probe based on GaAs. Microsyst. Technol. **14**, 1021 (2008), Figs. 7, 8. ©2008, Springer Publishing Ltd., Springer Science and Business Media.

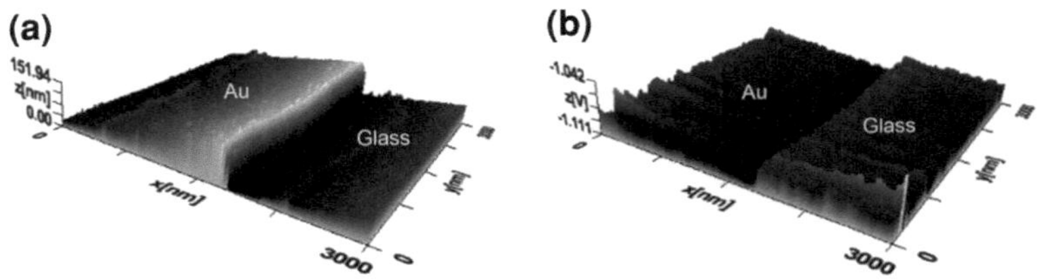

Fig. 16 M-AFM images: **a** AFM topography image of the sample (Au film coating on glass wafer substrate) obtained by the M-AFM probe, the working environment temperature at 24.5°C and the relative humidity 38.4%; **b** microwave image by converting the measured microwave signals into the voltage without calibration

the glass area is larger than that over the Au area, because the scanning started from the Au area with an initial offset of nulling operation.

By analyzing the scanning profile obtained from Fig. 16b, the outputting voltage measured over the Au area is −1093 mV around, and that over the glass area is −1059 mV around. The difference in the measured voltages between the Au and the glass areas is 34 mV. This value is sensitive enough for the evaluation of the electrical properties of materials. Therefore, the M-AFM should allow one to scan the electrical properties of materials in nanometer orders.

4.5 Measurement of Electrical Properties by M-AFM

The principle of the method described here is based on the reflection of microwave from materials. Since the reflection from the bottom surface can be neglected [23, 28], the measured reflection coefficient of the microwave signal can be expressed by considering the reflection only from the top surface as [48]

$$\Gamma = \frac{\eta - \eta_0}{\eta + \eta_0}, \tag{4}$$

where

$$\eta = \sqrt{\frac{\mu}{\varepsilon - j\frac{\sigma}{\omega}}} \tag{5}$$

and

$$\eta_0 = \sqrt{\frac{\mu_0}{\varepsilon_0}}. \tag{6}$$

In the above equations, Γ represents the reflection coefficient, η, σ, μ, and ε are the intrinsic impedance, conductivity, permeability, and permittivity of materials, respectively, and η_0, σ_0, μ_0, and ε_0 are those of free space. Symbolω denotes the angular frequency, and $j = \sqrt{-1}$.

For non magnetic materials, considering $\mu = \mu_0$, and using the above equations, the reflection coefficient, Γ, can finally be written as [28]

$$\Gamma = X + jY = \frac{1 - \sqrt{\frac{\varepsilon}{\varepsilon_0} - j\frac{\sigma}{\omega\varepsilon_0}}}{1 + \sqrt{\frac{\varepsilon}{\varepsilon_0} - j\frac{\sigma}{\omega\varepsilon_0}}}, \tag{7}$$

where X is the real part and Y the image part of the reflection coefficient. By solving the simultaneous equations for the real and imaginary parts of Eq. 7 and eliminating ε, the conductivity of the materials can be expressed as [28]

$$\sigma = \frac{4\omega\varepsilon_0 Y(1 - X^2 - Y^2)}{\left[(1+X)^2 + Y^2\right]^2}. \tag{8}$$

On the other hand, the amplitude, $|\Gamma_m|$, and the phase, θ_m, of the measured reflection coefficient, $\Gamma_{m,}$ can obtained by microwave measurement using following equation:

$$\Gamma_m = |\Gamma_m| e^{j\theta_m}. \tag{9}$$

Here, Γ_m is influenced by not only the electrical conductivity of metals but also the standoff distance, the reflection generated at the aperture part of the probe, the connection parts between the probe and coaxial line, and so on. Therefore, to obtain a correct value of σ, we must find the theoretical reflection coefficient Γ_t by calibrating the measured Γ_m.

Figure 17 shows the signal flow graph for the reflection measurement. Using S_{11}, S_{12}, S_{22} and Γ_t, Γ_m can expressed as [48]:

$$\Gamma_m = \frac{b_1}{a_1} = S_{11} + \frac{S_{12}^2 \Gamma_t}{1 - S_{22}\Gamma_t}, \tag{10}$$

where a_1 and b_1 are the input and output of microwave signal, S_{11}, S_{12} and S_{22} are the errors in the measurement system including losses and phase delays caused by the effects of the connectors, the cables and the M-AFM probe. By solving Eqs. 5–10, the theoretical reflection coefficient can expressed as

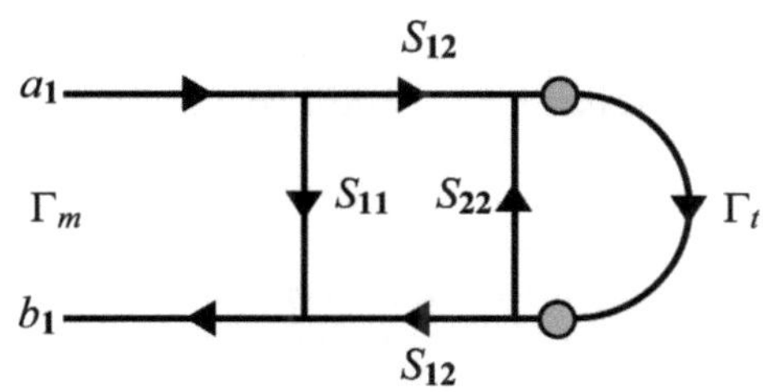

Fig. 17 Signal flow graph for the reflection measurement

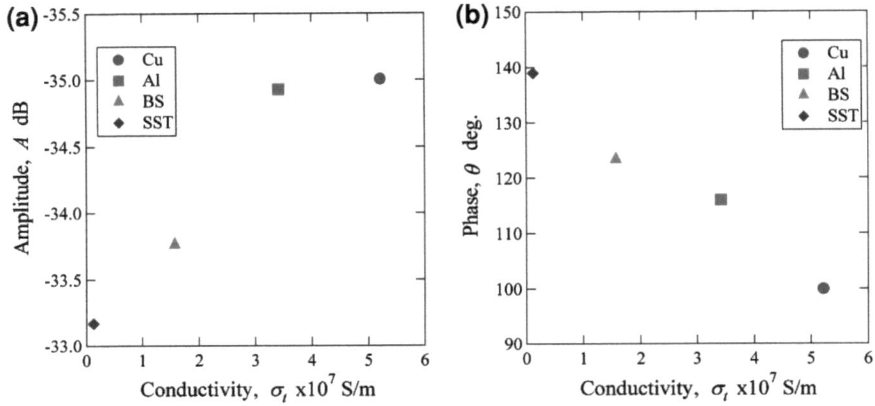

Fig. 18 Results of the microwave measurement:[12] **a** the relationship between the amplitude of reflection coefficient and the electrical conductivity, **b** the relationship between the phase of reflection coefficient and the electrical conductivity

$$\Gamma_t = \frac{\Gamma_m - S_{11}}{S_{12}^2 + S_{22}(\Gamma_m - S_{11})}. \tag{11}$$

Therefore, if S_{11}, S_{12} and S_{22} are determined, we can determine Γ_t from Γ_m using Eq. 11.

In order to measure the correct value of σ, the microwave measurement system must be calibrated. Three samples, of which the conductivity is known, are needed for determining the three unknown constants, S_{11}, S_{12} and S_{22}. Using three groups of the measured and theoretical reflection coefficients, $(\Gamma_{m1}, \Gamma_{t1})$, $(\Gamma_{m2}, \Gamma_{t2})$, and $(\Gamma_{m3}, \Gamma_{t3})$, S_{11}, S_{12} and S_{22} can be determined as follows:

$$S_{22} = \frac{(\Gamma_{t1} - \Gamma_{t2})(\Gamma_{m2} - \Gamma_{m3}) - (\Gamma_{t2} - \Gamma_{t3})(\Gamma_{m1} - \Gamma_{m2})}{\Gamma_{13}(\Gamma_{t1} - \Gamma_{t2})(\Gamma_{m2} - \Gamma_{m3}) - \Gamma_{t1}(\Gamma_{t2} - \Gamma_{t3})(\Gamma_{m1} - \Gamma_{m2})}, \tag{12}$$

$$S_{12}^2 = \frac{(\Gamma_{m1} - \Gamma_{m2})(1 - S_{22}\Gamma_{t1})(1 - S_{22}\Gamma_{t2})}{\Gamma_{t1} - \Gamma_{t2}}, \tag{13}$$

$$S_{11} = \Gamma_{m1} - \frac{S_{12}^2\Gamma_{t1}}{1 - S_{22}\Gamma_{t1}}. \tag{14}$$

Finally, the conductivity σ can be determined from Eqs. 5–8 and 5–11.

Figure 18 shows the measured amplitude and phase of the reflection coefficient verses the electrical conductivity, for a sample with copper (Cu), aluminum

[12] Reprinted with permission from Fujimoto, A., Zhang, L., Hosoi, A., Ju, Y.: Structure modification of M-AFM probe for the measurement of local conductivity. Proceedings of DTIP2010. Page 22. ©2010, EDA publishing.

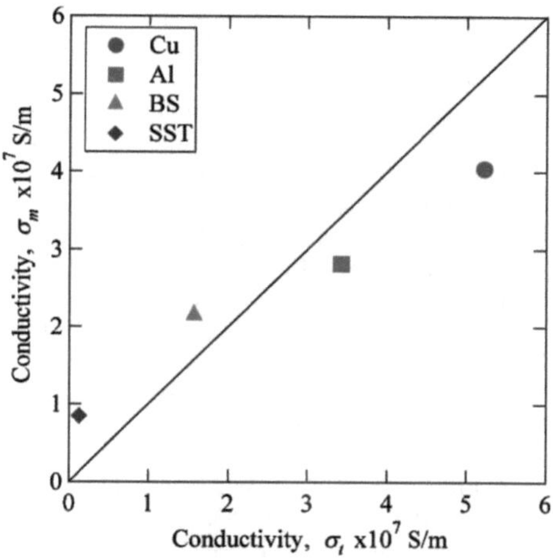

Fig. 19 The relationship between the conductivity, σ_m, obtained by the microwave measurement and the conductivity, σ_t, measured by a high-frequency conductometry (see footnote 13)

(Al), brass (BS) and stainless steel (SST) surfaces, by a M-AFM probe [12]. In the experiment, a network analyzer was used with the working frequency selected at 44.5 GHz. The AFM worked on the noncontact mode. The resonance frequency of the M-AFM probe was 368.5 kHz and the quality value was 615.5. Figure 19 shows the relationship between the conductivity σ_m, obtained by the microwave measurement and σ_t, measured by high-frequency conductometry. Here, each σ_m was obtained by carrying out the calibration as described above using the other three as references. These results indicate that microwave measurement can discriminate the conductivity of metallic micro and nano materials quantitatively.

Acknowledgments The author acknowledges partial support from the Japan Society for the Promotion of Science under Grant-in-Aid for Scientific Research (A) Grant No. 20246028, and Dr. M. Chen for his help in preparing the manuscript.

References

1. Bid, A., Bora, A., Raychaudhuri, A.K.: Temperature dependence of the resistance of metallic nanowires of diameter ≥15 nm: applicability of Bloch-Grüneisen theorem. Phys. Rev. B **74**(1–8), 035426 (2006)
2. Bøggild, P., Hansen, T.M., Kuhn, O., Grey, F., Junno, T., Montelius, L.: Scanning nanoscale multiprobes for conductivity measurements. Rev. Sci. Instrum. **71**, 2781–2783 (2000)
3. Boughton, R.I.: Size-dependent deviations from Matthiessens-rule in the resistivity of gallium. J. Phys. F **11**, L155–L156 (1981)
4. Braun, E., Eichen, Y., Sivan, U., Ben-Yoseph, G.: DNA-templated assembly and electrode attachment of a conducting silver wire. Nature **39**, 775–778 (1998)
5. Chambers, R.G.: The conductivity of thin wires in magnetic field. Proc. Roy. Soc. Lond. Ser. A **202**, 378–394 (1950)
6. Chaudhari, P., Habermeier, H.U.: Quantum localization in amorphous W-Re alloys. Phys. Rev. Lett. **44**, 40–43 (1980)

7. Cornelius, T.W., Toimil-Molares, M.E., Neumann, R., Karim, S.: Finite-size effects in the electrical transport properties of single bismuth nanowires. J. Appl. Phys. **100**(1–5), 114307 (2006)
8. Dingle, R.B.: The electrical conductivity of thin wires. Proc. Roy. Soc. Lond. Ser. A **201**, 545–560 (1950)
9. Duewer, F., Gao, C., Takeuchi, I., Xiang, X.D.: Tip-sample distance feedback control in a scanning evanescent microwave microscope. Appl. Phys. Lett. **74**, 2696–2698 (1999)
10. Durkan, C., Welland, M.E.: Size effects in the electrical resistivity of polycrystalline nanowires. Phys. Rev. B **61**, 14215–14218 (2000)
11. Fuchs, K., Wills, H.H.: The conductivity of thin films according to the electron theory of metals. Proc. Camb. Philos. Soc. **34**, 100–108 (1938)
12. Fujimoto, A., Zhang, L., Hosoi, A., Ju, Y.: Structure modification of M-AFM probe for the measurement of local conductivity. Proc. DTIP 22–26 (2010)
13. Giordano, N.: Experimental study of localization in thin wires. Phys. Rev. B **22**, 5635–5654 (1980)
14. Giordano, N., Gilson, W., Prober, D.E.: Experimental study of Anderson localization in thin wires. Phys. Rev. Lett. **43**, 725–728 (1979)
15. Gu, W., Choi, H., Kim, K.: Universal approach to accurate resistivity measurement for a single nanowire: theory and application. Appl. Phys. Lett. **89**(1–3), 253102 (2006)
16. Hasegawa, S., Shiraki, I., Tanikawa, T., Petersen, C.L., Hansen, T.M., Boggild, P., Grey, F.: Direct measurement of surface-state conductance by microscopic four-point probe method. J. Phys. Condens. Matter **14**, 8379–8392 (2002)
17. Heisig, S., Danzebrink, H.U., Leyk, A., Mertin, W., Münster, S., Oesterschulze, E.: Monolithic gallium arsenide cantilever for scanning near-field microscopy. Ultramicroscopy **71**, 99–105 (1998)
18. Hinode, K., Hanaoka, Y., Takeda, K., Kondo, S.: Resistivity increase in ultrafine-line copper conductor for ULSIs, Part 2. Jpn. J. Appl. Phys. **40**, L1097–L1099 (2001)
19. Huang, Q., Lilley, C.M., Bode, M.: Surface scattering effect on the electrical resistivity of single crystalline silver nanowires self-assembled on vicinal Si (001). Appl. Phys. Lett. **95**, 103112 (2009)
20. Iwata, N., Wakayama, T., Yamada, S.: Establishment of basic process to fabricate full GaAs cantilever for scanning probe microscope applications. Sens. Actuators A Phys. **111**, 26–31 (2004)
21. Josell, D., Burkhard, C., Li, Y., Cheng, Y.-W., Keller, R.R., Witt, C.A., Kelley, D.R., Bonevich, J.E., Baker, B.C., Moffat, T.P.: Electrical properties of superfilled sub-micrometer silver metallizations. J. Appl. Phys. **96**, 759–768 (2004)
22. Josell, D., Brongersma, S.H., Tökei, Z.: Size-dependent resistivity in nanoscale interconnects. Annu. Rev. Mater. Res. **39**, 231–254 (2009)
23. Ju, Y., Saka, M., Abé, H.: NDI of delamination in IC packages using millimeter-waves, IEEE. Trans. Instrum. Meas. **50**, 1019–1023 (2001)
24. Ju, Y., Inoue, K., Saka, M.: Contactless measurement of electrical conductivity of semiconductor wafers using the reflection of millimeter waves. Appl. Phys. Lett. **81**, 3585–3587 (2002)
25. Ju, Y., Ju, B.F., Saka, M.: Microscopic four-point atomic force microscope probe technique for local electrical conductivity measurement. Rev. Sci. Instrum. **76**(1–3), 086101 (2005a)
26. Ju, Y., Sato, H., Soyama, H.: Fabrication of the tip of GaAs microwave probe by wet etching. Proc interPACK 2005 (CD-ROM), 73140 (2005b)
27. Ju, Y., Hirosawa, Y., Soyama, H., Saka, M.: Contactless measurement of electrical conductivity of Si wafers independent of wafer thickness. Appl. Phys. Lett. **87**(1–3), 162102 (2005c)
28. Ju, B.F., Ju, Y., Saka, M.: Quantitative measurement of submicrometre electrical conductivity. J. Phys. D Appl. Phys. **40**, 7467–7470 (2007a)
29. Ju, Y., Kobayashi, T., Soyama, H.: Fabrication of a GaAs microwave probe used for atomic forcemicroscope. Proc interPACK 2007 (CD-ROM), 33613 (2007b)

30. Ju, Y., Kobayashi, T., Soyama, H.: Development of a nanostructural microwave probe based on GaAs. Microsyst. Technol. **14**, 1021–1025 (2008)
31. Ju, Y., Hamada, M., Kobayashi, T., Soyama, H.: A microwave probe nanostructure for atomic force microscopy. Microsyst. Technol. **15**, 1195–1199 (2009)
31. Keren, K., Krueger, M., Gilad, R., Ben-Yoseph, G., Sivan, U., Braun, E.: Sequence-specific molecular lithography on single DNA molecules. Science **297**, 72–75 (2002)
33. Kopanski, J.J., Marchiando, J.F., Loweny, J.R.: Scanning capacitance microscopy measurements and modeling: Progress towards dopant profiling of silicon. J. Vac. Sci. Technol. B **14**, 242–247 (1996)
34. Leunissen, L.H.A., Zhang, W., Wu, W., Brongersma, S.H.: Impact of line edge roughness on copper interconnects. J. Vac. Sci. Technol. B **24**, 1859–1862 (2006)
35. Liu, K., Chien, C.L., Searson, P.C.: Finite-size effects in bismuth nanowires. Phys. Rev. B **58**, 14681–14684 (1998)
36. MacDonald, D.K.C., Sarginson, K.: Size effect variation of the electrical conductivity of metals. Proc. Roy. Soc. Lond. Ser. A **203**, 223–240 (1950)
37. MacFadyen, D.N.: On the preferential etching of GaAs by H_2SO_4-H_2O_2-H_2O. J. Electrochem. Soc. **130**, 1934–1941 (1983)
38. Maîtrejean, S., Gers, R., Mourier, T., Toffoli, A., Passemard, G.: Experimental measurements of electron scattering parameters in Cu narrow lines. Microelectron. Eng. **83**, 2396–2401 (2006)
39. Marom, H., Mullin, J., Eizenberg, M.: Size-dependent resistivity of nanometric copper wires. Phys. Rev. B **74**(1–9), 045411 (2006)
40. Martin, Y., Abraham, D.W., Wickramasinghe. H.K.: High-resolution capacitance measurement and potentiometry by force microscopy. Appl. Phys. Lett. **52**, 1103–1105 (1988)
41. Mayadas, A.F.: Intrinsic resistivity and electron mean free path in aluminum films. J. Appl. Phys. **39**, 4241–4245 (1968)
42. Mayadas, A.F., Shatzkes, M.: Electrical-resistivity model for polycrystalline films: the case of arbitrary reflection at external surfaces. Phys. Rev. B **1**, 1382–1389 (1970)
43. Mayadas, A.F., Shatzkes, M., Janak, J.F.: Electrical resistivity model for polycrystalline films: the case of specular reflection at external surfaces. Appl. Phys. Lett. **14**, 345–347 (1969)
44. Nagase, M., Takahashi, H., Shirakawabe, Y., Namatsu, H.: Nano-four-point probes on microcantilever system fabricated by focused ion beam. Jpn. J. Appl. Phys. **42**:4856–4860 (2003)
45. Petersen, C.L., Grey, F., Shiraki, I., Hasegawa, S.: Microfour-point probe for studying electronic transport through surface states. Appl. Phys. Lett. **77**, 3782–3784 (2000)
46. Petersen, C.L., Hansen, T.M., Bøggild, P., Boisen, A., Hansen, O., Hassenkam, T., Grey, F.: Scanning microscopic four-point conductivity probes. Sens. Actuators A. Physics **96**, 53–58 (2002)
47. Petzold, M., Landgraf, J., Füting, M., Olaf, J.M.: Application of atomic force microscopy for microindentation testing. Thin Sol. Films **264**, 153–158 (1995)
48. Pozar, D.M.: Microwave Engineering, 2nd edn. Wiley, New York (1998).
49. Richter, J., Mertig, M., Pompe, W.: Construction of highly conductive nanowires on a DNA template. Appl. Phys. Lett. **78**, 536–538 (2001)
50. Richter, J., Mertig, M., Pompe, W., Vinzelberg, H.: Low-temperature resistance of DNA-templated nanowires. Appl. Phys. A **74**, 725–728 (2002)
51. Schindler, G., Steinlesberger, G., Engelhardt, M., Steinhögl, W.: Electrical characterization of copper interconnects with end-of-roadmap feature sizes. Sol. State Electron **47**, 1233–1236 (2003)
52. Sondheimer, E.H.: The mean free path of electrons in metals. Adv. Phys. **1**, 1–42 (1952)
53. Steinhögl, W., Schindler, G., Steinlesberger, G., Engelhardt, M.: Size-dependent resistivity of metallic wires in the mesoscopic range. Phys. Rev. B **66**(1–4), 075414 (2002)

54. Steinhögl, W., Schindler, G., Steinlesberger, G., Traving, M., Engelhardt, M.: Comprehensive study of the resistivity of copper wires with lateral dimensions of 100 nm and smaller. J. Appl. Phys. **97**(1–7), 023706 (2005)
55. Steinlesberger, G., Engelhardt, M., Schindler, G., Steinhögl, W., von Glasow, A., Mosig, K., Bertagnolli, E.: Electrical assessment of copper damascene interconnects down to sub-50 nm feature sizes. Microelectron. Eng. **64**, 409–416 (2002)
56. Steinlesberger, G., Engelhardt, M., Schindler, G., Kretz, J., Steinhögl, W., Bertagnolli, E.: Processing technology for the investigation of sub-50 nm copper damascene interconnects. Sol. State Electron **47**, 1237–1241 (2003)
57. Tabib-Azar, M., Akiwande, D.: Real-time imaging of semiconductor space-charge regions using high-spatial resolution evanescent microwave microscope. Rev. Sci. Instrum. **71**, 1460–1465 (2000)
58. Toimil-Molares, M.E., Höhberger, E.M., Schaeflein, C., Blick, R.H., Neumann, R., Trautmann, C.: Electrical characterization of electrochemically grown single copper nanowires. Appl. Phys. Lett. **82**, 2139–2141 (2003)
59. Vazquez-Mena, O., Villanueva, G., Savu, V., Sidler, K., van den Boogaart, M.A.F., Brugger, J.: Metallic nanowires by full wafer stencil lithography. Nano Lett. **8**, 3675–3682 (2008)
60. White, A.E., Tinkham, M., Skocpol, W.J., Flanders, D.C.: Evidence for interaction effects in the low-temperature resistance rise in ultrathin metallic wires. Phys. Rev. Lett. **48**, 1752–1755 (1982)
61. Williams, W.D., Giordano, N.: Experimental study of localization and electron-electron interaction effects in thin Au wires. Phys. Rev. B **33**(12), 8146–8154 (1986)
62. Wu, W., Brongersma, S.H., Van Hove, M., Maex, K.: Influence of surface and grain-boundary scattering on the resistivity of copper in reduced dimensions. Appl. Phys. Lett. **84**, 2838–2840 (2004)
63. Yan, H., Park, S.H., Finkelstein, G., Reif, J.H., LaBean, T.H.: DNA-templated self-assembly of protein arrays and highly conductivity nanowires. Science **301**, 1882–1884 (2003)
64. Zhang, L., Ju, Y., Hosoi, A., Fujimoto, A.: Microwave AFM imaging for the measurement of electrical properties on the nanometer scale (2010) (to be submitted)
65. Zhang, W., Brongersma. S.H., Li, Z., Li, D., Richard, O., Maex, K.: Analysis of the size effect in electroplated fine copper wires and a realistic assessment to model copper resistivity. J. Appl. Phys. **101**(1–11), 063703 (2007)
66. Zhang, Z.B., Sun, X.Z., Dresselhaus, M.S., Ying, J.Y., Heremans, J.: Electronic transport properties of single-crystal bismuth nanowire arrays. Phys. Rev. B **61**, 4850–4861 (2000)

Modification of Nano/Micromaterials

Hironori Tohmyoh and Mikio Muraoka

Abstract Although nano/micromaterials have attracted considerable attention due to their excellent physical properties and geometrical merits, these should be modified for specific purposes or be assembled in systems for many fields of application. The cutting and welding of materials must be the principle operation for this purpose. The welding and cutting technology utilizing Joule heat is first described together with some experiments and applications. First, heat transfer problem in thin wires are treated. And then, two Pt wires with diameters of about 800 nm are shown to successfully be welded by Joule heating. Melting and solidification at the point contact of thin wires occurred continuously under a constant current supply and the welding of wires is completed within several seconds in self-completed manner. Moreover, the welding technology for low-dimensional materials have been found to be effective for manipulating materials and for generating functional elements, e.g., electromagnetic rings and thermo-electric elements. A unique technique for creating nanocoils from straight nano-wires is described.

H. Tohmyoh (✉)
Department of Nanomechanics, Tohoku University, Aoba 6-6-01, Aramaki, Aoba-ku, Sendai 980-8579, Japan
e-mail: tohmyoh@ism.mech.tohoku.ac.jp

M. Muraoka
Department of Mechanical Engineering, Akita University, 1-1 Tegatagakuen-machi, Akita 010-8502, Japan
e-mail: muraoka@ipc.akita-u.ac.jp

M. Saka (ed.), *Metallic Micro and Nano Materials*, Engineering Materials,
DOI: 10.1007/978-3-642-15411-9_6,

1 Welding and Cutting of Materials

1.1 Introduction

Recently, low-dimensional materials, such as nanowires (NWs), nanotubes, nanoparticles and so on, have attracted considerable attention due to their excellent mechanical, electrical, optical and thermal properties, and these are expected to be used as functional elements, especially in nanoelectromechanical systems (NEMS) or microelectromechanical systems (MEMS). The importance of cutting and welding technologies for low-dimensional materials, e.g., metallic or semiconductor nanowires (NWs), carbon nanotubes (CNTs) [56], polymer nanofibres [19] and nanoparticles [24], has been increasing day by day because these technologies are indispensable for modifying these materials for specific purposes or assembling them in systems in many fields of application.

Li et al. [31] successfully cut a Ag NW that had a diameter of a few tens of nanometers by applying a high load to the NW on a glass slide using a nanoindenter. Wu and Yang [64] joined 20 nm diameter Ge NWs end to end by thermal heating (Fig. 1). These techniques may be used for generating small-scale structures with various functions. However these techniques require that the indenter or heater is in physical contact with the object, and the geometry of the indenter or heater puts limitations on the capability of the techniques for cutting and joining nanomaterials. Techniques that use an electron beam [28] or laser beam [24] can provide local non-contact heating of small-scale materials, and these are capable of welding two objects at a specified point without problems arising from the geometry of the objects. Xu et al. [65] joined Au NWs by irradiating the point at

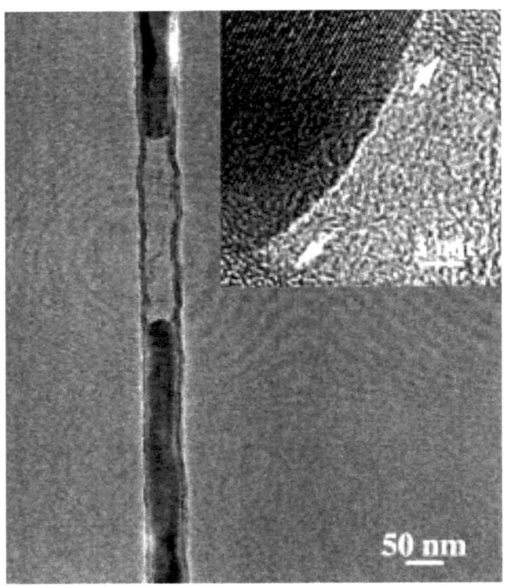

Fig. 1 TEM image of two Ge nanowires separated by a vacuum gap, created by the thermal cutting process. The *inset* shows a high-resolution TEM image at the wire-tube interface. The (111) lattice fringe of the Ge nanowire can clearly be seen. Reprinted with permission from Wiley, from [64]

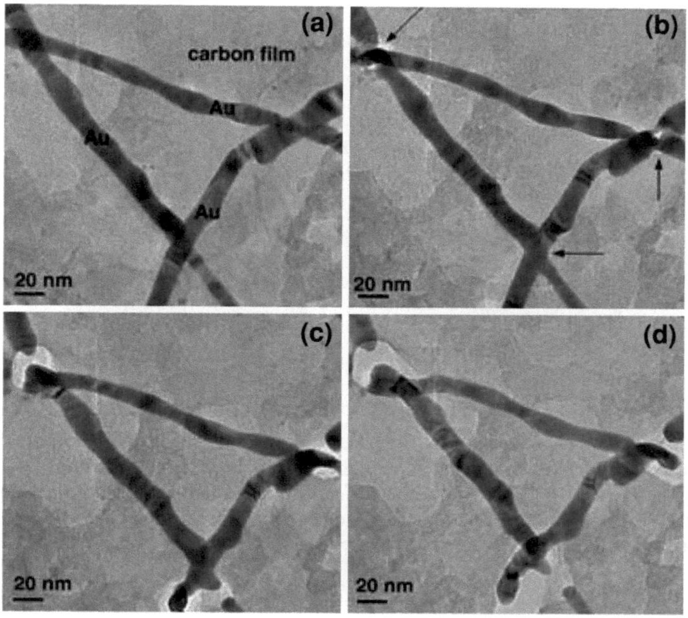

Fig. 2 Welding of three single-crystalline Au NWs with a 10-nmdiameter HIEB. **a** The original Au NWs. **b–d** A closed loop is formed after welding the NWs together at their overlapping regions (highlighted by *arrows*), and then the loop is cut off from the rest of the NWs. Holes are seen in the supporting carbon film under the joints. The whole process from **a** to **d** takes 30 min. Reprinted with permission from Wiley, from [65]

which two single-crystalline Au NWs crossed with a high-intensity electron beam (Fig. 2). These techniques, however, are difficult to apply at shadow points that are obscured from the electron or laser beam.

In contrast with the above-mentioned techniques, cutting and welding techniques for conductive materials utilizing Joule heating can be used for a wide range of applications because heat can easily be induced by a current supply. Hirayama et al. [18] reported the joining of CNTs in a transmission electron microscope (TEM). The tip of a straight CNT was brought into contact with a looped CNT in a TEM, and the two were joined at the point of contact by applying a constant bias voltage. Jin et al. [22] also demonstrated the connection of two CNTs in a TEM. Two CNTs with the same diameter were connected together by Joule heating and the associated electro-migration effect. Moreover, they discovered that any two CNTs with different diameters could be joined with the assistance of a tungsten metal particle [22]. A metal nanovolume solder have recently been found to be effective for Joule heat welding of low-dimensional materials (Fig. 3) [36].

Welding is accomplished by phase transition of the material at the contact point. Firstly, melting at the point of contact occurs and then the molten part solidifies. Low-dimensional materials show unusual thermal behavior due to their

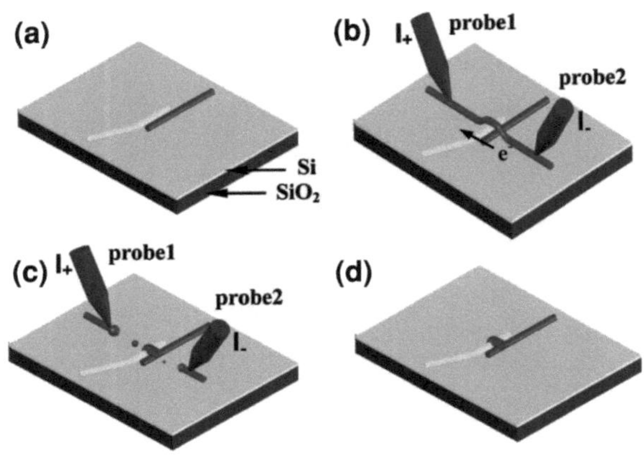

Fig. 3 Schematic of nanoconstruction by the welding of individual nanoobjects (nanowires) using nanoscale solder. **a** Assembly of individual nanoobjects into a desired pattern using a nanomanipulator probe. **b** Placement of a sacrificial nanowire in contact with the nanostructure to be welded. **c** Nanowelding the nanoobjects together by an electrical signal. **d** Completed nanoweld. Reprinted with permission from [36]. Copyright (2009) by the American Chemical Society

specific geometry and properties [30]. Moreover, the melting point of low-dimensional materials is found to be lower than their bulk equivalents [12, 17, 59]. Consequently, controlling the melting and solidification phenomena for low-dimensional materials is not an easy task.

In this chapter, Joule heat welding technology of thin metallic wires (Fig. 4) is described together with some experiments [50]. The welding technology has two specialties for creating future NEMS or MEMS. One specialty is that melting and solidification at the point of contact of the wires occurs continuously over a period of several seconds in a self-completed manner under constant current supply [49]. Another specialty is that the melting phenomenon at the nanocontact of two thin wires can be controlled by a governing parameter [48].

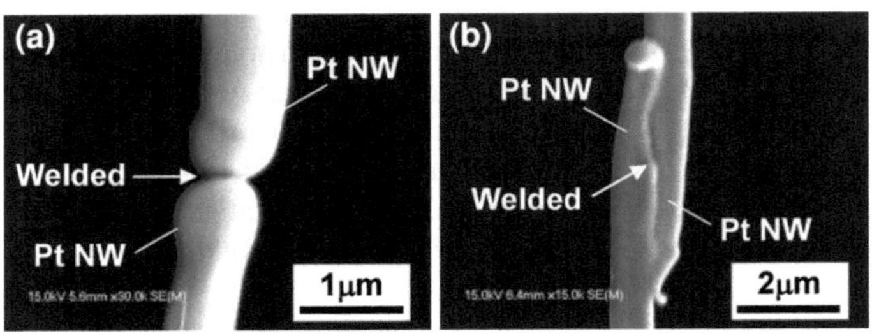

Fig. 4 FE-SEM micrographs of welded Pt wire. **a** Straight welding. **b** Lap welding. Reprinted from [50], copyright (2007), with permission from Elsevier

1.2 Heat Conduction in Thin Wires

Let us consider the electro-thermal problem, where the current I flows through a thin wire of length l and cross-sectional area A [48]. The wire is assumed to be so thin that the temperature is uniform at all points of the cross section. The temperature dependencies of the electrical conductivity, σ, and the heat conductivity, K, are ignored for simplicity. The Thomson effect, which is related to the supplied current and the temperature gradient, can be ignored because Thomson heating is usually very small compared with Joule heating. The conduction of heat in the wire is governed by the following one-dimensional differential equation [9]:

$$\frac{\partial T}{\partial t} = k\frac{\partial^2 T}{\partial x^2} - v(T - T_0) + \frac{q}{\rho c}, \tag{1}$$

where T is the temperature and is a function of time, t, and the position in the wire, x. The x-axis is taken to be along the length of the wire and its origin is located at one end of it. The symbol ρ is density, c specific heat, $k = K/(\rho c)$, $v = Hp/(\rho cA)$, H surface conductance, p perimeter, T_0 the ambient temperature surrounding the wire and $q = I^2/(A^2\sigma)$. The solution of Eq. (1) is given by $T = T_1 + T_2$, where T_1 is the steady-state temperature, and T_2 the transient temperature.

For simplicity, we assume that no heat transfer from the surface of the wire to the ambient occurs, i.e., the second term on the right hand side of Eq. (1) can be ignored. The temperature at both ends of the wire is denoted by T_0, and the temperature at any point in the wire at $t = 0$ is T_0. The quantities T_1 and T_2 are given by

$$T_1 = \frac{q}{2K}\left(lx - x^2\right) + T_0, \tag{2}$$

and

$$T_2 = \frac{2ql^2}{K\pi^3}\sum_{m=1}^{\infty}\frac{(-1)^m - 1}{m^3}\exp\left[-k\left(\frac{m\pi}{l}\right)^2\right]t\,\sin\frac{m\pi}{l}x. \tag{3}$$

We consider the electro-thermal system, in which a current, I, flows through a thin wire. The diameter of the wire d is set to 800 nm. The temperature at the both ends of the wire, T_0, is set to 293 K, see Fig. 5a. Figure 5b shows the temperature changes at the middle of the section of the wire ($x = l/2$) against t for a system with $l = 100$ μm and $I = 6$ mA. The material properties used in the calculation were: $\sigma = 9.45 \times 10^6$ S/m, $K = 72$ W/mK, $\rho = 2.15 \times 10^4$ kg/m^3 and $c = 134$ J/kg K, respectively. Immediately after the current supply, the temperature reaches a steady state value, T_1, at around 0.2 ms. The distributions of T_1 against x for $I = 4$, 6 and 8 mA are shown in Fig. 5c. In each case, the maximum value of temperature, $T_{1,\max}$, is at the middle of the section of the wire ($x = 50$ μm), and this increases with increasing I. The relationships between

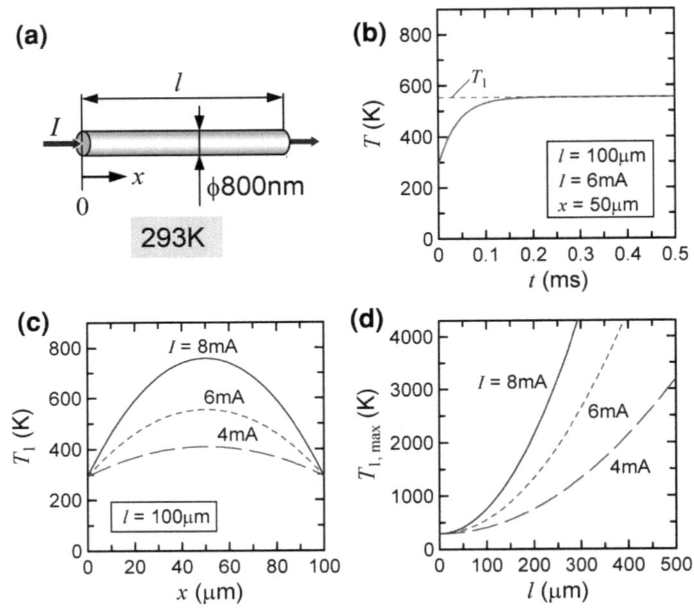

Fig. 5 **a** One-dimensional electro-thermal analysis. Here we consider the problem for a wire with a diameter of 800 nm and a length l. **b** Temperature variation at the middle of the section of the wire against t for the system with $l = 100$ μm and $I = 6$ mA. **c** Distributions of T_1 against x for the supplied currents of 4, 6 and 8 mA. (d) The relationships between $T_{1,\max}$ and l. Reprinted with permission from [48]. Copyright (2009), American Institute of Physics

$T_{1,\max}$ and l for $I = 4$, 6 and 8 mA are shown in Fig. 5d, and the value of $T_{1,\max}$ is given by

$$T_{1,\max} = \frac{1}{8K\sigma}\left(I\frac{l}{A}\right)^2 + T_0. \tag{4}$$

The maximum value, $T_{1,\max}$, is proportional to l^2; thus a smaller current is needed for longer wires to get the same value of $T_{1,\max}$.

Let us consider the effect of thermal boundary conditions on $T_{1,\max}$. For simplicity, let us consider the electro-thermal problem where the current flows through a continuous wire. In Cases 1 and 2, the current I flows through the entire length, l, of the wire. On the other hand, in Cases 3 and 4, the current is supplied locally to just the middle part of the wire, where the total length of the wire is L and the length over which current is supplied is l. In Cases 1 and 3, there is no heat transfer from the surface of the wire to the ambient, whereas in Cases 2 and 4, heat transfer at the wire surface is considered. The values of T_1 and $T_{1,\max}$ for Case 1 are given by Eqs. (2) and (4). On the other hand, the values of T_1 and $T_{1,\max}$ for Case 2 are given by

$$T_1 = \frac{q}{v\rho c}\left[1 - \frac{e^{bx} + e^{b(l-x)}}{1 + e^{bl}}\right] + T_0, \tag{5}$$

and

$$T_{1,\max} = \frac{q}{v\rho c}\left(1 - \frac{2e^{0.5bl}}{1 + e^{bl}}\right) + T_0, \tag{6}$$

where b is given by $\sqrt{v/k}$. In Cases 3 and 4, the length of the wire is L and the current I is supplied locally to a segment of length l located at the center of the wire. The origin of x is located at an end of the wire. In Case 3, the values of T_1 at the segment with flowing current and $T_{1,\max}$ are given by

$$T_1 = \frac{q}{2K}\left(Lx - x^2 - a^2\right) + T_0, \quad a \leq x \leq a + l, \tag{7}$$

and

$$T_{1,\max} = \frac{q}{8K}\left(2Ll - l^2\right) + T_0, \tag{8}$$

where $a = (L - l)/2$. Heat transfer from the wire surface is considered for Case 4, and the values of T_1 and $T_{1,\max}$ are given by

$$T_1 = \frac{q}{v\rho c}\left[1 - \frac{e^{ba} + e^{-ba}}{2}\frac{e^{bx} + e^{b(L-x)}}{1 + e^{bL}}\right] + T_0, \quad a \leq x \leq a + l, \tag{9}$$

and

$$T_{1,\max} = \frac{q}{v\rho c}\left(1 - \frac{e^{b(L-0.5l)} + e^{0.5bl}}{1 + e^{bL}}\right) + T_0. \tag{10}$$

Figure 6 shows the relationships between $T_{1,\max}$ and slenderness, d/l, for 4 different thermal conditions. In all the cases considered, the temperature at each end of the wire is constant at T_0. The material properties used in the calculation were: $\sigma = 9.45 \times 10^6$ S/m, $K = 72$ W/mK, $\rho = 2.15 \times 10^4$ kg/m^3, $c = 134$ J/kgK and $H = 30$ W/m^{2}K, respectively. The values of L and T_0 considered are 0.5 mm and 293 K, respectively. The diameter of the wire, d, is 800 nm and the length l at $d/l = 0.0016$ is 0.5 mm. Therefore, the values of $T_{1,\max}$ for Cases 3 and 4 are the same as those for Cases 1 and 2 at $d/l = 0.0016$. The value of $T_{1,\max}$ for Case 1 was close to that for Case 2, and the value of $T_{1,\max}$ for Case 3 was close to that for Case 4 except at smaller values of d/l. These results indicate that heat transfer from the wire surface to the ambient has little effect on the changes in $T_{1,\max}$ at larger values of d/l for the wire geometries considered. On the other hand, Fig. 6 shows that the values of $T_{1,\max}$ for Cases 3 and 4 are higher than those for Cases 1 and 2. This shows that $T_{1,\max}$ is highly affected by the thermal boundary conditions at the points at which current is supplied. The heat generated by Joule heating in the region of the wire carrying current diffuses toward the lower

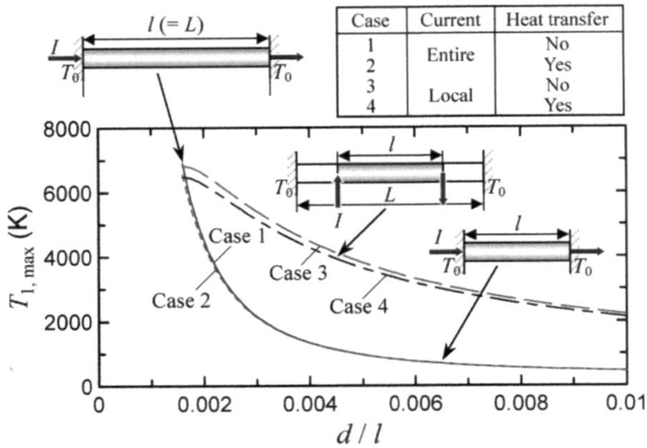

Fig. 6 The relationships between $T_{1,\max}$ and d/l for Cases 1–4. Reprinted with permission from [48]. Copyright (2009), American Institute of Physics

temperature region in the wire where no current flows. As a result, the temperature on each side of the segment in the wire carrying current increases, and the value of $T_{1,\max}$ in this case becomes higher than the case in which the wire ends are at a constant temperature.

1.3 Melting Phenomenon at Nanocontacts

1.3.1 Analysis of Temperature at Nanocontacts

Welding of objects can be realized by the melting at the contacts of two objects and the following solidification of the molten part of the contacts. In this section, let us consider the melting phenomenon at the nanocontacts of two small-scaled objects.

The electro-thermal problem, where the current flows in two thin wires in contact with each other, is considered. The diameter and the length of each wire are 800 nm and 50 μm, respectively. Three different contact conditions are considered [48]; first, two wires in perfect contact (Model 1); secondly, only a small part at the tips of the wires are in contact, and *I* flows through a contact area with a diameter of 8 nm (Model 2). In this case, the local current density at the point of contact is higher than that for Model 1; thirdly, two wires with 45 degree angle cone-shaped tips and a tip diameter of 8 nm are in contact (Model 3). In all the models considered, the temperature at one end of the wire was assumed to be 293 K, as was the ambient temperature. The temperature at all points in the wires at $t = 0$ was 293 K. Heat transfer from the surfaces of the wires to the ambient was considered, and the contact plane was thermally insulated by considering the symmetry of the thermal problem. These models were analyzed using the three-dimensional finite element modeling program. By considering the symmetry, a 1/8 size model was employed. Currents of 4, 6 and 8 mA were supplied to the wire systems.

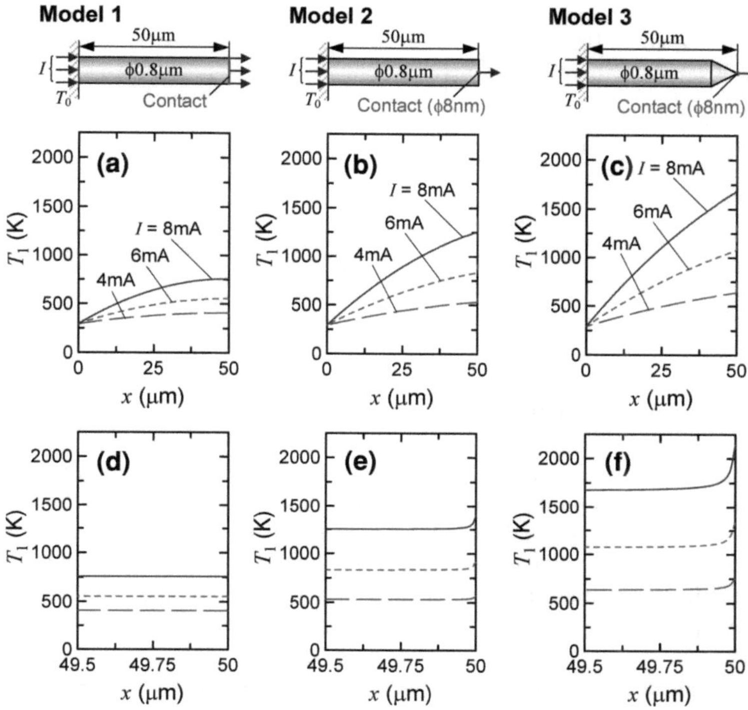

Fig. 7 Distributions of T_1 against x at the point of the contact for different wire systems under current supplies of $I = 4$, 6 and 8 mA. The results for Model 1 are shown in **a**, **d**, those for Model 2 in **b**, **e**, and Model 3 in **c**, **f**. The distributions shown in **d–f** are enlarged from those shown in **a–c**, respectively. Reprinted with permission from [48]. Copyright (2009), American Institute of Physics

The material properties used in the calculation were the same as those used in Eq. (1). Distributions of T_1 against x at the center of the wire for Models 1–3 are shown in Fig. 7a–c, respectively, and the distributions over a smaller range of x are shown in Fig. 7d–f, respectively. Figure 7 clearly shows that T_1 for Model 2 is higher than Model 1 except at the wire end ($x = 0$) and that Model 3 has the highest value of T_1 among the three models considered. These tendencies are obvious in the vicinity of the contacts. Figure 8 shows the difference in temperature between the positions $x = 50$ and 49.5 μm, i.e., $\Delta T = T_1$ ($x = 50$ μm) $-$ T_1 ($x = 49.5$ μm), against I for each of the models considered. It is clear that ΔT increases with increasing supplied current, particularly for Model 3. These results indicate that a specific temperature field forms in the contact region between the two wires, with a higher current density and a smaller volume around the contact region inducing a higher temperature field. Doelling et al. [13] also reported that a nanospot weld was repeatedly performed at the contact between gold tip and substrate in the microswitch and the higher temperature was realized at the contact.

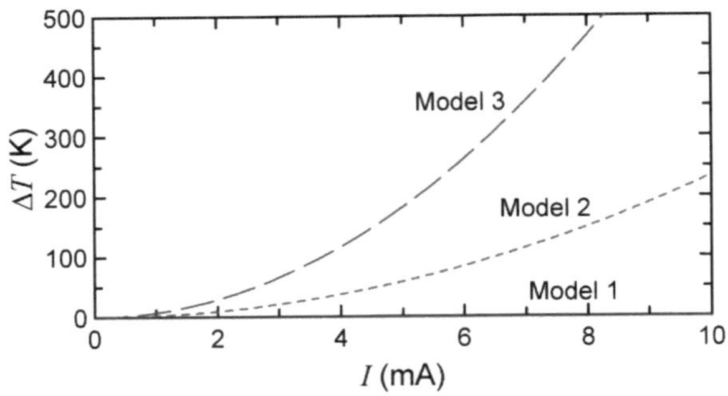

Fig. 8 The values of ΔT for Models 1 to 3 against I. Here, ΔT is given by T_1 ($x = 50$ μm) − T_1 ($x = 49.5$ μm). Reprinted with permission from [48]. Copyright (2009), American Institute of Physics

1.3.2 Governing Parameter

The temperature at the contact between two wires governs the melting phenomenon and the contacts melt provided that the temperature at the contact reaches the melting point of the material. However, because it is very difficult to determine the temperature at the contact due to its complex geometry, such as the shapes of the contacting asperities, the actual area and location of the contacts, etc., an alternative parameter, which governs the melting phenomenon of the contacts, is required.

The actual area of the contact between two thin wires is considerably smaller than the cross sectional areas of the wires, and at the contact point, the current density becomes much greater than that in the wire. If the local current density, in other words, the Joule heating, governs the temperature in the contact region, a smaller current should be sufficient to push the temperature above the melting point of the material. However, the temperature always has a finite value even if the current density becomes infinite [38]. This suggests that the local current density does not govern the temperature at the point of contact.

Here, let us recall linear elastic fracture mechanics (LEFM). The stress around a crack tip always becomes infinite, but no crack growth occurs under a lower applied load. In LEFM, fracture is discussed based on the stress intensity factor, which is determined from the crack length, the applied stress and the geometry of the specimen, and the fracture occurs when the stress intensity factor reaches a critical value, which is called the fracture toughness. Although there is no mathematical analogy between the melting at the point of contact and fracture mechanics, we notice that there is a physical analogy between these phenomena as shown in Table 1.

Based on the knowledge of fracture mechanics and the similarities between this and the melting phenomenon, an alternative parameter, which governs the melting phenomenon of the contacts, has been proposed [48]. The parameter U is described by the following equations

$$U = u_0 f, \tag{11}$$

Table 1 Physical analogy between the melting phenomenon and the case of two-dimensional cracks considered in linear elastic fracture mechanics. Reprinted with permission from [48]. Copyright (2009), American Institute of Physics

Phenomenon	Fracture (LEFM)	Melting at contacts
Geometrical origin of singularity	Crack (Length = $2a$)	Contact point
Applied quantity	Applied stress, S	Applied current density, I/A
Quantity showing concentration	Local stress	Local current density
Governing parameter	Stress intensity factor, $K = (S\sqrt{\pi a} \times F^{b})$	U
Critical value of parameter[a]	Fracture toughness, K_c	U_c

[a] Critical value of the governing parameter for the phenomenon concerned. [b] F is a function of a and other geometrical quantities.

and

$$u_0 = I\frac{l}{A}, \tag{12}$$

where f is a function of the geometrical quantities used to represent the difference in thermal boundary conditions between the actual situation and in ideal conditions, and l is the total length of the wire system. The function f has the value unity in ideal conditions. In ideal conditions no heat transfer from the wire surface occurs and the temperature of both ends of the wire is kept at room temperature, T_0. The temperature at the middle of a wire system without a contact, T, depends on the parameter U defined in Eq. (4) and is given by

$$T = \frac{1}{8K\sigma}U^2 + T_0, \tag{13}$$

where K is the heat conductivity and σ the electrical conductivity, respectively. Moreover, for simplicity, if σ is considered to be independent of temperature, u_0/σ corresponds to the voltage between the ends of the wire system. It is verified from later experiments that U governs the melting phenomenon at the contact point between two thin wires and gives the successful welding conditions whether the thin wires are joined or not [48, 49].

1.3.3 Thermal Function

In LEFM, the difference in the mechanical boundary conditions between the actual conditions and ideal situation, whereat the crack is located in the infinite body, is calibrated by a function F for determining the stress intensity factor, see Table 1. The difference in boundary conditions can be described by a function f and the function is determined from the experiments for cutting thin wires [48]. If we supply a current to the wire so that the maximum temperature on the wire is above the melting point of the wire due to Joule heating and apply a force to shear the molten part of the wire, the wire can be cut at its middle position of the wire. The current required to cut a wire of length l, for the case where the temperature of both sides of the wire is constant at T_0 and no heat transfer from the wire surface occurs, can analytically be determined to be

$$(I_C)_0 = \sqrt{(T_M - T_0) 8 K \sigma \frac{A}{l}}, \tag{14}$$

where T_M is the melting point of the wire. On the other hand, in the experiments the current required to cut a wire is described by

$$(I_C)_{\text{exp}} = \sqrt{(T_M - T_0) 8 K \sigma \frac{A}{l} \frac{1}{f}}. \tag{15}$$

Therefore, the function f is given by

$$f = \frac{(I_C)_0}{(I_C)_{\text{exp}}}. \tag{16}$$

The function f represents the difference in the heat transfer properties of the wire system.

1.4 Cutting of Ultrathin Pt Wires

1.4.1 Experimental

Extremely thin Pt wires with a diameter of 800 ± 200 nm were used in the experiments. The cutting experiments were conducted to discover the capability of

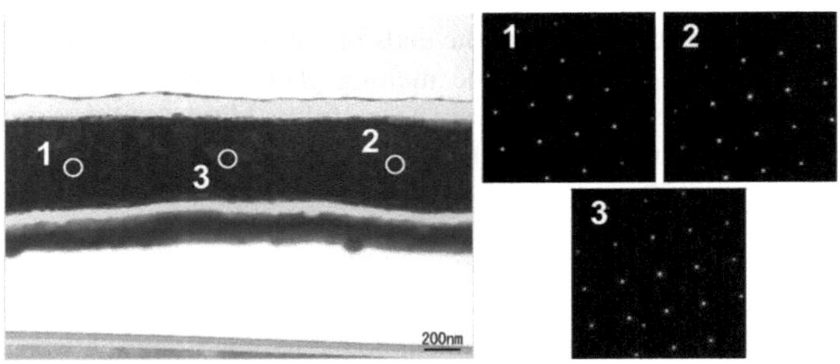

Fig. 9 TEM images of the examined Pt thin wire. Selected area diffraction patterns inserted were obtained at the different positions on the wire where the electron beam direction was parallel to the [001] of the Pt, and those showed that the examined points existed at the same grain. Reprinted with permission from [49]. Copyright (2009) by the American Physical Society

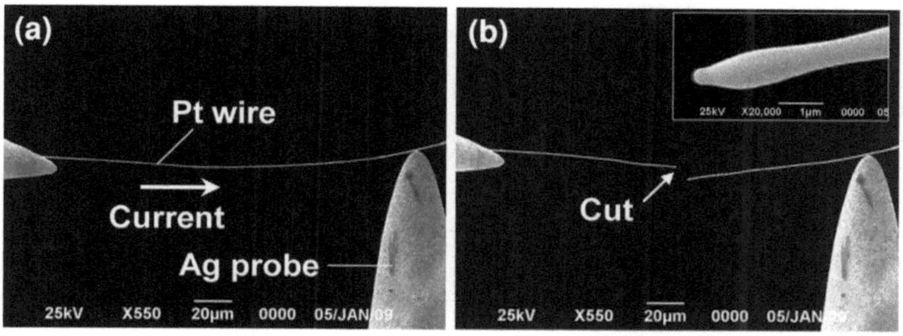

Fig. 10 SEM micrographs obtained in the cutting experiment in the case of $l = 200$ μm before **a** and after a current supply of $I = 2.7$ mA **b**. Inserted micrograph in **b** shows the swelled tip of the cut wire. Reprinted with permission from [49]. Copyright (2009) by the American Physical Society

Joule heat in cutting the wire and to determine the thermal function in a scanning electron microscope (SEM) and in air. The wires were Ag-coated, making the overall diameter about 75 μm. This type of wire is known as Wollaston wire. The Ag-coated Pt wires were cut into 10 mm lengths, and each was soldered to a Cu electrode on the surface of a chip. The Ag around the tips of the wires was then removed by HNO_3 to expose the Pt wire. Figure 9a shows a transmission electron microscope (TEM) micrograph of a cross-section of the examined Pt wires. No clear grain boundary was observed in the TEM image. The selected area diffraction patterns obtained at any point on the wire showed the same pattern, demonstrating that the wire was of an elongated grain structure [37].

As well as the thin Pt wires, cone-shaped Ag probes were also attached to the Cu electrode chips and these were used to locally supply current to a segment of the Pt wires, see Fig. 10a. The length of the Pt wire with current flowing, i.e., the distance between the root of the wire and the contact point of the Ag probe, l, was 50, 100, 200, 300 or 400 μm. A small force, which was needed to shear the molten part of the wire and thus cut the wire, was applied by bending the wire with the Ag probe [50]. Current was supplied to the wire system for 20 s. If the wire remained intact, the supply was stopped. The current was increased by 0.1 mA and the experiment repeated. This process was repeated until the wire was cut (Fig. 10b). For all values of l examined, the wires were cut at the center. The SEM micrograph inserted in Fig. 10b shows the detail around the tip of the wire after cutting, which shows a bead formed at the tip. This indicates that the wire had been locally melted by current-induced heating at the cutting point. The surface tension in the liquid phase of the molten Pt would have contributed to formation of the bead at the tip.

1.4.2 Results and Thermal Function

Figure 11a shows the current required to cut wires in the SEM and in air for various values of l [49]. The current decreases as l increases, and the current

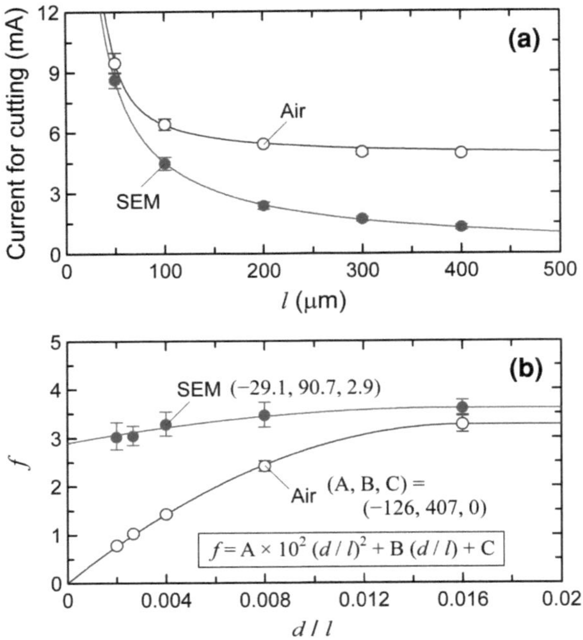

Fig. 11 **a** Current required for cutting thin Pt wires in SEM and air for various wire lengths. **b** The relationships between f and d/l for SEM and air environments obtained from the cutting experiments. Reprinted with permission from [49]. Copyright (2009) by the American Physical Society

required for cutting the wire in the SEM are lower than those in air. The experimental measurements are in good agreement with the calculated data shown in Fig. 5. Figure 11b shows the relationship between f and d/l for the SEM and air environments. Here we assume that the melting point of the Pt wire is $T_M = 2{,}042$ K [60] and that $T_0 = 293$ K. Because the value of d is constant in the present study, smaller values of d/l indicate longer l. By nonlinear fitting, we obtained the following f for each environment [49];

$$f_1 = -29.1 \times 10^2 \left(\frac{d}{l}\right)^2 + 90.7\frac{d}{l} + 2.9, \tag{17}$$

for the SEM environment, and

$$f_2 = -126 \times 10^2 \left(\frac{d}{l}\right)^2 + 407\frac{d}{l}, \tag{18}$$

for the air environment.

Values of f greater than unity indicate that the temperature at the middle of the wire in a realistic situation is higher than that for the ideal conditions considered, in which there is no heat loss from the wire, and this is due to the difference in temperature at the ends of the wire. Although the temperature of both ends of the wire was considered to be constant at T_0 for the ideal conditions, it is higher than T_0 in a realistic situation. On the other hand, if the temperature at the middle of the wire in a realistic situation becomes less than that of the ideal conditions,

f becomes less than unity. This situation is brought about by the difference in heat loss from the wire surface. In a realistic situation, heat loss occurs from the wire surface and this becomes significant for longer wires. On the other hand, there is no heat loss from the wire surface for the ideal conditions.

In the examined range of d/l, f_1 is larger than f_2. This indicates that the temperature of the wire easily increases in the SEM compared to in air. This is due to the difference in heat loss from the wire surface, and the significant effect of insulation in the SEM. Actually, this tendency was more remarkable for longer wires (smaller d/l). The larger surface area of longer wires gave greater heat exchange between the wire and the environment, due to the larger surface to volume ratio. Needless to say, if the temperature distribution in the wire changes, heat conduction in the wire also changes. As a result, the temperature at both ends of the wire would change. For shorter wires (larger d/l), f_1 approaches f_2. This is because heat loss from the wire surface is ineffective for shorter wires, which have insufficient surface area, and the difference between the environments is minimal. In both SEM and air, the value of f approaches 3.5 for shorter wires. This value of f depends on the balance of heat conduction in the wire and indicates that the temperature at each end of the wire is greater than T_0.

Let us consider the difference in the limit of $(I_C)_{\text{exp}}$ for longer wires. From Eqs. (15) and (17), the value of $(I_C)_{\text{exp}}$ for longer wires in the SEM is given by

$$\lim_{l\to\infty}(I_C)_{\text{exp}} = \lim_{l\to\infty} A\sqrt{(T_M - T_0)8K\sigma}\,\frac{1}{-29.1\times 10^2\frac{d^2}{l} + 90.7d + 2.9l} \to 0. \quad (19)$$

In the case of the SEM, a very small current, i.e. almost zero, is enough for cutting extremely long wires. This is due to the greater insulation in the SEM. On the other hand, from Eqs. (15) and (18), the value of $(I_C)_{\text{exp}}$ for longer wires in air is

$$\lim_{l\to\infty}(I_C)_{\text{exp}} = \lim_{l\to\infty} A\sqrt{(T_M - T_0)8K\sigma}\,\frac{1}{-126\times 10^2\frac{d^2}{l} + 407d} \to 4.8\text{mA}. \quad (20)$$

In the limit of f_2 for longer wire, a critical current for cutting the wire exists. Because of the heat loss from the wire surface, the temperature at the middle of the wire never reaches the melting point unless the current is greater than this critical value. In this experiment, the critical value for cutting the wire is about 4.8 mA, and wires of any length would not be cut under 4.8 mA.

1.5 Welding of Ultrathin Pt Wires

1.5.1 Experimental

Thin Pt wires with a diameter of 800 ± 200 nm were used in the experiments. The thin wires were the same as that used for cutting experiments in 1.4. The wires were fixed onto chips containing Cu electrodes, and two chips each carrying a Pt

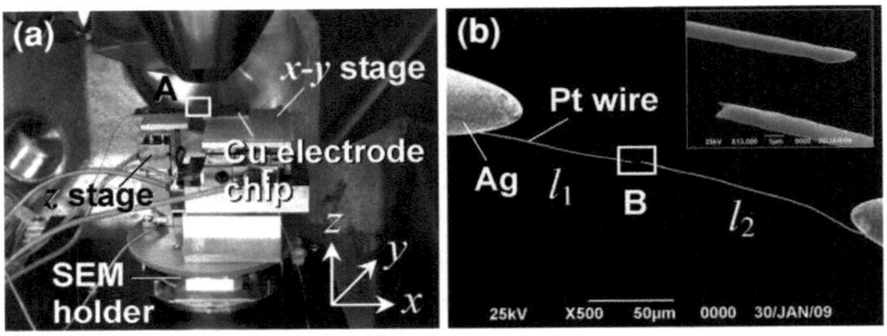

Fig. 12 **a** The SEM holder with three-axis piezoelectric manipulators developed for welding two thin wires. **b** The magnification view of A in **b**. The Pt wires to be joined are of lengths l_1 and l_2. *Inset* shows the details around the tips of the wires. All the experiments were conducted in SEM under vacuum condition. Reprinted with permission from [49]. Copyright (2009) by the American Physical Society

wire were placed on piezoelectric manipulators. The piezoelectric manipulators were assembled on the scanning electron microscope (SEM) holder for welding the Pt wires in SEM (Fig. 12a) [49].

Using the piezoelectric manipulators, the tips of the wires were brought into contact with each other (Fig. 12b). Here, a small compressive force was applied to the wire system to obtain a stable electrical contact. This was indispensable to keep the wire straight during the melting phase. After contact was confirmed, a constant direct current was applied to the wire system for 20 s. A power source placed outside the SEM was used to supply power and measure the voltage, and current was carried into the SEM by an electrical feedthrough. The experiments were conducted for combinations of wires with different lengths. The lengths of the exposed Pt wires are denoted by l_1 and l_2, as shown in Fig. 12b. After the current supply had been on for 20 s, it was determined by SEM observation whether welding was successful or not. The same process was performed repeatedly with increasing current until the wires were successfully welded. The whole process for welding two Pt wires was conducted *in-situ* while under observation in the SEM. The variation of voltage with time was also measured. The same experiments were also conducted in air, where a high-resolution digital microscope was employed and the magnification of the microscope was sufficient for welding experiments.

1.5.2 Monitoring of Welding Process

Figure 13 shows SEM micrographs during the welding process of two thin wires [49]. Figure 13a–d correspond to the points labeled in the circuit voltage history shown in the lower part of Fig. 13. The circuit voltage slightly decreases immediately after the current is supplied [50]. In the SEM micrographs shown in Fig. 13a and b, the shape of the parts in contact changed during this period, i.e. several seconds after the supply was switched on. On the other hand, as shown

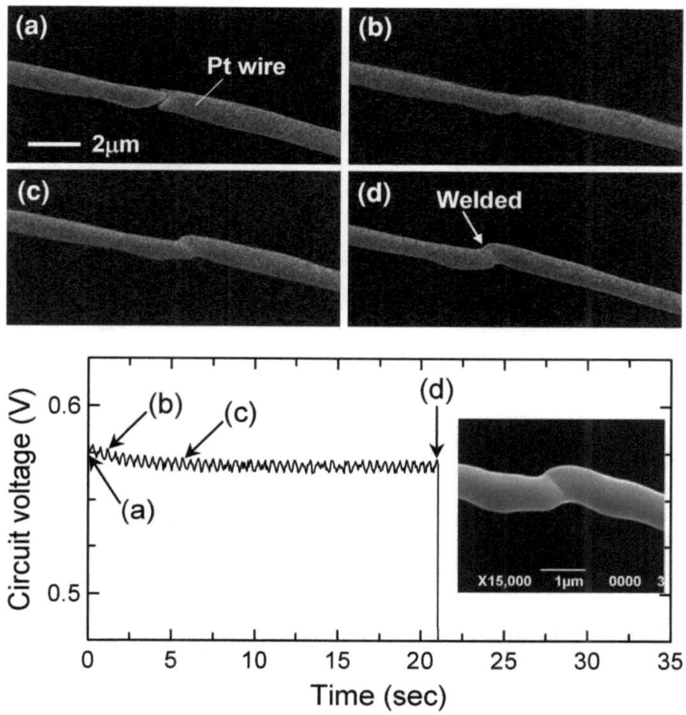

Fig. 13 *In-situ* observation of the welding process of two Pt wires. **a** Before current supply. **b–d** The snap shots obtained from the SEM movie at 1, 6 and 21 s after starting the current supply. The corresponding time history for circuit voltage is shown in the lower figure. The applied current was constant at 2.2 mA for 21 s. The welding process, that consisted of the melting and solidification at the contact, was completed over several seconds after the current supply. Reprinted with permission from [49]. Copyright (2009) by the American Physical Society

Fig. 13c and d, with current still flowing through the wire, from points c–d the circuit voltage remained constant and no visible change in the contact region was observed. These experimental results show that a changing voltage indicates melting in the contact region, whereas a constant voltage indicates that the contact region has already solidified.

From the SEM observations, the welding mechanism of two thin wires is considered to be as follows (Fig. 14). First, the contact region of the wires melts locally by current-induced heating and the contact area increases. The increase in contact area brings about a decrease in circuit resistance and the circuit voltage decreases under the constant current supply. The reduction in Joule heating reduces the temperature in the contact region and the material there solidifies as the temperature falls below the melting point. Note that this welding process is a self-completed mechanism and that melting and solidification in the contact region are achieved within several seconds of turning on the constant current supply.

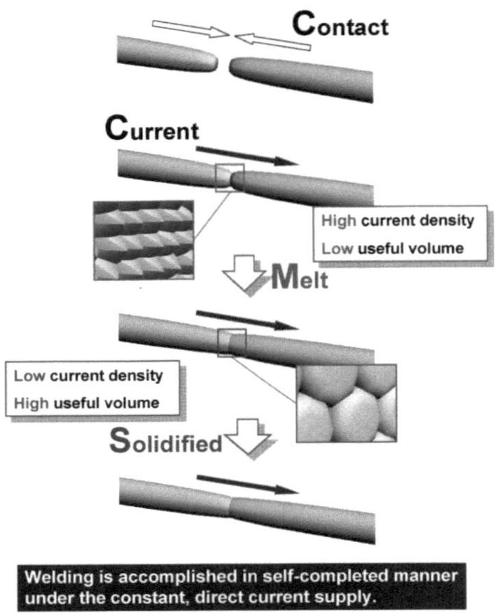

Fig. 14 Schematic of the self-completed welding process

1.5.3 Welding Condition

The results for joining Pt wires in air are summarized in Table 2 [48]. Attempts were made to weld two Pt wires of lengths l_1 and l_2 using a current supply of I. A total of 19 samples with different combinations of wires were welded in air, of which 15 samples were successfully welded. The samples are numbered in order of the total length of the wires. The results are classified into three categories. Successful welding is denoted by S, and incomplete welding by N. The symbol M indicates that the welding failed after joining. In this case, the welding was at first successful, but a point in the wire was cut immediately due to higher Joule heating. After the welding was completed, a small force was applied in the vicinity of the joint with the Ag probe. If the wires separated under this force, the joint was classified as N.

Figure 15 shows examples of SEM micrographs of the joints. Figure 15a and b shows successful joints for Sample Nos. 2 and 6, respectively. It is clear that the wires are joined end to end and the boundary between them is very smooth. Figure 15c shows an unsuccessful joint for Sample No. 13 and this was classified into M. In this case, although the wire was cut at a point away from the joint, the joint was very smooth, similar to the successful joint. Note that joints were not achieved at lower currents, see Table 2. From the experimental results, we found that the current applied to the wire system does not directly govern the melting at the point of contact between the two wires in the present system, with wires of different lengths. In fact, in the case of Sample No. 17, a current of 4.5 mA was insufficient to join the wires, but was sufficient to join the wires for the cases of Sample Nos. 2, 6, 9 and 10.

Table 2 Results of welding experiments in air. Reprinted with permission from [48]. Copyright (2009), American Institute of Physics

Sample No.	Symbol[a]	l_1 (μm)	l_2 (μm)	l^b (μm)	I (mA)
1	M	411.5	82.5	494.0	5.0
2	S	383.8	94.0	477.8	4.5
3	M	165.2	213.4	378.6	6.0
4	S	263.6	83.1	346.8	4.7
5	S	118.1	191.7	309.8	4.2
6	S	214.5	76.4	290.9	4.5
7	M	180.5	99.8	280.2	6.0
8	S	135.4	97.1	232.5	5.0
9	S	160.2	45.8	206.0	4.5
10	N	59.5	141.2	200.7	3.5
	N	…	…	…	4.0
	S	…	…	…	4.5
11	S	86.5	98.2	184.7	4.8
12	S	99.0	80.7	179.8	4.2
13	M	88.1	76.2	164.3	5.7
14	S	110.0	42.5	152.4	5.3
15	S	109.4	37.2	146.6	4.0
16	S	89.6	49.5	139.1	4.3
17	N	104.7	31.5	136.2	4.5
	S	…	…	…	5.0
18	N	57.2	68.2	125.4	3.5
	N	…	…	…	4.0
	S	…	…	…	4.5
19	S	73.1	42.0	115.1	4.8

[a] S indicates the successful joining and N indicates incomplete joining. The case of the joint, which was cut during the current supply due to higher Joule heating, is classified into M. [b] The diameter of all wires was 800 nm, and l is given by $l_1 + l_2$.

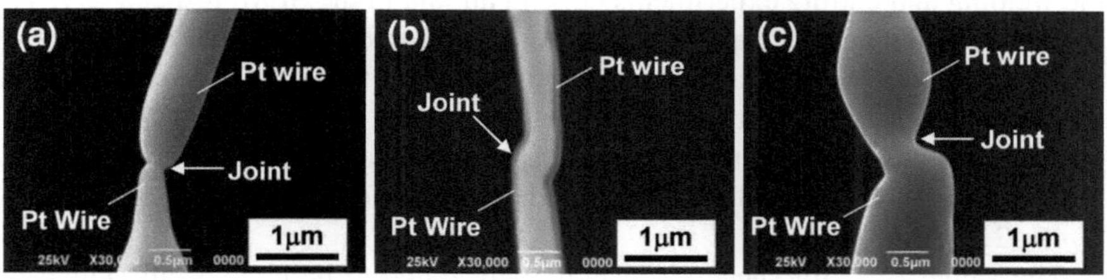

Fig. 15 SEM micrographs of joints. The successful joints for Sample Nos. 2 and 6 are shown in **a** and **b**, respectively. The trial for joining Sample No. 13 was judged to be unsuccessful, even though the joint is smooth **c**. Reprinted with permission from [48]. Copyright (2009), American Institute of Physics

Experiments were also conducted in SEM on 15 samples (S1–S15), and of these, 11 combinations were welded. Note that welds in the SEM were achieved at lower currents than in air. This suggests that the temperature reached at the contact

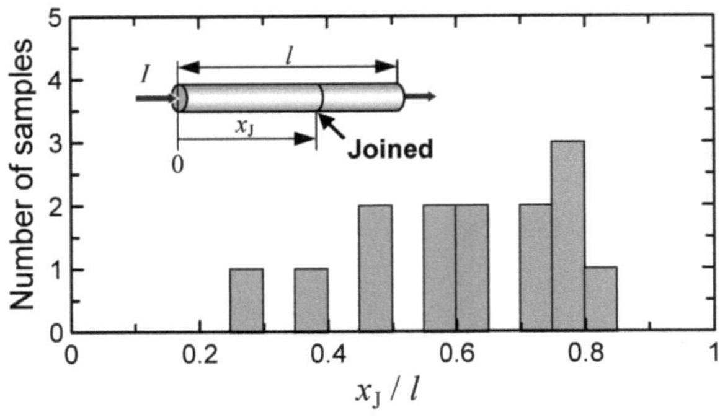

Fig. 16 The position of the joint in the wire system. Note that the position of the joint farthest from the middle of the wire system is more than 0.4 *l*. Reprinted with permission from [48]. Copyright (2009), American Institute of Physics

point of the wires in the SEM was higher than that in air for the same constant current. This is most likely due to the difference in insulation provided by the vacuum. Heat transfer from the wire surface is restricted under vacuum conditions, and this would cause the temperature to rise in the wire system.

Figure 16 shows the successful welding position observed in air [48]. It was shown experimentally confirmed that two wires could be connected by Joule heating at points some distance from the center of the wire system. This was achieved with the aid of a specific temperature field in the contact region of the two wires described in Sect. 1.3.1. Actual wires have some surface roughness, and the contact between two wires is considered to consist of a group of small contacts. In this case, the specific temperature field may be formed throughout the contact region of two wires due to higher current density and a smaller contact volume, and the points of contact may be melted by the current supply provided that the temperature at the contacts exceeds the melting point of the material. Therefore, joining two wires is possible due to the specific temperature field in the region of contact.

Figures 17 and 18 display the relationships between the values of U and d/l for the welding and cutting experiments done in air and in the SEM, respectively. The parameter U defined by Eq. (11) corresponds to the temperature at the middle of the wire system without a contact and it governs the melting phenomenon at the nanocontact between two thin wires. Two wires were found to be successfully welded provided that the value of U is in a specific range, i.e., $U_C < U < U_U$. The values of U_U was 3.1 MA/m, and U_C was found to be about 70% of U_U (= 2.2 MA/m) [48].

The values of U for successful welding are found to be in a specific range for both air (Fig. 17) and SEM (Fig. 18) environments, and the welding conditions defined by U were found to be independent of environment. Because the difference in thermal boundary conditions was calibrated in U by f, the difference in welding current between the SEM and air environments was due to the difference in the thermal boundary conditions, e.g., the heat loss from the wire surface and the temperature at both ends of the wires. Inserting $U = 2.2$ MA/m (U_C) into Eq. (13), we get $T = 1{,}180$ K. On the other hand, because the contact region of two wires melts under this condition, the temperature at the contacts is considered to be

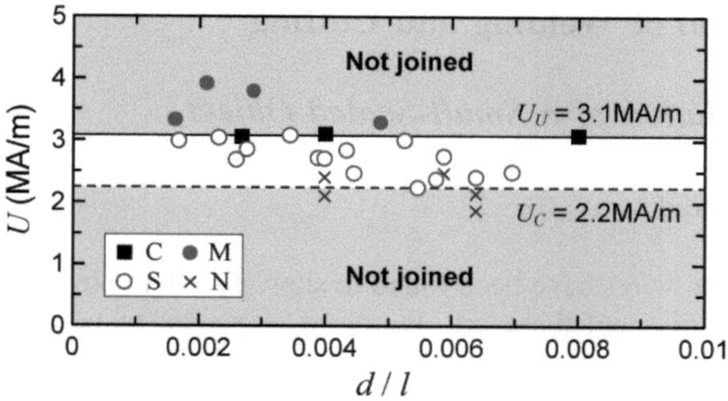

Fig. 17 The relationship between the values of U for successful welding and d/l for the welding experiments in air. Here the open circles show successful welding (S) and the crosses mark incomplete welding (not joined) (N). The solid circles also show welding where the wires were immediately cut after joining due to higher Joule heating (M). The data for the cutting experiments are plotted with solid squares (C). Reprinted with permission from [48]. Copyright (2009), American Institute of Physics

Fig. 18 The relationship between the values of U for successful welding and d/l for the welding experiments in SEM. Reprinted with permission from [49]. Copyright (2009) by the American Physical Society

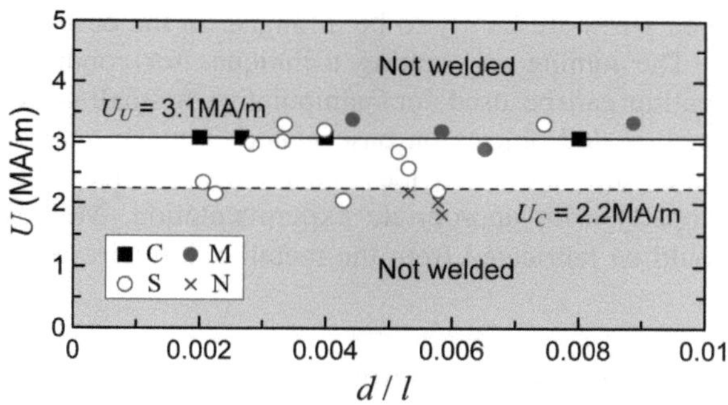

$T_M = 2{,}042$ K. The ratio T_M/T is about 1.7. In other words, the specific temperature distribution at the nanocontact between two thin wires, that gives roughly 70% higher temperature than that at the center of a wire system without a contact is preserved independently of the thermal boundary conditions of the wires. This finding is important for understanding welding of thin wires under various environments and thermal boundary conditions. The current to achieve successful welding can be decided provided that f for the environment is known. Moreover, longer thin wires may possibly be fabricated by repeatedly welding wires. In this case, the current should be decided by considering the total length of the wires to be joined and the environmental situation.

The parameter U governs the melting phenomenon at the contact point between two thin wires, and two wires can be successfully joined provided the value of U is in a specific range, i.e., $U_C < U < U_U$. The wide margin for U for successful joining gives this joining technique utilizing Joule heating a wide applicability.

2 Application of Welding and Cutting

2.1 Manipulation of a Small-Scaled Object

2.1.1 Background

The metallic thin wires have been found to show excellent physical properties and geometrical merits, and these are, therefore, expected to be used as structural, electrical, optical or thermal components in future NEMS or MEMS. However, it is not an easy task to manipulate thin wires for measuring physical properties or assembling them in NEMS/MEMS because the objects are too small for handling. Wu et al. [63] successfully determined mechanical properties of Au nanowire. In this trial, a large number of Au nanowires were scattered on the Si substrate having grooves, and the wires which crossed the grooves were selected for testing after the ends of the wires were welded at the groove edges by laser. The novel technique, however, includes random factor in the sample preparation, and the testing may have limitation on the sample situation. For example, the thin wires that are prepared in chemical or physical ways usually stand on the substrate [39, 58], and such wires are hardly to be arranged on the desired positions.

The joining and cutting techniques for conductive thin wires utilizing Joule heating can be used for manipulating a small-scaled object. The rigid joints of small-scaled objects on base materials make easy of their flexible manipulation, which ultimately provides the facility to check their utility by measuring their property with appropriate experimentation. Moreover, the functional elements could be fabricated from the metallic thin wires by the manipulation technique.

2.1.2 Concept of Manipulation

The term "manipulation" is defined as that the objects are flexibly moved in three-dimensional space with enough force. This includes griping, carrying, and positioning of the objects for getting their functional use. The basic idea of the proposed manipulation technique is to use substrate material that is placed on the 3-axis manipulator, and the thin metallic wires as objects are to be joined on the substrate. Then the thin wires become possible to be flexibly manipulated by the 3-axis manipulator.

The Ag coated Pt wire was used as objective material. The nominal diameter of this Pt wire was 650 nm. The coated Pt wire was cut into 10 mm length, and Ag around the tip of the wire was removed by HNO_3 to expose Pt. Indium (In) wire with the diameter of 500 μm was used as low melting point substrate, and the end of In wire was cut to have sharp tip. The Pt wire with the diameter of 5 μm was used for supplying current to the 650 nm-diameter Pt wire. All the wires were soldered to Cu electrode chips. A schematic of circuit diagram for fabricating the Pt wire sample is shown in Fig. 19a. The tips of the 650 nm-diameter Pt wire and

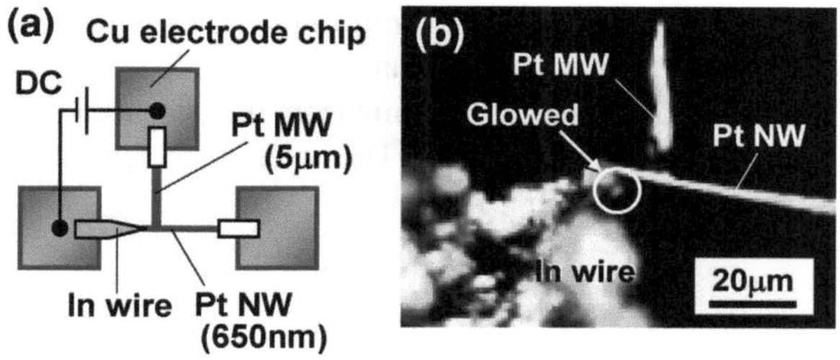

Fig. 19 **a** Schematic of electrical circuit. **b** Microscope image during joining operation. Reprinted with permission of the Society of Materials Science, Japan, from [51]

In wire were brought in contact each other. And then the Pt wire of 5 μm thick was contacted with the 650 nm-diameter Pt wire to supply the current from the 5 μm thick. Pt wire to In wire through a small portion of the 650 nm-diameter Pt wire. Melting point of In is lower than that of Pt. Therefore, the In was melted at the contact of the Pt NW and a joint between the In and the Pt NW was obtained, see Fig. 19b. The 650 nm-diameter Pt wires on In substrate were used for electrical test and mechanical testing of chapter titled Evaluation of Mechanical Properties.

2.1.3 Electrical Test

For electrical test, very-small circuits of the Pt wires were fabricated by joining the Pt wires and the In substrate to Cu electrodes. The diameters of the tested Pt wires were 650 nm and 5 μm. An example of the Pt wire with the diameter of 650 nm is shown in Fig. 20a. Figure 20b shows the magnified view of the joint between the Pt wire and the In wire which was confirmed to be a good electrical conductive

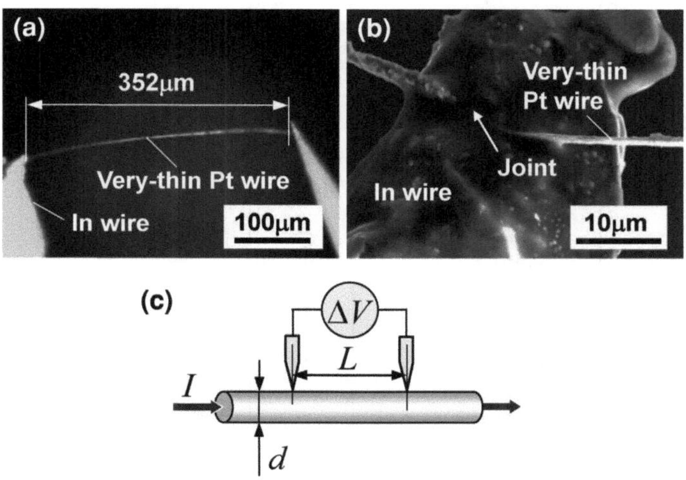

Fig. 20 **a** An example of the very-thin Pt wire circuit prepared by joining. **b** Magnified view of the joint. **c** Schematic of potential drop test for thin wires. Reprinted with permission of the Society of Materials Science, Japan, from [51]

path. The length of the 650 nm-diameter Pt wire was 352 μm and that of the 5 μm-diameter Pt wire was 350 μm.

A voltmeter was used to measure the potential drop (ΔV) across a distance (s) of the Pt wire, see Fig. 20c. The average electrical resistivity (ρ_e) in length L is given by

$$\rho_e = \frac{\pi d^2}{4L}\frac{\Delta V}{I}, \tag{21}$$

where d is the diameter of the Pt wire, I supplied current. The values of ΔV for determining ρ_e were measured at $L = 82$, 125 and 180 μm for the 650 nm-diameter Pt wire and at $L = 85$, 121 and 181 μm for the 5 μm-diameter Pt wire.

Figure 20a and b represents the values of ΔV against I for the 650 nm-diameter Pt wire and for the 5 μm diameter Pt wire under the various values of L, respectively. Figure 21a–d shows the relationships between ρ_e and I^2 for Pt wires with diameter of 650 nm and 5 μm, respectively. Figure 21c clearly shows that ρ_e is proportional to I^2. On the other hand, no dependency of I was observed for the 5 μm-diameter Pt wire. The increase of ρ_e with I^2 is due to the increase in temperature of the ultrathin Pt wire. By fitting the relation between ρ_e and I^2 with least-square method, the value of ρ_e at room temperature ($I^2 = 0$) was determined

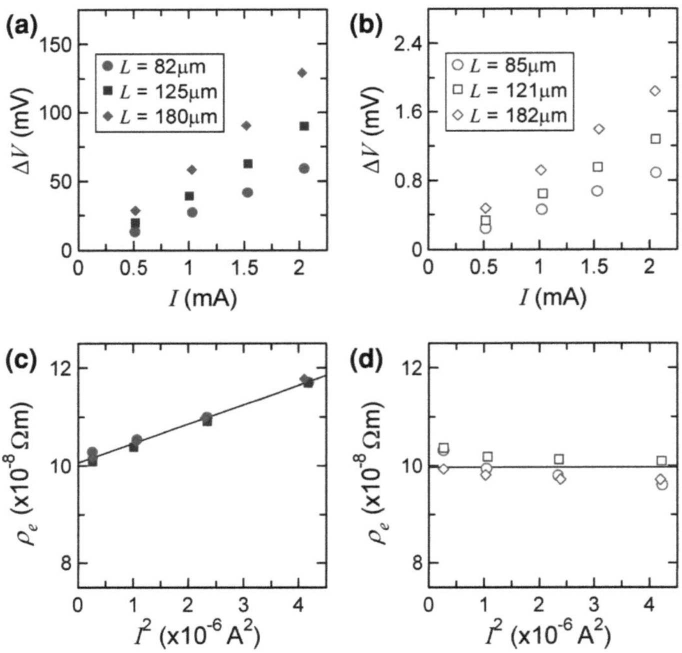

Fig. 21 Potential drop (ΔV) as a function of I and L for **a** very-thin Pt wire and **b** thin Pt wire. Relationships between ρ_e and I^2 for **c** very-thin Pt wire and **d** thin Pt wire. Reprinted with permission of the Society of Materials Science, Japan, from [51]

as 10.2×10^{-8} Ωm. On the other hand, the average value of ρ_e of the 5 μm-diameter Pt wire was 10.0×10^{-8} Ωm. It was confirmed that the both Pt wires had almost same ρ_e and the values were close to the reported value of ρ_e for bulk Pt (10.6×10^{-8} Ωm).

2.2 Fabrication of a Small-Scaled Electromagnetic Ring

2.2.1 Background

Loop-shaped structures such as micro-rings or micro-coils have attracted considerable attention recently because such structures are expected to be used as functional elements in nano-electromechanical systems (NEMS) and micro-electromechanical systems (MEMS), especially for the electromagnetic elements. Put simply, a micro-ring/coil can behave as an inductor in a small-scale integrated circuit [23, 66]. Also, these elements can generate small magnetic fields when they are supplied with a current. Lee et al. [29] utilized the electromagnetic interaction between a micro-coil through which a current was flowing and a magnet to realize a pumping action. Sutanto et al. [45] utilized the electromagnetic coil interaction for electromagnetic bistable latching microvalves. Watson and Chan [61] successfully generated microwaves by using a micro-ring supplied with a high-frequency alternating current. Many types of sensors, such as strain sensors [2], temperature sensors [69], proximity sensors [42], mechanical sensors [3], and so on, have been realized by the use of loop-shaped structures.

There are two main approaches to fabricate small looped structures. The first involves a bottom-up approach utilizing various chemical reactions, such as the focused-ion-beam-assisted chemical vapor deposition [35] and microwave chemical vapor deposition [57]. Although various bottom-up approaches can be used to realize the mass production of smaller rings or coils, it is usually difficult to control the geometrical shapes of the structures. Moreover, it is difficult to connect the electrodes and electrical lines for supplying the current to the looped structures when they are prepared by using a bottom-up approach. On the other hand, an alternative top-down approach can be used to realize the variety of ring designs that can more readily include electrodes. For example, rings with sub-micron diameters have been prepared on substrates by using electron-beam lithography [8] and nano-imprint lithography [10]. However, the presence of the substrate behind the ring may be an obstacle in many applications, e.g., antenna applications. At the macroscopic level, the looped structures are obtained by winding the wire onto a core, which causes local plastic deformation of the wire. However, at the microscopic level, it is difficult to manipulate thin wires and to generate sufficiently high stresses for inducing the local plastic deformation in the wire. A free-standing Pt micro-ring (MR) with electrodes is successfully fabricated on a chip by using a welding technique that utilizes Joule heating [53].

2.2.2 Fabrication

Figure 22 shows the set-up that was used to fabricate the MRs. Two types of thin Pt wires with diameters of 650 nm and 5 μm were used. All of the Pt wires were Ag coated, making the overall diameter about 75 μm. The Ag coatings around the tips of the Pt wires were removed by dipping the wires in HNO_3 to expose the underlying metal, see the inset scanning electron microscope (SEM) image in Fig. 22. One Pt NW (Pt NW A) was positioned on a Cu electrode chip, and a second Pt NW (Pt NW B) was positioned on a 3-axis nano-manipulator. A third Pt wire with a diameter of 5 μm was also positioned on the Cu electrode chip. The Pt wire and Pt NW A were insulated electrically with each other because the air gap existed between the Cu electrodes for both Pt wires.

By moving Pt NW B with the manipulator, Pt NW A and Pt NW B were brought into contact at the tips of both NWs, and by supplying a direct current (SW 1 was closed), the two Pt NWs could be successfully welded together due to Joule heating. The rigid joint between the two thin Pt wires enabled us to twist the wires around the Ag core by using the nano-manipulator. Figure 23a shows a digital microscope image of Pt NW A after it has been twisted 180 degrees around the Ag core. Further images in which Pt NW A is wound further (by 360 and 720 degrees) are shown in Fig. 23b and c, respectively. After Pt NW A was twisted by 2 full turns, it was then twisted 180 degrees in the opposite direction to create a space between Pt NW A and the Ag core. This space was indispensable in removing the wrapped Pt NW from the Ag core; see Fig. 23d. Pt NW A, which was now in the form of a MR, was then contacted to the Pt wire on the same electrode chip, and a current was supplied (SW2 was closed) at the contact point between Pt NW A and the thicker Pt wire. In this case, no current flowed through the Pt MR (Fig. 23e). The contacting point was first welded by Joule heating. After that, the wire system with flowing current was cut out by also Joule heating at a point of the Pt NW with the cross-sectional area smaller than Pt wire; see Fig. 23f. At the end of this process, a free-standing Pt MR had been successfully prepared on a Cu electrode chip.

Figure 24a shows the SEM micrograph of the Cu electrode chip with the Pt MR. The Pt MR is located in ‘region A’ in the micrograph, and the MR is simply

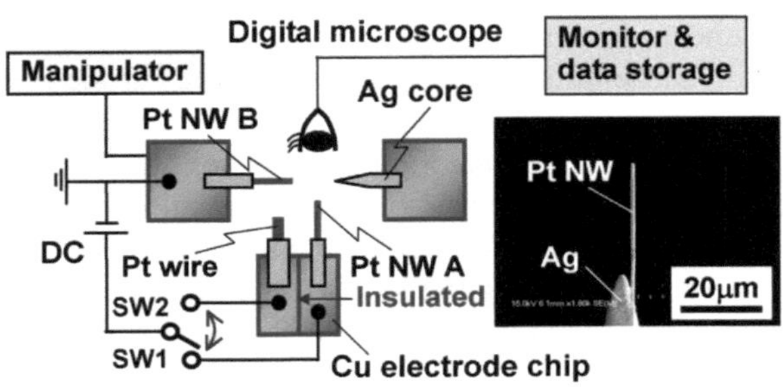

Fig. 22 Set-up for fabricating the micro-rings. An SEM image of the Pt NW is inserted. Reprinted with permission from the Institute of Physics, from [53]

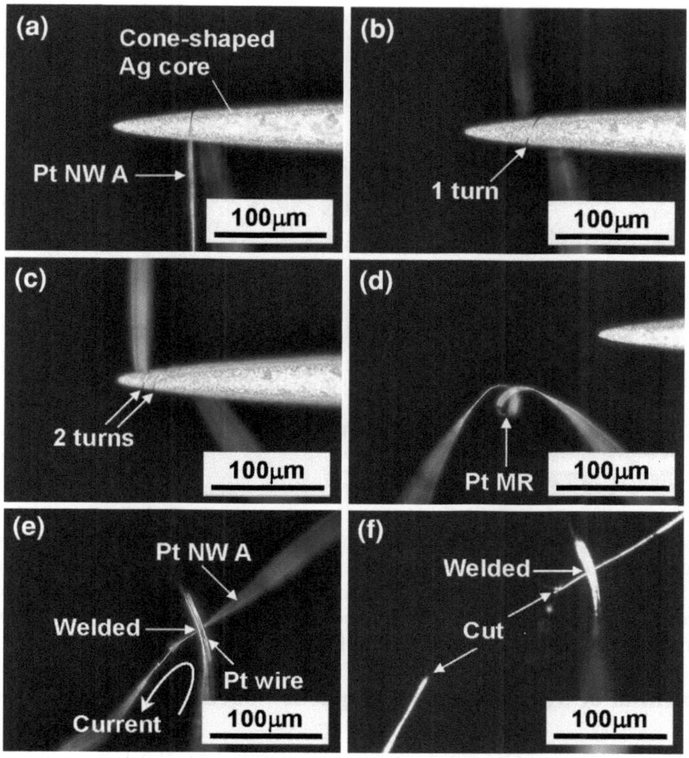

Fig. 23 Digital microscope images taken during the fabrication of the micro-ring. Images show Pt NW A twisted around the Ag core at 180 (**a**), 360 (**b**) and 720 degrees (**c**), respectively. The turned Pt NW A was then pulled off the Ag core (**d**). Pt NW A with the MR was then contacted with the Pt wire and a current was supplied at the contact point between Pt NW A and the thicker Pt wire (**e**). The contact point was first jointed, and the circuit was opened at a point in Pt NW B (**f**). Reprinted with permission from the Institute of Physics, from [53]

supported by two Ag beams. A magnified view of the Pt MR is shown in Fig. 24b, and the fabricated MR can be seen to be an elliptical-shaped ring. The long axis of the elliptical MR was 30 μm, while its short axis was 15 μm, respectively. A smaller Pt MR was also fabricated by replacing the Ag core with the 5 μm-diameter Pt core. Figure 24c and d shows SEM micrographs of the electrode chip and a magnified view of the fabricated Pt MR. The diameter of the MR in this case was 11 μm. Moreover, it was confirmed that currents of up to 3 mA were able to be supplied to both of the Pt MRs that we fabricated. The presence of several deformation points in both of the Pt MRs indicated that local yielding of the Pt thin wires had occurred at these points; see Fig. 24b and d. The thinner Ag core may allow the wire to attain the required yield-strain more readily, and many yield-points can be observed in the smaller Pt MR (Fig. 24d).

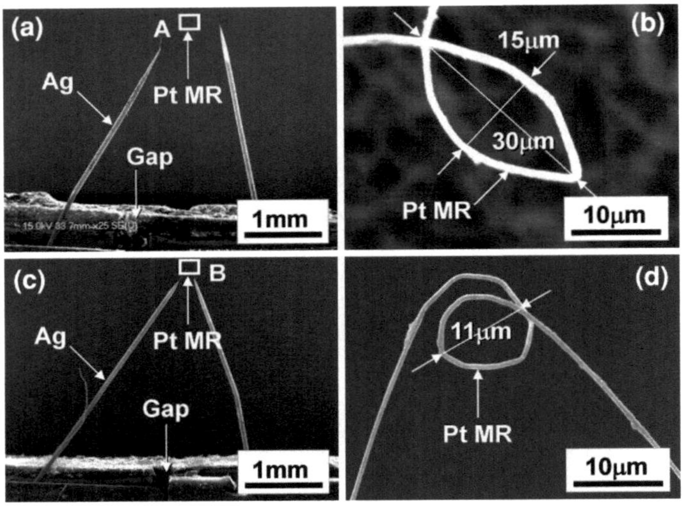

Fig. 24 SEM micrograph of a Cu electrode chip carrying a Pt MR fabricated by rolling a NW around a cone-shaped Ag core (**a**). The Pt MR is located in region A in the micrograph, and a magnified view of the Pt MR is shown in (**b**). Equivalent images of a Pt MR fabricated by rolling an NW around a 5 μm-diameter Pt wire are shown in (**c**) and (**d**), respectively. Reprinted with permission from the Institute of Physics, from [53]

2.2.3 Potential of Micro-Rings as Magnetic Sources

To confirm that a magnetic field could be generated by the fabricated Pt MR, the electromagnetic interaction between the MR with flowing current and a permanent magnet was observed. The diameter and the thickness of the permanent magnet were 15 and 4 μm, respectively, and the magnetic flux density of the magnet was 0.085 T. A fabricated Pt MR with a diameter of 11 μm was located at a position 200 μm above the surface of the magnet. Figure 25a and b shows digital microscope images of the Pt MR without and with supplied current ($I = 100$ μA), respectively. The Pt MR was initially at the focal point of the microscope (Fig. 25a), but the focal point was then shifted away from its original position when the current was supplied (Fig. 25b). This result indicated that the Pt MR moved in the vertical direction due to an electromagnetic interaction between the permanent magnet and the Pt MR when a current was flowing. The relationship between the vertical displacement of the Pt MR in terms of δ and I is shown in Fig. 25c. The values of δ were determined from the changes in the focal point of the microscope. The (+) current indicates the case where the current flowed in a counter clock-wise direction, while the (−) current is the case where the current flowed clock-wise. The Pt MR moved in the 'upwards' direction in the case of the (+) current, whereas it moved in a 'downwards' direction in the case of the (−) current. The absolute values of δ were almost the same for a given value of I if the direction of the current was ignored, and these were proportional to I. Because the Pt MR was supported by two straight beams, the values of δ were proportional to the vertical forces acting on the Pt MR.

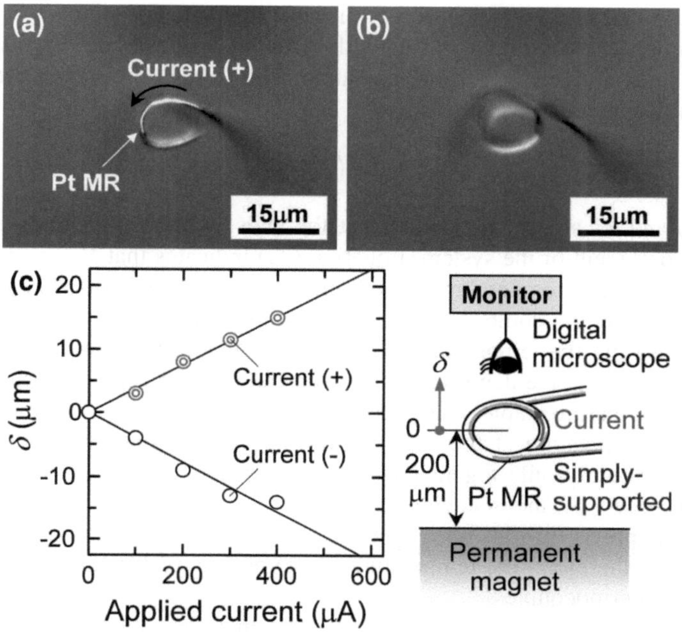

Fig. 25 Digital microscope images of a Pt MR without (**a**) and with (**b**) applied current ($I = 100$ μA), respectively. The relationship between the vertical displacements of the Pt MR supported by simple beams in terms of δ and I are shown in (**c**) together with the experimental setup. Here, the (+) current indicates the case where the current flowed in the counter clock-wise direction, and the (−) current is the case where the current flowed clock-wise. Reprinted with permission from the Institute of Physics, from [53]

The current that is flowing the wire is also generates the magnetic field around the wire. However, because the total magnetic field in the vertical direction generated by the wire becomes zero, no electromagnetic interaction between the wire and the magnet occurs. Therefore, the observed electromagnetic interaction was due to the presence of the magnetic field emitted from the Pt MR with flowing current. The present design of a micro-ring supported by two simple beams, which can be linearly deflected against the supplied current, can be expected to be used for micro switches in small-scale circuits.

2.3 Fabrication of a Small-Scaled Thermoelectric Element

2.3.1 Background

Seebeck effect refers to the generation of voltage as a result of a temperature gradient and its efficiency is represented by the Seebeck coefficient, *S*. The effect can be exploited in thermoelectric-power generators and for temperature sensing, by connecting two dissimilar metals with different Seebeck coefficients which is well-known as a principle for thermocouple. The voltage difference between the two dissimilar metal, V_{12}, is given by

$$V_{12} = S_{12}(T_H - T_L), \tag{22}$$

where T_H and T_L are the temperature at the ends of dissimilar joints and S_{12} is Seebeck coefficient of the system. Equation (22) indicates that V_{12} is proportional to the difference in temperature at both ends of dissimilar joints. The coefficient S_{12} is decided by the combination of metals to be jointed.

For measuring the temperature at very-small region, the combination of thin, dissimilar films that are deposited on the substrate can be used. For example, this type of film thermocouple can successfully be used for measuring the rapid changes in temperature at the local region induced by a short laser pulse [11]. Also, temperature profile together with the topographyic data at the local region can be recorded by the scanning probe microscopy that has a dissimilar film structure on probe tip [44, 62]. However, the substrate sometimes becomes an obstacle, and therefore, the temperature of 3D objects is difficult to be measured by the film thermocouple.

So far, recently, various types of thin wires including nanowires are successfully fabricated and synthesized by chemical and physical methods, and the giant thermoelectric effect of nanowires is observed [4]. The combination of dissimilar nanowires may initiate an effective thermoelectric element which leads to realize sensitive temperature measurement for 3D objects. Here the welding technology for dissimilar metals is a key factor for realizing the above trial. A thin, free-standing Pt/W thermocouple was successfully fabricated on a small electrode chip using the Joule heat welding technology [52].

2.3.2 Fabrication of Thermocouple

Thin Pt wire of 5 μm diameter and two types of W wires with diameters of 20 and 5 μm were used to construct the thermocouple. The Pt wire was coated with Ag having the overall diameter of 200 μm. Ag around the tips of the Pt wires was removed by HNO_3 solution and was soldered on the surface of a Cu electrode chip and thin W wire was also soldered to the same Cu electrode with an isolation. The Cu electrode chip was fixed onto a sample base. Ag probe with cone-shaped was also attached to another Cu electrode chips and was positioned on a three axis manipulator to supply the current through a segment of the desired wire. To make the thermocouple, the thick W wire was located upon another manipulator with Cu electrode chip. By moving the manipulator, tip of the thin W wire was brought into contact at the tip of Pt wire and a direct current was applied to the system with the support of Ag probe where current passed through a part of Pt wire and thin W wire. With the increasing of the current the tip of Pt wire was melted due to the

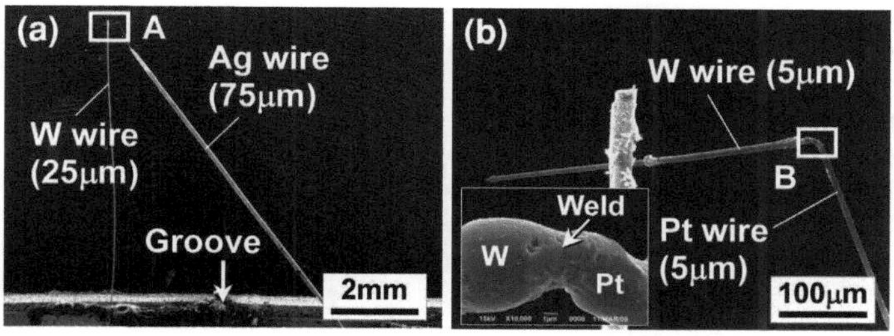

Fig. 26 SEM micrographs of fabricated Pt/W thermoelectric element. **a** Whole view. **b** Details around the Pt/W dissimilar weld (from [52], © (2010) IEEE)

Joule heating. Then current supply was being constant for 10 s before switched off and the wires were successfully welded together. After then by moving the manipulator, tip of the thick W wire was brought into contact at the W wire and was also welded by supplying a direct current with the Ag probe. The fabrication of the thermocouple was completed under the direct observation of a high-resolution optical microscope.

The SEM micrograph of the fabricated thermocouple element on the electrode chip is shown in Fig. 26 as an example. The tips of the dissimilar metal welds were less than 10 μm.

2.3.3 Seebeck Effect

Figure 27a shows the experimental setup for observing Seebeck effect of the fabricated thermoelectric elements. The Pt wire having the diameter of 5 μm was used as heat source and the temperature of the wire was controlled by changing the current supplied through the wire. Both ends of wire were fixed on In and Ag wire substrates, and the length of the wire was 350 μm. The temperature at the middle of a wire system was determied from Eq. (13). The dissimilar joint of thermocouple was contacted to the middle part of the heater wire and the voltage was successfully observed in the thermoelectric circuit, see Fig. 27b. Immediately after the contacts, the voltage took the constant value within several seconds. The stable

Fig. 27 Measurement of Seebeck coefficient. **a** Experimental setup. Electrodes 1 and 2 were connected to power supply and the electrode 3 and 4 were connected to a volt meter. **b** Current supplied to the Pt wire and the voltage observed by volt meter (from [52], © (2010) IEEE)

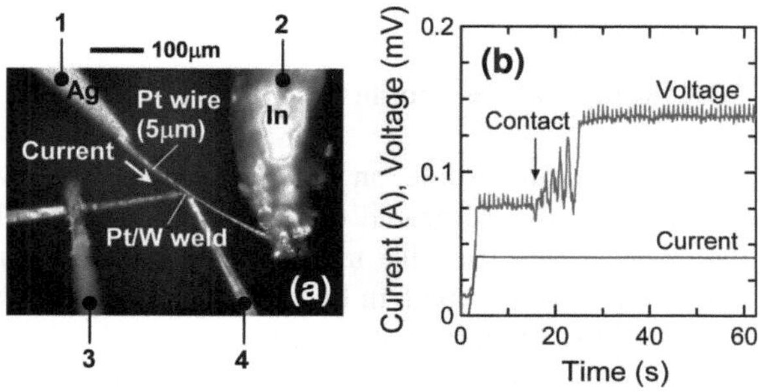

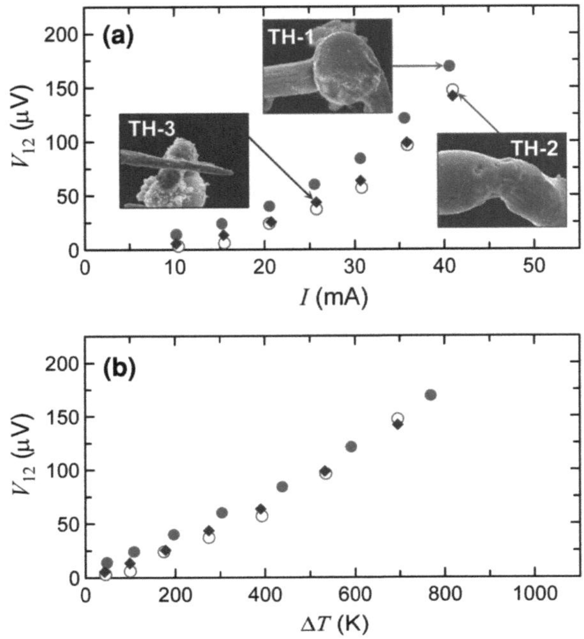

Fig. 28 Experimental results. **a** V_{12} versus I. **b** V_{12} versus ΔT. The details of weld for three thermocouple (TH-1, TH-2 and TH-3) are inserted (from [52], © (2010) IEEE)

voltage is denoted by V_{12} in this paper and we measured V_{12} with the supplied current range of 10–40 mA.

Relations between the observed voltage (V_{12}) and supplied current through the heater wire (I) are presented in Fig. 28a. Total three thermocouples (TH-1, TH-2 and TH-3) were fabricated by the same procedure as described in the previous section, and the photographs of the details around the dissimilar metal tips are inserted in Fig. 28a. It was observed that V_{12} increased with the increasing of I. Figure 28b shows the relationships between V_{12} and the temperature at the contacting portion ($\Delta T = T - T_0$). The voltage due to Seebeck effect was proportional to the difference in the temperature between the tip of the dissimilar joint and the electrode chip. The values of S_{12} for the Pt/W joints of TH-1, TH-2 and TH-3 were 0.211, 0.217 and 0.206 μV/K, respectively.

2.3.4 Temperature Profile on Thin Pt Wire

The temperature profile on the thin wire with flowing current of 40 mA was measured by the fabricated thermocouple element (TH-3). The wire system was the same as that used for measuring Seebeck coefficient. In the measured temperature profile as shown in Fig. 29, the temperature at the middle of the wire was highest. Also, the temperature at one end of the heater wire that was connected to the Ag wire was lower than the temperature at another end of the wire that was

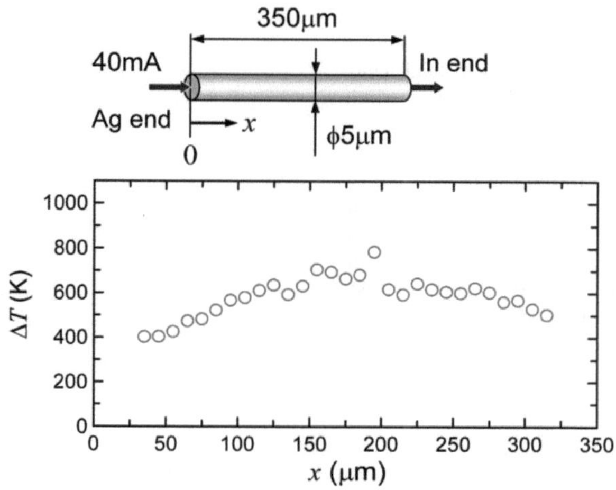

Fig. 29 The relationship between ΔT and x measured by the fabricated Pt/W thermoelectric element (from [52], © (2010) IEEE)

connected to the In wire. This interesting result suggested that the heat was easily transported toward the Ag end than the In end.

3 Permanent Bending of Brittle Nanowires for Formation of Nanocoils

3.1 Introduction

Nanocoils (NCs), also referred as nanohelixes or nanosprings elsewhere, are the most attractive, complex nanowire (NW) structures that have ever been demonstrated. They are candidates for nanoscale sensors and actuators, in addition to acting as mechanical springs. The well studied technique for assembling NCs is the use of polar-surface-dominated nanobelts (NBs) or NWs of ZnO [27], SnO_2 [67], InP [43] and AlN [15]. The electrostatic surface energy imposes spontaneous self-assembly into NCs during synthesis. Figure 30 shows the SEM image of polarization-induced NCs of ZnO NBs created in the pioneering work by Kong and Wang [27]. Some syntheses of nonpolar NCs, such as SiC ([68], BC [33], and $NaFe_4P_{12}$ [32] have also been reported, where the mechanisms are believed to originate from the anisotropic effects of the catalyst on SiC and BC, and the anisotropic crystal growth due to the specific crystal structure of $NaFe_4P_{12}$.

Inductance NCs necessitate good conductivity. However the above techniques are restricted to the specific non-metallic materials: oxide (ZnO, SnO_2), semiconductors (InP, $NaFe_4P_{12}$), and insulating ceramics (AlN, SiC, BC). Focused-ion-beam-assisted chemical vapor deposition (FIB-CVD) has been applied to the assembly of conducting carbon NCs [35], but was unable to create metallic NCs. Additionally, FIB-CVD is time/cost-consuming. Further, the misfit strain due to

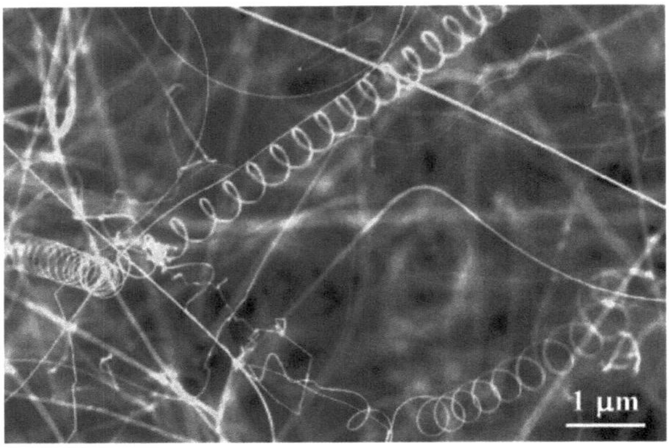

Fig. 30 SEM image of the as-synthesized ZnO nanobelts, showing helical nanostructure. The typical width of the nanobelt is ~30 nm, and pitch distance is rather uniform. The helixes are right-handed. Reprinted with permission from [27]. Copyright 2003 American Chemical Society. Image courtesy of Prof. Z. L. Wang, Georgia Tech

lattice mismatch of a semiconducting bilayer epitaxially grown on a substrate has also been utilized to mechanically roll the bilayer after removing the substrate [41]. Patterning the bilayer on a substrate allows the design of nanotubes (NTs) and NCs [20].

Recently, Muraoka et al. [34] presented a simple method for mechanically bending a NW. In this method, a metallic film is sputtered onto a straight NW with a circumferentially nonuniform film thickness. The residual strain in the coating metal results in the permanent bending of the NW, thereby forming NCs. This technique does not necessarily require the epitaxial growth of a film to induce film strain and extends the range of materials for NCs. As is well known, macroscopic coils or springs are fabricated through plastic deformation by bending. However it is somewhat difficult to achieve this in the nanoworld. The thinness of NWs significantly reduces the bending stress against a bending moment. Thus, NWs are quite flexible but can hardly be deformed plastically, even if they are metallic.

Section 3 deals with the permanent bending technique of NWs by using the residual strain of the deposited film [34] together with an enhanced bending technique by creeping the NW to remove its constraint on the bending deformation [54]. In advance to them, mechanisms of residual stresses in thin films are briefly discussed together with an analytical treatment of bending deformation induced by the residual stresses.

3.2 Residual Stress in Deposited Thin Films

Residual stresses in thin films deposited on substrates are divided into two types: extrinsic and intrinsic stresses. Extrinsic stress is due to post-deposition processing such as cooling or due to external influences such as impurity adsorption within film structure. The former is thermal stress. When the temperature during the film deposition differs from that at measurement of the residual stress, thermal stress is induced in the film whose thermal expansion mismatches that of the substrate. The

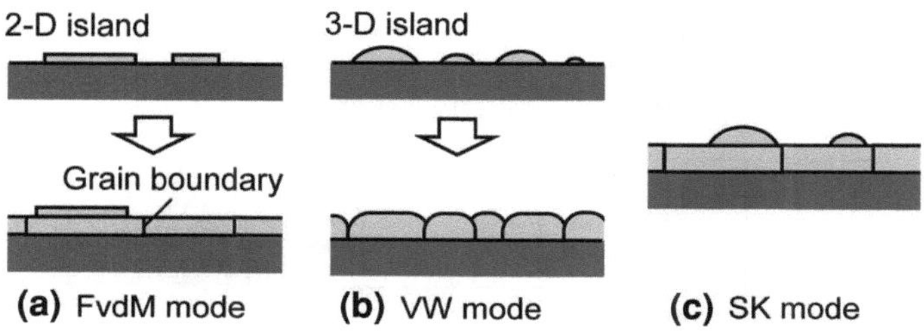

Fig. 31 Film growth modes. **a** Frank-van der Merwe (FvdM) growth mode, where the film atoms wet the subtrate surface. The film grows atomic layer by layer. **b** Volmer-Weber (VW) growth mode, where the film atoms do not wet the substrate surface. If the grain sizes are preserved, the film grows in columnar form, otherwise, increase in lateral size of grains due to recrystallization. **c** Stranski–Krastanov (SK) growth mode, which is a mixed mode between FvdM and VW modes and induced by extrinsic factors such as film stress.

latter is extra stress introduced by interaction of polar molecules adsorbed within porous films.

The film formation depends in part on the thermodynamics of the film-substrate interface; in particular, it would depend on how well the film wets the substrate. However, the deposited film is not in thermodynamic equilibrium. The film deposition process is strongly influenced by kinetics parameters (such as the deposition rate and the surface mobility of the atoms), defects, etc., which determine when and how the nucleation takes place and when the film becomes continuous. Basically, the film formation is divided into the following three growth modes [25]. Frank–van der Merwe (FvdM) growth or Volmer–Weber (VW) growth modes (Fig. 31) are observed when the film atoms wet the substrate surface or not, respectively. The other case where it is between them (mixed mode) is called Stranski–Krastanov (SK) growth mode (Fig. 31). The film formation typically follows VW mode for lower substrate temperatures, and epitaxial growth follows FvdM mode. In FvdM mode, two-dimensional (2D) islands of atomic monolayer are grown on the substrate, the atoms fill the remaining channels, and then form continuous layer before growing next layer. VW mode is basically similar to FvdM mode, where islands are 3D instead of 2D. SK mode is due to extrinsic factors such as residual stress and interfacial alloy.

Intrinsic stress is defined as stress in the as-deposited film, i.e., stress that develops during the deposition, see Fig. 32. It is mainly caused by microstructure of the film, which includes interaction of grains, impurities (gas atoms) entrapped during the deposition, lattice mismatch between the film and substrate, solid state reactions that change the film volume, and interfacial phase formation [14, 25]. During the early stages of deposition (before the grain interactions become dominant), it is possible for the surface stress to generate intrinsic stress, where the stress is initially compressive and increases in magnitude with increasing the thickness up to a certain thickness for metallic films [7]. This surface stress effect is equivalent to the lattice expansion mechanism [25].

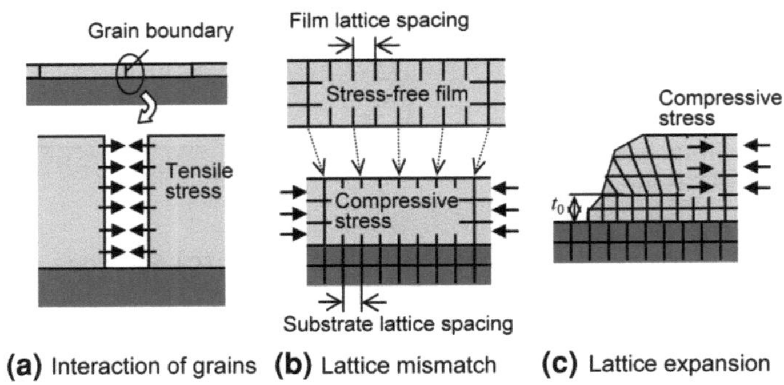

Fig. 32 Examples of mechanisms of intrinsic stress. **a** Grain boundaries are of reduced density because the grains of differing orientations laterally touch. Therefore, interatomic forces try to close the gaps at grain boundaries and stretch grains, resulting in tensile stress. **b** For a strong film-substrate interaction and small lattice-mismatch, the interface becomes coherent and induces compressive or tensile stresses in the epitaxially grown film. **c** For a weak film-substrate interaction, the interface becomes noncoherent. The atomic bonding of the film during the earlier stage of deposition (the thickness < a specific value t_0) energetically relaxes to contract its bond length (its lattice spacing) in stress-free state due to the high surface-to-volume ratio. When increasing the thickness, the lattice spacing tries to expand and to approach to that of bulk, more precisely to that of bulk stressed under effect of surface stress. The expansion is constrained by the underlying layer of the thickness t_0 that initially was stress-free and produces compressive stress in the overlying layer.

Bending deformation of film-deposited substrates due to intrinsic stress can be analytically dealt with by using a misfit strain e_0. If the film deposited on the substrate of the lateral length l_S is nondestructively removed from the substrate, the film deforms and recovers its lateral length l_F in the stress-free state. The misfit strain is defined as $e_0 = (l_S - l_F)/l_F$. For epitaxially grown films with coherent interfaces, the misfit strain e_0 coincides with the lattice misfit parameter $e_0 = (c_S - c_F)/c_F$, where c_S and c_F are the lattice constant (in bulk) of the substrate and film materials. When the misfit strain varies along the film thickness, the removed film may bend. For simplicity, assume the uniform misfit strain in the film and negligible surface and interface stresses. Figure 33a shows a 2D model of film-deposited substrates that are elastically bent by the misfit strain. The equilibrium state is obtained by an equivalent process as follows: the film in the stress-free state (Fig. 33b) is elastically stretched with the elastic strain equal to e_0 (Fig. 33c). It is firmly glued on the substrate surface (Fig. 33d), and then the tensile stress applied to the film is removed (Fig. 33a).

In the 2D model, the film under an intrinsic stress bends the substrate as an arch with radius R (= $1/\kappa$) of the neutral surface. The strain $e_B(\eta)$ of a layer in the substrate at a distance η from the bottom of the substrate is given by

$$e_B(\eta) = -(\eta - c_0)/R = -(\eta - c_0)\kappa, \tag{23}$$

where c_0 is the position of the neutral surface (the strain-free layer) in the substrate. The strain in the film is a combination of the bending strain $e_B(d_S)$ and the

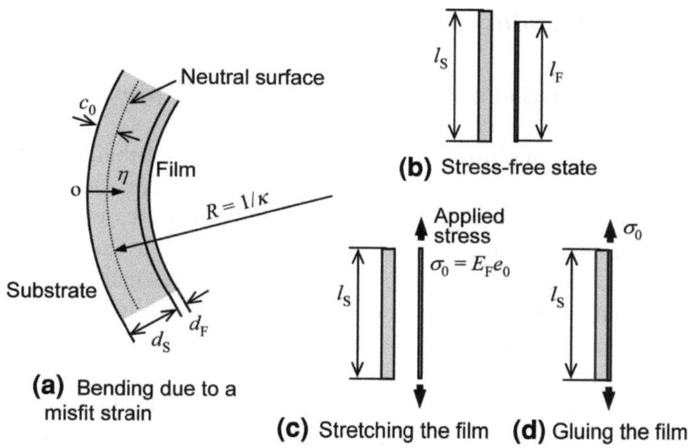

Fig. 33 A 2D continuum model of film-deposited substrates

misfit strain e_0, where $d_F << d_S$ is assumed. The total strain energy of a unit area of the substrate and film is given by

$$U = \frac{E_S}{2} \int_0^{d_S} e_B(\eta)^2 d\eta + \frac{E_F}{2} [e_B(d_S) + e_0]^2 d_F, \tag{24}$$

where E_S and E_F are the Young's moduli of the substrate and film. In the case of 3D substrates such as wafers and plates, their Young's moduli are replaced by the corresponding biaxial moduli $E_S/(1 - \nu_S)$ and $E_F/(1 - \nu_F)$ (ν_S, ν_F: Poisson's ratios), the total energy is duplicated, and R denotes the radius of the spherically bent substrate.

Values κ and c_0 are found by minimizing the energy with respect to those quantities. This gives $c_0 = d_S/3$ [55] and an expression of κ:

$$\kappa = \frac{6 r_E r_d e_0}{(1 + 4 r_E r_d) d_S}, \tag{25}$$

where $r_E = E_F/E_S$ and $r_d = d_F/d_S$ are ratios of the Young's modulus and thickness, respectively. From Eqs. (23) and (25), the intrinsic stress σ_F is determined as

$$\sigma_F = E_F e_0 / (1 + 4 r_E r_d). \tag{26}$$

Meanwhile, thermal stress is also dealt with by using a misfit strain. Equating the misfit strain e_0 with $e_{0T} = (\alpha_S - \alpha_F)\Delta T$ in Eqs. (25) and (26), we obtain the bending curvature and thermal stress, where α_S, α_F, and ΔT ($= T_B - T_A$) are thermal expansion coefficients of the substrate and film, and the change in temperature from at the deposition (T_A) to at the measurements (T_B) of deformation and stress, respectively.

3.3 Misfit-Strain-Induced Bending of Nanowires

3.3.1 Nanocoil Formation of Chromium-Deposited Copper Nanowires

A circumferentially nonuniform thickness of an overlayer unbalances the residual strains and produces the bending deformation of a NW. Test samples of straight NWs of copper (Cu), standing on a multilayered substrate, were created by a stress-induced migration process [39]. The substrate was a $Cu_2O/Cu/Ta/SiO_2/Si$ system, fabricated as described in Sect. 3.3.1. The Cu NWs had diameters ranging from 20 to 50 nm with lengths of several microns, as shown in Fig. 34a. The number of NWs per a unit area of the substrate surface, (the sample density), was about $10/\mu m^2$. Most of NWs were somewhat tilted off the vertical.

After the substrate was diced into chips of about 10 mm in size, several were tested for the applicability of the bending process. The deformational process of NWs relied on a modification of their surfaces by coating that resulted in emergence of residual bending-strains in the NWs. Coating was carried out with physical vapor deposition (PVD) at low energy to prevent any considerable damage to the NWs. The electron-beam evaporation technique (EBET) was used as a low energy PVD, at a pressure of 1.1×10^{-3} Pa. The coating materials used were Cr, Pt, Ni or Al, all being good conductors. When a sputtering technique at a

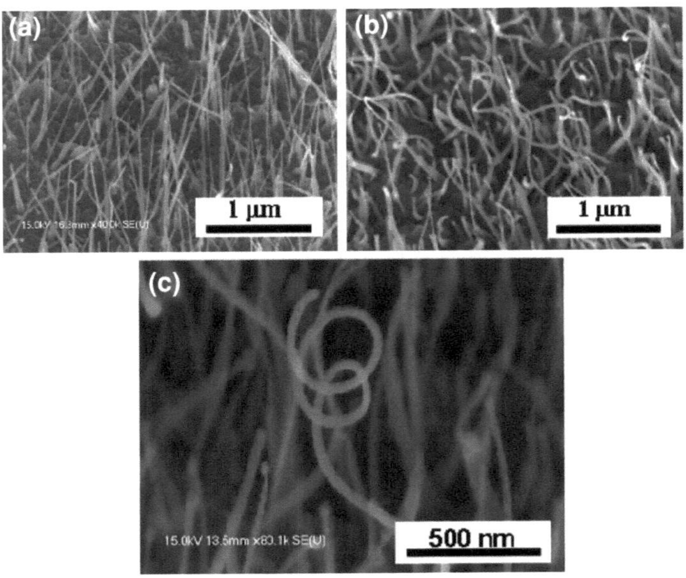

Fig. 34 FE-SEM image of Cu NWs and Cr/Cu NWs. The image was taken with the substrate inclined at 40 degrees. **a** Cu NWs before deposition of Cr. **b** Cr/Cu NWs formed into helixes after the deposition of Cr by EBET. **c** A two-turn Cr/Cu nanocoil. Reprinted with permission from [34]. Copyright 2008 American Scientific Publishers

relatively high energy was applied as an experiment, a large fraction of Cu NWs on a chip were broken away from their trunks. In contrast, EBET rarely produced such damage. A thickness monitor, built into the EBET equipment, was set for a film thickness of 20 nm on a flat surface. Even when the evaporated atoms were vertically incident on the chip surface, the Cu NWs were not uniformly coated, due to their oblique orientation. The thickness of the overlayer was expected to be biased between the exposed side and the shadowed side of an oblique Cu NW. After the coating, the NWs were cooled to room temperature, and were analyzed by field-emission scanning electron microscopy (FE-SEM).

FE-SEM images of the Cr-coated Cu NWs (Cr/Cu NWs) are shown in Figs. 34b and c. The Cr coating was the most successful. We can see that some of Cr/Cu NWs are bent into helical structures. Figure 34c depicts a particularly bent NW: a two-turn NC with a coil diameter of about 300 nm. The wire fineness of the NCs was about 70 nm. Such NCs infrequently existed and their coil diameters ranged from 300 to 500 nm. The production rate is estimated less than 1% of the total NWs. This is a sample density of less than 0.1/μm^2. Despite the low production rate, it is adequate for the achievement of the creation of the smallest conducting (metal-coated) NCs in the world. In addition, FE-SEM analysis also showed that Pt coating generated one-turn NCs, but none of more than one turn. Al coating did not even yield one-turn NCs. The coating effect on bending seems to weaken in order of Cr, Pt, Ni and Al.

Both of the misfit strains of thermal expansion and of generating intrinsic stress are candidates for the driving force to bend the NW. Let us estimate their contributions to the bending process, using the simple model of Fig. 35a. The cross-section of a coated NW is modeled as a circle (radius: a_{NW}) for a Cu NW surrounded by an eccentric circle (radius: a_{OUT}, eccentricity: y_{off}). The bending radius (R) takes a positive value when the thicker-coated side of the NW is concave. In the same manner as in Fig. 33, the equilibrium shape is determined as [34]:

$$\kappa = \frac{4y_{\mathrm{off}} r_E r_a{}^2[(\alpha_{\mathrm{NW}} - \alpha_{\mathrm{F}})\Delta T + e_0]}{a_{\mathrm{NW}}{}^2(r_E r_a{}^2 - r_E + 1)(r_E r_a{}^4 - r_E + 1) - 4y_{\mathrm{off}}{}^2 r_E r_a{}^2(r_E - 1)}, \tag{27}$$

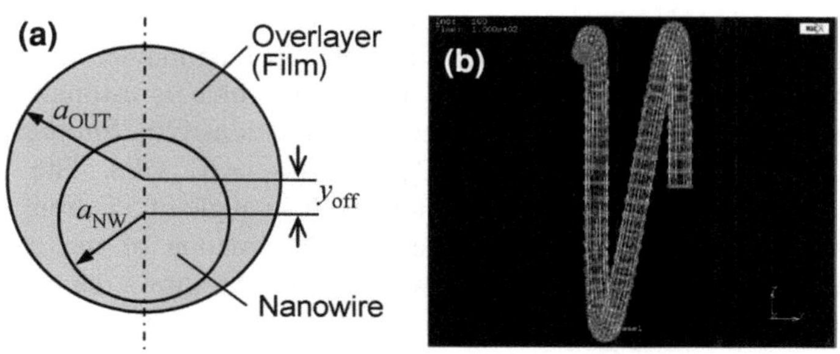

Fig. 35 **a** Cross-sectional model of a coated nanowire and **b** The finite element analysis of misfit-strain-induced bending

where $r_E = E_F/E_{NW}$ (E_{NW} instead of E_S in Sect. 3.2) and $r_a = a_{OUT}/a_{NW}$. The thermal expansion coefficients are assumed to be constant for the simplicity. As an example, a Cr/Cu NW with $a_{NW} = 25$ nm and $a_{OUT} = 35$ nm ($r_a = 1.4$) is evaluated. The maximum eccentricity $y_{off} = a_{OUT} - a_{NW}$ is also assumed, then the maximum thickness 20 nm and the minimum thickness 0. The evaporation process transforms a fraction of the kinetic energy of the metal vapor into thermal energy in the NWs. Because of the small volume of a NW, the temperature of NWs is expected to rise easily. Assuming $T_A = 600$ K, $T_B = 293$ K, and the material properties of bulk samples: $r_E = 2.5$, $\alpha_{NW} = 18 \times 10^{-6}$/K, $\alpha_F = 8 \times 10^{-6}$/K as averages over the temperature variation, we obtain the bending radius $R = 24$ μm from Eq. (27), when ignoring the misfit strain (e_0) that produces intrinsic stress. This value of R is about 100 times larger than the experimental results, namely the radii of the Cr/Cu NCs (150–250 nm). Thus the thermal strain is a minor contribution. The minor contribution is not changed even if different values are used for r_E because the bending radius, estimated from Eq. (27), has the minimum around the assumed value ($r_E = 2.5$). Consequentially, we propose that the formation of NCs relies mainly on the misfit strain (e_0) of the overlayer. The misfit strain (e_0), resulting in $R = 200$ nm, is estimated to be 0.36 from Eq. (27). This value of e_0 is huge, and would be unrealistic. This result suggests other factors encountered in the bending process in addition to the misfit strain. The additional factors will be discussed in the following section.

The highly curved surface of Cu NWs disadvantages the crystal growth of overlayers and, thus, easily yields lattice defects such as grain boundaries in overlayers. Therefore, relatively large misfit strains are expected in the overlayers. The experiments roughly showed that the coating effect on bending weakens in order of Cr, Pt, Ni and Al. This can be explained by the intrinsic stress. In deposition process, some of the lattice defects are naturally annealed by atomic diffusion. The lower the melting point of film material, the easier the diffusion of atoms and then stronger the annealing effect [47]. Actually the melting points are lower in order of Cr, Pt, Ni and Al.

The elastic anisotropy of Cu NWs explains why bending the NWs is coupled with twist into helical structures with a large pitch-to-radius ratio (Fig. 34b) or with a small pitch-to-radius ratio (NC, Fig. 34c). The growth of Cu NWs is expected in the [111] crystal direction because the close-packed plane (111) has the maximum surface energy. Young's modulus in longitudinal direction of Cu NWs also takes a maximum: $E_{[111]} = 191$ GPa, which is estimated by using crystal constants of elasticity [1] and theoretical relations [26]. Similarly, $E_{[100]}$ is estimated to be about one-third of $E_{[111]}$: $E_{[100]} = 67$ GPa. The equivalent directions [100], [010] and [001] are oblique with an angle of 55° from the axis of Cu NWs. The elastic anisotropy makes the deformation of the coated NWs complex, namely a bend coupled with twist. Figure 35b shows the result of finite element analysis (FEA) on the equilibrium shape of a Cr-coated Cu NW, where the elastic anisotropy of Cu is taken account into, the same parameters as in the analysis by using Eq. (27) are used, and the misfit strain is set to be 0.36.

The equilibrium shape is a coil with a radius of about 200 nm. The FEA result verifies the coupling of the bending with twist.

3.3.2 Enhanced Bending by Viscous Flow of Core Material

Here we present an efficient method for the helical formation of coated NWs [54]. In this method, the temperature is elevated in an inert gas to induce creep deformation only in the core material, i.e., NWs. This is done with the NWs coated so that the melting point of the NW material is less than that of the coating. The heat treatment releases the constraint of the core material on the bending deformation, and enhances the bending. The driving force of the bending is the misfit strain induced by the size mismatch between the coating film and NW. A single NW picked up by using a micro-cantilever is examined under well-defined experimental conditions.

Figure 36 shows the experimental procedure. The starting material (i.e., a straight NW) was a single-crystalline CuO NW. Numerous CuO NWs were grown from the surface of a Cu wire (ϕ 0.1 mm) by heating the wire in air at 873 K for 4 h [21, 46]. The fabrication technique of NWs is based on a self-assembly method called the vapor–solid mechanism, which differs from that in the Sect. 3.3.1. The AFM cantilever that attaches the NW was set in the sputtering chamber so that the NW was parallel to a Cr target; it was then sputtered with Cr (Fig. 36b). Conditions used in the sputtering ensured a film thickness of 1 nm on a flat surface, and the film on the NW had a maximum thickness of considerably less than the NW diameter (about 100 nm). The sputtering process resulted in a small amount of bending curvature for the coated NW. The metal-coated NW on the cantilever was then heated in Ar gas at 873 K for 30 min to induce creep deformation only in the CuO NW (Fig. 36c). The melting point of CuO NWs is expected to be considerably lower than that of the coating metal (Cr) since the melting points of bulk CuO and Cr are 1,338–1,501 and 2,180 K, respectively [5, 16, 40].

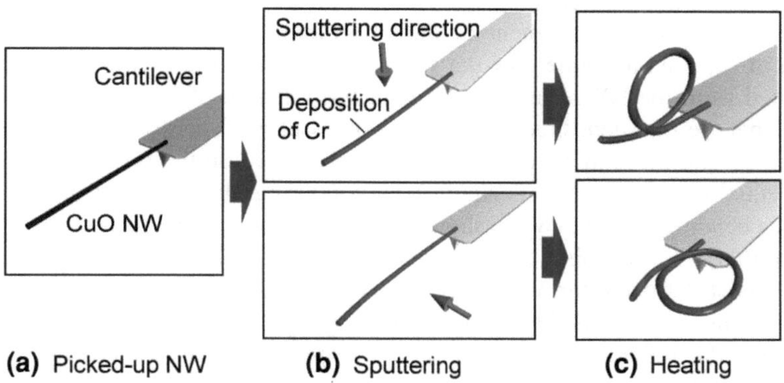

Fig. 36 Experimental procedure for enhanced bending of coated nanowires

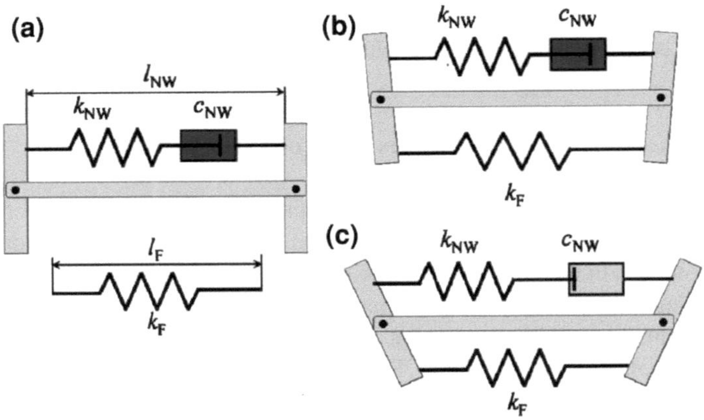

Fig. 37 Simple rheological model for explanation of enhanced bending: **a** before sputtering, **b** after sputtering, and **c** after heating. Reprinted with permission from [54]. Copyright 2010 American Scientific Publishers

The mechanism of enhanced bending is explained by means of a simple rheological model shown in Fig. 37. The model shows a small segment along the axial direction of the NW or coated NW. The upper spring with a damper and the lower spring represent rheological properties of the NW and coating film, respectively. The spring constant k_{NW} of the NW is much larger than the spring constant k_F of the coating film because of the difference in thickness. The coefficient c_{NW} of viscous damping, which represents creep deformation of the NW, depends on temperature. At room temperature, the damper is frozen ($c_{NW} = 0$). For the coating film, the damper is neglected because of the high melting point. Mechanical interaction between the NW and film due to film deposition is expressed by a hinged link for simplicity. Both of the vertical levers roughly correspond to two cross-sections separated by a small distance along the coated NW. The horizontal lever corresponds to a neutral surface. Inclination of both vertical levers means a bending deformation of the coated NW.

Figure 37a shows a segment of the stress-free NW before sputtering with a length of l_{NW} and a segment of the stress-free coating film that is hypothetically separated from the NW. The film segment has a length of l_F less than l_{NW}. This means that the film deposited on the NW possesses a misfit strain of $e_0 = (l_{NW} - l_F)/l_F$ (>0), and the film becomes elastically elongated on the NW with an elastic strain equal to e_0 if the NW is rigid. After sputtering, the misfit strain of the film imposes bending on the NW, as shown in Fig. 37b, where the linkage model is in self-equilibrium after joining the elastically elongated film spring to the link. The damper is still frozen during sputtering because the average temperature for the process is not as high as that inducing marked creep. The stiff spring of the NW constrains the restoring motion of the film spring, which reduces the bending curvature. In the final process, i.e., heating in inert gas, the damper of the NW is activated as shown in Fig. 37c. The residual restoring energies of the NW and film springs drive the damper to viscous elongation. Finally, all of the residual stresses are released, and the film spring recovers its natural length l_F. This means that the misfit strain of the film fully contributes to the bending deformation.

In the experiment, several sputtering directions were examined to check on bending direction of NW. Figures 38a and b show scanning electron microscopy (SEM) images of a CuO NW (sample No. 1) before and after sputtering, respectively. Figures 38c and d show SEM images of helical formation samples after the heating treatment, which have coil diameters of about 4 μm (sample No. 2) and 900 nm (sample No. 3), respectively, and are comprised of two turns.

The efficiency of coil formation was about 40%. The improved efficiency was obtained compared with that in Sect. 3.3.1 (<1%), but the formation failed for 60% of the samples. The estimated film thickness was about 1 nm on the impingement

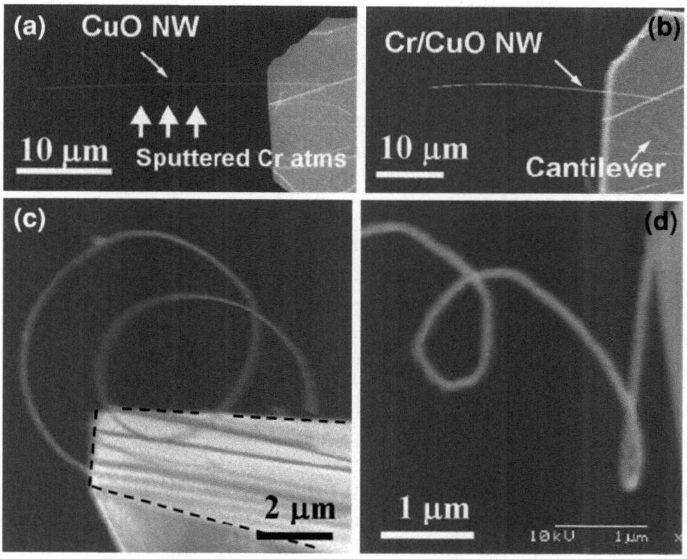

Fig. 38 SEM images of CuO NW and Cr-deposited CuO NW attached to microcantilever: **a** and **b** Sample 1 before and after sputtering, respectively. **c** Sample 2 after heating in Ar gas (the broken lines indicate the edge of the cantilever back-surface) and **d** Sample 3 after heating in Ar gas. Sample 2 (image **c**) was sputtered in the direction normal to the cantilever back-surface, while Sample 1 (image **a**) was rotated by 90° around the longitudinal axis of the cantilever and then sputtered. Therefore, the direction of impingement of Cr atoms was that shown by the arrows in the image **a**. The back surface of the cantilever in the image **c** was viewed from an oblique angle. Because the impingement direction was normal to the back surface, the image **c** gives the evidence that the coated NW bends toward the target side even after creeping, and the deposited side of the NW becomes concave. The sputtering direction in the image **d** was parallel to the back surface, but was opposite to that in the image **a**. The image **d** was taken from the direction normal to the back surface. The supported end of the NW is located lower out of the image. The image **d** also provides the evidence of the bending direction predicted by our proposed mechanism. Reprinted with permission from [54].

side of NWs and much less than 1 nm on the back side (shadow side). Uncompleted coating on the back side might have been encountered in the experiments. During creep deformation of the CuO NW, the surface tension of bare CuO surface on the back side could give an opposite influence on the bending direction.

As mentioned, CuO NWs were synthesized in air at a temperature of 873 K without creep deformation of the NWs. In contrast, the NWs in Ar gas actually showed creep at the same temperature, see the report by Toku and Muraoka [54]. This discrepancy in material phases is due to the effect of NW surface free energy on the melting point. Thermodynamics for small or thin solids states that a melting point depends on the surface free energy and surface curvature [6]. The surface free energy of the NWs in Ar gas is larger than that in air, which indicates a lower NW melting point in Ar gas.

Making some modifications of the 3D model (Fig. 35a) for thinner films in this experiment yielded simple, theoretical formulas on the curvature κ and neutral surface of elastically bent film-coated NWs, see the report by Toku and Muraoka [54]. The creep of the core (the NW) increases the permanent strain (creep strain) and reduces the stress due to relaxing the elastic strain. The creep strain itself is driven by the stress. Therefore, no creep strain is expected at the neutral axis for a while after onset of creeping. Here, we assume that the neutral axis is kept to be at the same position as that before creeping. This is correct for the 2D model discussed in Sect. 3.2 because c_0 contains no parameters related to material properties. The complete stress-relaxation of the core after creeping may be expressed by the limiting case of $E_{\mathrm{NW}} \rightarrow 0$. The enhanced curvature κ_{c} after creeping is obtained by taking the limit with the neutral axis kept stationary in the theoretical formulas on the elastically bending of film-coated NWs [54]:

$$\kappa_{\mathrm{c}} = \frac{8\pi e_0}{(\pi^2 + 4)d_{\mathrm{NW}}}. \tag{28}$$

Equation (28) states that the curvature after heating depends only on the NW diameter d_{NW} and the misfit strain (e_0), and hence the coil diameter does not explicitly depend on the Cr film thickness. In preliminary experiments, we tried some values of the film thickness and confirmed the dependence of the coil diameter on the film thickness. Thinner films resulted in smaller coil diameters. This fact suggests that the misfit strain of sputtered Cr film is tensile and increases with decreasing the film thickness.

For comparison of the curvature after heating with that after sputtering, the ratio of κ_{c} and κ was calculated by using the experimental conditions of $d_{\mathrm{F}} = 1$ nm (the maximum thickness), $d_{\mathrm{NW}} = 100$ nm, $E_{\mathrm{F}} = 40$ GPa (See Table 1 in Evaluation of Mechanical Properties), and $E_{\mathrm{NW}} = 200$ GPa (See Fig.21a in Evaluation of Mechanical Properties). Because of size effects, the value of E_{F} is quite smaller than bulk Cr value (278 GPa). The ratio $\kappa_{\mathrm{c}}/\kappa$ was predicted to be 114, i.e., the curvature after heating increases 114 times more than that after sputtering. In addition, these values must contain some errors. Taking 30% error of these values into account, the maximum of $\kappa_{\mathrm{c}}/\kappa$ is predicted to be 390 and the minimum to be 34.

It is seen from Eq. (28) that the coil diameter ($\approx 2R = 2/\kappa_c$) is proportional to the NW diameter and is in a reciprocal proportion to the misfit strain. The curvature radius is estimated to be $R \approx 150$ μm from SEM images of coated NWs before creeping, e.g., Fig. 38b. The misfit strain is estimated to be $e_0 = 0.009 - 0.19$ from the theoretical formula of κ (see the report by Toku and Muraoka [54]) together with the value R and the values E_{NW}, d_{NW}, E_F and d_F, whose error of 30% is taken into account. Equation (28) together with the values of e_0 predicts coil diameters of 770 nm–8.9 μm. The prediction almost coincides to the observed coil diameters (Fig. 38c 4 μm and d 900 nm).

From the discussion on the enhanced bending, we expect that the coil diameter can be downsized by using a thinner NW, as stated by Eq. (28). Furthermore, it would be plausible that the NCs observed by using EBET (e.g., Fig. 34) may be due to creeping of the core material at locally elevated temperatures during the deposition.

Acknowledgments H. Tohmyoh acknowledges partial support from the Ministry of Education, Culture, Sports, Science and Technology (MEXT), Japan under Grant-in-Aid for Young Scientists (A) Grant No. 21686012 and S. Fukui for his help in preparing the manuscript. M. Muraoka acknowledges partial support from the Japan Society for the Promotion of Science (JSPS), through the Grant-in-Aid for Scientific Research (B) Grant No. 20360049 and Ms. Y. Ishigami and Y. Toku for their help in preparing the manuscript.

References

1. Anderson, O.L.: Determination and some uses of isotropic elastic constants of polycrystalline aggregates using single-crystal data. In: Mason, W.P. (ed.) Physical Acoustics III—Part B. Academic Press, New York (1965)
2. Bhola, B., Song, H.C., Tazawa, H., Steier, W.H.: Polymer microresonator strain sensors. IEEE Photon Technol. Lett. **17**, 867–869 (2005)
3. Bintoro, J.S., Papania, A.D., Berthelot, Y.H., Hesketh, P.J.: Bidirectional electromagnetic microactuator with microcoil fabricated on a single wafer: Static characteristics of membrane displacements. J. Micromech. Microeng. **15**, 1378–1388 (2005)
4. Boukai, A.I., Bunimovich, Y., Tahir-Kheli, J., Yu, J.-K., Goddard III, W.A., Heath, J.R.: Silicon nanowires as efficient thermoelectric materials. Nature **451**, 168–171 (2008).
5. Brady, G.S., Clauser, H.R., Vaccari, J.A.: Materials Handbook, 15th edn. McGraw-Hill, New York (2002)
6. Cahn, J.W.: Surface stress and the chemical equilibrium of small crystals – I. the case of the isotropic surface. Acta Metall. **28**, 1333–1338 (1980)
7. Cammarata, R.C.: Surface and interface stress effects in thin films. Prog. Surf. Sci. **46**, 1–38 (1994)
8. Carl, A., Lohau, J., Kirsch, S., Wassermann, E.F.: Magnetization reversal and coercivity of magnetic-force microscopy tips. J. Appl. Phys. **89**, 6098–6104 (2001)
9. Carslaw, H.S., Jaeger, J.C.: Conduction of Heat in Solids, 2nd edn. Oxford University Press, London (1959)
10. Chen, Y., Lebib, A., Li, S.P., Natali, M., Peyrade, D., Cambril, E.: Nanoimprint fabrication of micro-rings for magnetization reversal studies. Microelectron. Eng. **57–58**, 405–410 (2001)
11. Day, G.W., Gaddy, O.L., Iversen, R.J.: Detection of fast infrared laser pulses with thin film thermocouples. Appl. Phys. Lett. **13**, 289–290 (1968)

12. Dick, K., Dhanasekaran, T., Zhang, Z., Meisel, D.: Size-dependent melting of silica-encapsulated gold nanoparticles. J. Am. Chem. Soc. **124**, 2312–2317 (2002)
13. Doelling, C.M., Vanderlick, T.K., Song, J., Srolovitz, D.: Nanospot welding and contact evolution during cycling of a model microswitch. J. Appl. Phys. **101**, 124303(1–7) (2007)
14. Doerner, M.F., Nix, W.D.: Stress and deformation processes in thin films on substrates. Crit. Rev. Solid State Mater. Sci. **14**, 225–268 (1988)
15. Duan, J., Yang, S., Liu, H., Gong, J., Huang, H., Zhao, X., Tang, J., Zhang, R., Du, Y.: AlN nanorings. J. Cryst. Growth **283**, 291–296 (2005)
16. García-Labiano, F., de Diego, L.F., Adánez, J., Abad, A., Gayán, P.: Temperature variations in the oxygen carrier particles during their reduction and oxidation in a chemical-looping combustion system. Chem. Eng. Sci. **60**, 851–862 (2005)
17. Goldstein, A.N., Echer, C.M., Alivisatos, A.P.: Melting in semiconductor nanocrystals. Science **256**, 1425–1427 (1992)
18. Hirayama, H., Kawamoto, Y., Ohshima, Y., Takayanagi, K.: Nanospot welding of carbon nanotubes. Appl. Phys. Lett. **79**, 1169–1171 (2001)
19. Huang, J., Kaner, R.B.: Flash welding of conducting polymer nanofibres. Nature Mater. **3**, 783–786 (2004)
20. Huang, M., Boon, C., Roberts, M., Savage, D.E., Lagally, M.G., Shaji, N., Qin, H., Blick, R., Nairn, J.A., Liu, F.: Nanomechanical architecture of strained bilayer thin films: from design principles to experimental fabrication. Adv. Mater. **17**, 2860–2864 (2005)
21. Jiang, X., Herricks, T., Xia, Y.: CuO nanowires can be synthesized by heating copper substrates in air. Nano. Lett. **2**, 1333–1338 (2002)
22. Jin, C., Suenaga, K., Iijima, S.: Plumbing carbon nanotubes. Nature Nanotechnol. **3**, 17–21 (2008)
23. Kim, J.W., Jung, M.H., Park, N.K., Yun, E.J.: Microfabrication of solenoid-type RF SMD chip inductors with an Al_2O_3 core. Curr. Appl. Phys. **8**, 631–636 (2008)
24. Kim, S.J., Jang, D.-J.: Laser-induced nanowelding of gold nanoparticles. Appl. Phys. Lett. **86**, 033112(1–3) (2005)
25. Koch, R.: The intrinsic stress of polycrystalline and epitaxial thin metal films. J. Phys. Condens. Matter **6**, 9519–9550 (1994)
26. Kollár, L.P., Springer, G.S.: Mechanics of Composite Structures. Cambridge University, Cambridge (2003)
27. Kong, X.Y., Wang, Z.L.: Spontaneous polarization-induced nanohelixes, nanosprings, and nanorings of piezoelectric nanobelts. Nano. Lett. **3**, 1625–1631 (2003)
28. Krasheninnikov, A.V., Nordlund, K., Keinonen, J., Banhart, F.: Ion-irradiation-induced welding of carbon nanotubes. Phys. Rev. B. **66**, 245403(1–6) (2002)
29. Lee, C.Y., Chang, H.T., Wen, C.Y.: A MEMS-based valveless impedance pump utilizing electromagnetic actuation. J. Micromech. Microeng. **18**, 035044(1–9) (2008)
30. Li, D., Wu, Y., Kim, P., Shi, L., Yang, P., Majumdar, A.: Thermal conductivity of individual silicon nanowires. Appl. Phys. Lett. **83**, 2934–2936 (2003)
31. Li, X., Gao, H., Murphy, C.J., Caswell, K.K.: Nanoindentation of silver nanowires. Nano. Lett. **3**, 1495–1498 (2003)
32. Liu, H., Cui, H., Wang, J., Gao, L., Hang, F., Boughton, R.I., Jiang, M.: Growth of $NaFe_4P_{12}$ skutterudite single crystalline nanosprings synthesized through a hydrothermal-reduction-alloying method. J. Phys. Chem. B **108**, 13254–13257 (2004)
33. Mcllroy, D.N., Zhang, D., Kranov, Y., Norton, M.G.: Nanospring. Appl. Phys. Lett. **79**, 1540–1542 (2001)
34. Muraoka, M., Settsu, N., Saka, M.: Residual-strain-induced nanocoils of metallic nanowires. J. Nanosci. Nanotechnol. **8**, 439–442 (2008)
35. Nakamatsu, K., Nagase, M., Igaki, J., Namatsu, H., Matsui, S.: Mechanical characteristics and its annealing effect of diamondlike-carbon nanosprings fabricated by focused-ion-beam chemical vapor deposition. J. Vac. Sci. Technol. B **23**, 2801–2805 (2005)
36. Peng, Y., Cullis, T., Inkson, B.: Bottom-up nanoconstruction by the welding of individual metallic nanoobjects using nanoscale solder. Nano. Lett. **9**, 91–96 (2009)

37. Sacharoff, A.C., Westervelt, R.M.: Physical properties of ultrathin drawn Pt wires. Phys. Rev. B **29**, 6411–6418 (1984)
38. Saka, M., Sun, Y.X., Ahmed, S.R.: Heat conduction in a symmetric body subjected to a current flow of symmetric input and output. Int. J. Thermal. Sci. **48**, 114–121 (2009)
39. Saka, M., Yamaya, F., Tohmyoh, H.: Rapid and mass growth of stress-induced nanowhiskers on the surfaces of evaporated polycrystalline Cu films. Scr. Mater. **56**, 1031–1034 (2007)
40. Schmid, R.: A thermodynamic analysis of the Cu–O system with an associated solution model. Metall. Trans. B **14**, 473–481 (1983)
41. Schmidt, O.G., Eberl, K.: Thin solid films roll up into nanotubes. Nature **410**, 168 (2001)
42. Seidemann, V., Büttgenbach, S.: Closely coupled micro coils with integrated flux guidance: fabrication technology and application to proximity and magnetoelastic force sensors. IEEE Sensors J. **3**, 615–621 (2003)
43. Shen, G.Z., Bando, Y., Zhi, C.Y., Yuan, X.L., Sekiguchi, T., Golberg, D.: Single-crystalline cubic structured InP nanosprings. Appl. Phys. Lett. **88**, 243106(1–3) (2006)
44. Shi, L., Plyasunov, S., Bachtold, A., McEuen, P.L., Majumdar, A.: Scanning thermal microscopy of carbon nanotubes using batch-fabricated probes. Appl. Phys. Lett. **77**, 4295–4297 (2000)
45. Sutanto, J., Hesketh, P.J., Berthelot, Y.H.: Design, microfabrication and testing of a CMOS compatible bistable electromagnetic microvalve with latching/unlatching mechanism on a single wafer. J. Micromech. Microeng. **16**, 266–275 (2006)
46. Tan, E.P.S., Zhu, Y., Dai, L., Sow, C.H., Tan, V.B.C., Lim, C.T.: Crystallinity and surface effects on Young's modulus of CuO nanowires. Appl. Phys. Lett. **90**, 163112(1–3) (2007)
47. Thornton, J.A., Hoffman, D.W.: Stress-related effects in thin films. Thin Solid Films **171**, 5–31 (1989)
48. Tohmyoh, H.: A governing parameter for the melting phenomenon at nanocontacts by Joule heating and its application to joining together two thin metallic wires. J. Appl. Phys. **105**, 014907(1–9) (2009)
49. Tohmyoh, H., Fukui, S.: Self-completed Joule heat welding of ultrathin Pt wires. Phys. Rev. B **80**, 155403(1–7) (2009)
50. Tohmyoh, H., Imaizumi, T., Hayashi, H., Saka, M.: Welding of Pt nanowires by Joule heating. Scr. Mater. **57**, 953–956 (2007)
51. Tohmyoh, H., Takeda, H., Akanda, M.A.S.: Evaluation of mechanical and electrical properties of very-thin Pt wires by utilizing joining technique with Joule heating. J. Soc. Mater. Sci. **58**, 847–851 (2009) (in Japanese)
52. Tohmyoh, H., Takeda, H., Khan, M.N.I., Saka, M.: Fabrication of freestanding thin Pt/W thermocouple by Joule heat welding. Proc. EuroSimE. 2010, 1–4 (2010)
53. Tohmyoh, H., Takeda, H, Saka, M.: Fabrication of a free-standing Pt MR on an electrode chip as a small magnetic source. J. Micromech. Microeng. **19**, 085013(1–5) (2009)
54. Toku, Y., Muraoka, M.: Helical formation of coated nanowires by viscous flow of core material. Nanosci. Nanotechnol. Lett. (2010) (in press)
55. Townsend, P.H., Barnett, D.M., Brunner, T.A.: Elastic relationships in layered composite media with approximation for the case of thin films on a thick substrate. J. Appl. Phys. **62**, 4438–4444 (1987)
56. Tsai, S.H., Shiu, C.T., Jong, W.J., Shih, H.C.: The welding of carbon nanotubes. Carbon **38**, 1899–1902 (2000)
57. Varadan, V.K., Hollinger, R.D., Varadan, V.V., Xie, J., Sharma, P.K.: Development and characterization of micro-coil carbon fibers by a microwave CVD system. Smart Mater. Struct. **9**, 413–420 (2000)
58. Wagner, R.S., Ellis, W.C.: Vapor-liquid-solid mechanism of single crystal growth. Appl. Phys. Lett. **4**, 89–90 (1964)
59. Wang, J., Chen, X., Wang, G., Wang, B., Lu, W., Zhao, J.: Melting behavior in ultrathin metallic nanowires. Phys. Rev. B **66**, 085408(1–5) (2002)
60. Wang, Z.L., Petroski, J.M., Green, T.C., El-Sayed, M.A.: Shape transformation and surface melting of cubic and tetrahedral platinum nanocrystals. J. Phys. Chem. B **102**, 6145–6151 (1998)

61. Watson, C.C., Chan, W.K.: High-spatial-resolution semiconductor characterization using a microwave eddy current probe. Appl. Phys. Lett. **78**, 129–131 (2001)
62. Williams, C.C., Wickramasinghe, H.K.: Scanning thermal profiler. Appl. Phys. Lett. **49**, 1587–1589 (1986)
63. Wu, B., Heidelberg, A., Boland, J.J.: Mechanical properties of ultrahigh-strength gold nanowires. Nature Mater. **4**, 525–529 (2005)
64. Wu, Y., Yang, P.: Melting and welding semiconductor nanowires in nanotubes. Adv. Mater. **13**, 520–523 (2001)
65. Xu, S., Tian, M., Wang, J., Xu, J., Redwing, J.M., Chan, M.H.W.: Nanometer-scale modification and welding of silicon and metallic nanowires with a high-intensity electron beam. Small **1**, 1221–1229 (2005)
66. Yamaguchi, M., Suezawa, K., Arai, K.I., Takahashi, Y., Kikuchi, S., Shimada, Y., Li, W.D., Tanabe S, Ito, K.: Microfabrication and characteristics of magnetic thin-film inductors in the ultrahigh frequency region. J. Appl. Phys. **85**, 7919–7922 (1999)
67. Yang, R.S., Wang, Z.L.: Springs, rings, and spirals of rutile-structured tin oxide nanobelts. J. Am. Chem. Soc. **128**, 1466–1467 (2006)
68. Zhang, D., Alkhateeb, A., Han, H., Mahmood, H., Mcllroy, D.N.: Silicon carbide nanosprings. Nano. Lett. **3**, 983–987 (2003)
69. Zhang, X., Li, X.: Design, fabrication and characterization of optical microring sensors on metal substrates. J. Micromech. Microeng. **18**, 015025(1–7) (2008)

Errata to: Metallic Micro and Nano Materials

Masumi Saka

Errata to: EngMat
DOI: 10.1007/978-3-642-15411-9

Preface (pp. v–vi)

Location	Printed	Should be	Comment
p. vi	Akita	Sendai	City name for H. Tohmyoh

Introduction (pp. 1–14)

Location	Printed	Should be	Comment
None			

Basis of Atomic Diffusion (pp. 15–52)

Location	Printed	Should be	Comment
p. 16, line 13	results	result	Delete "s"
p. 23, line 19	acceleration test of EM	acceleration test of EM for IC Al lines	Add "for IC Al lines"
p. 26, line 9	j_{x}	j_x	Put "$_{\mathrm{x}}$" in italics
p. 26, line 10	J_{x}	j_x	Decapitalize "*J*" and put "$_{\mathrm{x}}$" in italics
p. 27, line 4	,	.	Change comma to period after Eq. (33)
p. 30, line 17	0.00442	0.00442.	Add period

(continued)

The online version of the original book can be found under doi:10.1007/978-3-642-15411-9.

M. Saka (✉)
Department of Nanomechanics, Graduate School of Engineering, Tohoku University, Aoba 6-6-01, Aramaki, Aoba-ku, Sendai 980-8579, Japan
e-mail: saka@ism.mech.tohoku.ac.jp

(continued)

Location	Printed	Should be	Comment
p. 34, line 6	$-2\frac{\kappa\Omega\partial^2 N}{N_0\partial x\partial y}\sin 2\theta\Big\}$	$-2\frac{\kappa\Omega}{N_0}\frac{\partial^2 N}{\partial x\partial y}\sin 2\theta\Big\}$	Divide into two fractions
p. 37, line 5	Printed $+\frac{\sqrt{3}}{4}\pi d\Big[-\frac{\kappa\Omega}{N_0}\Big(\frac{\partial^2 N}{2}+\frac{\partial^2 N}{\partial y^2}\Big)-\frac{\kappa\Omega/N_0}{kT}\Big(D_x\frac{\partial N}{\partial x}+D_y\frac{}{\partial y}\Big)$ Should be $+\frac{\sqrt{3}}{4}\pi d\Big[-\frac{\kappa\Omega}{N_0}\Big(\frac{\partial^2 N}{\partial x^2}+\frac{\partial^2 N}{\partial y^2}\Big)-\frac{\kappa\Omega/N_0}{kT}\Big(D_x\frac{\partial N}{\partial x}+D_y\frac{\partial N}{\partial y}\Big)$ Comment Insert "∂x" into denominator and "∂N" into numerator		
p. 37, line 14	comparing	comparing void formation in	Add "void formation in"
p. 38, line 3	Printed $=\frac{\sqrt{3}d^2}{4\delta}\Big[AFD^*_{\text{gen}}\lvert\text{end}+\frac{\partial AFD^*_{\text{gen}}\lvert\text{end}}{\partial x}(l_d-l^*)\Big]$ Should be $=\frac{\sqrt{3}d^2}{4\delta}\Big[AFD^*_{\text{gen}}\lvert_{\text{end}}+\frac{\partial AFD^*_{\text{gen}}\lvert_{\text{end}}}{\partial x}(l_d-l^*)\Big]$ Comment Put "end" in inferior		
p. 38, line 11	Printed $v_d=\Big[AFD^*_{\text{gen}}\lvert\text{end}+\frac{\partial AFD^*_{\text{gen}}\lvert\text{end}}{\partial x}(l_d-l^*)\Big]l^*\Omega$ Should be $v_d=\Big[AFD^*_{\text{gen}}\lvert_{\text{end}}+\frac{\partial AFD^*_{\text{gen}}\lvert_{\text{end}}}{\partial x}(l_d-l^*)\Big]l^*\Omega$ Comment Put "end" in inferior		
p. 39, line 4	Printed $F^*_{\text{gb}}\lvert\text{end}=\sum_i\sum_j\Big\{v_d\lvert_{ij}-\Big[(AFD^*_{\text{gen}}\lvert\text{end})_i+\Big(\frac{\partial AFD^*_{\text{gen}}\lvert\text{end}}{\partial x}\Big)_i(l_d\lvert_{ij}-l^*)\Big]l^*\Omega\Big\}^2$ Should be $F^*_{\text{gb}}\lvert_{\text{end}}=\sum_i\sum_j\Big\{v_d\lvert_{ij}-\Big[(AFD^*_{\text{gen}}\lvert_{\text{end}})_i+\Big(\frac{\partial AFD^*_{\text{gen}}\lvert_{\text{end}}}{\partial x}\Big)_i(l_d\lvert_{ij}-l^*)\Big]l^*\Omega\Big\}^2$ Comment Put "end" in inferior		
p. 45, line 3 of figure caption of Fig. 14	[71]	[72]	Change reference number

Fabrication of Micro and Nano Metallic Materials (pp. 53–92)

Location	Printed	Should be	Comment
p. 55, line 17	'the electro-thermal'	the 'electro-thermal'	Change position of single quotation marks
p. 72, line 4	EM failure	EM failure in IC	Add "in IC"
p. 72, line 12	Eqs. (55), (57)	Eqs. (55)–(57)	Replace a comma with an en dash
p. 72, line 13	Eq. (58) in Basis of Atomic Diffusion given in Sect. 1.7 in Basis of Atomic Diffusion	Eq. (58) given in Sect. 1.7 in Basis of Atomic Diffusion	Delete the first "in Basis of Atomic Diffusion"
p. 73, line 2	Assuming	Assume	Be in imperative form
p. 81, line 15	chpater	chapter	Misspelling
p. 89, line 17	Acknowledgments	Acknowledgment	Change to singular

Evaluation of Mechanical Properties (pp. 93–142)

Location	Printed	Should be	Comment
p. 99, line 4	called as	called the	Change "as" to "the"
p. 101, in 3rd line from bottom	large value.	large value (see Sect. 2.4).	Insert "(see Sect. 2.4)"
p. 101, in 2nd line from bottom	called as	called the	Change "as" to "the"
p. 102, line 6	called as	called	Delete "as"
p. 105, in 4th line after Eq. (12)	rotary inertia)	rotary inertia	Delete a parenthesis
p. 113, line 4	Sect. 3.3.1	Sect. 3.1 in the preceding chapter	Replace "3.3.1" with "3.1 in the preceding chapter"
p. 124, in 1st line before Eq. (62)	in Eq. (53)	in Eq. (42)	Change 53 to 42
p. 130, line 16	spuattering	spattering	Misspelling
p. 133, line 1	Printed Nano/Micromaterials Should be Nano/Micromaterials		Delete indent
p. 136, line 1 of figure caption of Fig. 31	$\mathbf{a}E\ldots\mathbf{b}\sigma_{Y}\ldots\mathbf{c}E'\ldots$	$\mathbf{a}\ E\ldots\mathbf{b}\ \sigma_{Y}\ldots\mathbf{c}\ E'\ldots$	Insert a space for each
p. 137, in Acknowledgments	M.Muraoka. H.Tohmyoh.	M. M. H. T.	Change to initials

Evaluation of Electrical Properties (pp. 143–172)

Location	Printed	Should be	Comment
p. 145, line 15	nanowires	nanowire	Change to singular
p. 146, line 28	diameter a the only lateral dimension	diameter is the only lateral dimension	Change "a" to "is"
p. 151, the end of the caption of Fig. 4		(scale bar 2μm)	Add (scale bar 2μm)
p. 152, line 24	[24, 29]	[25, 28]	Change reference numbers
p. 152, line 37; p. 154, lines 4, 8 and 18; p. 156, line 14	[29]	[28]	Change reference number
p. 157, the last line of caption of Fig. 9	the mean	the variation of them	Replace "mean" with "variation of them"
p. 157, line 11	is on the nanometer	on the nanometer	Delete "is"
p. 158, line 17	[27]	[23]	Change reference number
p. 158, line 29	[25, 26, 30, 31]	[26, 29, 30, 31]	Change 25 to 29
p. 159, lines 9, 26 and 34	[25]	[26]	Change reference number
p. 159, line 16	strong	strongly	Change to adverb
p. 159, line 36; p. 161, line 7	[26]	[29]	Change reference number
p. 163, line 11	the step height of	the step height is	Change "of" to "is"
p. 165, line 5	simultaneous	simultaneously	Change to adverb
p. 166, line 11–12	[23, 28]	[24, 27]	Change reference numbers
p. 167, line 1	Symbolω	Symbol ω	Insert a space
p. 167, lines 4 and 8	[28]	[24]	Change reference number
p. 168, line 12	Eqs. 5–8 and 5–11	Eqs. 8 and 11	Delete "5-" for each
p. 169, the last line of caption of Fig. 19	13	12	Change footnote number
p. 170, ref. 28	(2007a)	(2007)	Delete "a"
p. 170, ref. 29	(2007b)	(2007)	Delete "b"
p. 171, line 5	31.	32.	Change reference number

Modification of Nano/Micromaterials (pp. 173–220)

Location	Printed	Should be	Comment
p. 173, line 7 in Abstract	in thin wires are	in thin wires is	Change "are" to "is"
p. 175, line 1 of figure caption of Fig. 2	10-nmdiameter	10-nm diameter	Insert a space
p. 177, line 6 from bottom	$K = 72$ W/mK	$K = 72$ W/(mK)	"mK" should be parenthesized
p. 177, line 5 from bottom	$c = 134$ J/kg K	$c = 134$ J/(kg K)	"kg K" should be parenthesized
p. 179, in 13th line from bottom	$K = 72$ W/mK	$K = 72$ W/(mK)	"mK" should be parenthesized
p. 179, in 13th line from bottom	$c = 134$ J/kg K	$c = 134$ J/(kg K)	"kg K" should be parenthesized
p.179, in 12th line from bottom	$H = 30$ W/m^{2}K	$H = 30$ W/(m^{2}K)	"m^{2}K" should be parenthesized
p. 181, line 2	Eq. (1).	1.2.	Change Eq. (1) to 1.2
p. 183, Table 1	I/A	*I*/*A*	Variables *I* and *A* should be italic
p. 188, line 2 of figure caption of Fig. 12	of A in **b**	of A in **a**	Change **b** to **a**
p. 188, in 5th line from bottom	Figure 13a–d correspond to	Figure 13a–d corresponds to	Change "correspond" to "corresponds"
p. 192, line 6	shown experimentally	experimentally	Delete "shown"
p. 196, line 10	Figure 20a and b represents	Figure 21a and b represents	Change 20 to 21
p. 196, line 12	Figure 21a–d shows	Figure 21c and d shows	Change "a–d" to "c and d"
p. 203, line 6 of figure caption of Fig. 27	electrode 3 and 4	electrodes 3 and 4	Change "electrode" to "electrodes"
p. 204, line 5	in the previous section,	in Sect. 2.3.2,	Change "the previous section" to "Sect. 2.3.2"

(continued)

(continued)

Location	Printed	Should be	Comment
p. 205, in 9th-10th lines from bottom	SiC ([68], BC [33], and $NaFe_4P_{12}$ [32] have	SiC [68], BC [33], and $NaFe_4P_{12}$ [32], have	Delete a parenthesis and insert a comma after [32]
p. 210, line 7	Sect. 3.3.1	Sect. 3.1 in Fabrication of Micro and Nano Metallic Materials	Replace "3.3.1" with "3.1 in Fabrication of Micro and Nano Metallic Materials"
p. 216, in 7th line from bottom	(See Table 1	(see Table 1	Change to lowercase
p. 216, in 6th line from bottom	(See Fig.21a	(see Fig. 21a	Change to lowercase and insert a space before 21a

Index (pp. 221–225)

Location	Printed	Should be	Comment
p. 221	Acceleration test, 23, 25, 33, 38, 40, 42, 45, 49	Acceleration test, 23, 25, 33, 38, 40, 42	Delete 45 and 49
p. 221	Anodic alumina, 5, 6, 145	Anodic alumina, 5, 6, 146	Change 145 to 146
p. 221	Aspect ratio, 46, 57, 59, 159, 160	Aspect ratio, 46, 57, 59, 159, 162	Change 160 to 162
p. 221	Printed Atomic diffusion, 11, 15, 16, 18, 19, 35, 36, 43, 46, 54, 82, 212 flux, 15, 17-21, 33, 36-38, 47, 48, 63, 66 Should be Atomic diffusion, 11, 15, 16, 18, 19, 36, 43, 46, 54, 82, 212 flux, 15, 17-21, 33, 36-38, 47, 66		Delete 35 for atomic diffusion Delete 48 and 63 for atomic flux
p. 221	Printed Bond-order-bond-length-bond strength (bond-OLS) correlation, 129 Should be Bond-order-bond-length-bond-strength (bond-OLS) correlation, 128		Insert a hyphen before "strength" and change 129 to 128
p. 221	Brass (BS), 168	Brass (BS), 169	Change 168 to 169
p. 221	Cantilever, 94, 98, 105, 107, 110, 121, 129, 133, 152, 159, 162, 213	Cantilever, 94, 98, 105, 107, 110, 121, 129, 133, 152, 160, 162, 213	Change 159 to 160
p. 221	Printed Characteristic constant, 15, 34, 35, 37, 39, 41, 42 Should be Characteristic constant, 15, 35, 37, 39, 41, 42		Delete 34

(continued)

(continued)

Location	Printed	Should be	Comment
p. 221	Coaxial line, 163, 164, 167	Coaxial line, 164, 167	Delete 163
p. 221	Printed Crystalline, 99, 112, 213 plane, 159, 160 Should be Crystalline, 99, 112, 213 plane, 159, 161		Change 160 to 161 for crystalline plane
p. 222	Printed Elastic modulus, 2, 12, … Should be Elastic modulus, 2, 12, …		Indent “modulus”
p. 222	Printed Electron, 17, 35, 43, 48 -beam-evaporation technique (EBET), 210 Should be Electron, 17, 35, 43, 48 -beam evaporation technique (EBET), 210		Replace a hyphen with a space before “evaporation”
p. 223	Printed Least-squares method (least-squaresfit), 38, Should be Least-squares method (least-squares fit), 38,		Insert a space before “fit”
p. 223	Lifetime, 16, 19, 20, 34, 62, 75, 78, 79, 80	Lifetime, 16, 19, 20, 35, 62, 75, 78, 79, 80	Change 34 to 35
p. 223	Printed Lower-coordinated surface-atom, 128 Should be Lower-coordinated surface atom, 128		Replace a hyphen with a space before “atom”
p. 223	Printed Mechanical diode (MD), 101 property, 45, 93, 126, 130, 136 Should be Mechanical diode (MD), 101, 102 property, 45, 93, 126, 130		Add 102 for mechanical diode and delete 136 for mechanical property
p. 223	Misfit strain, 206, 208, 214	Misfit strain, 205, 208, 214	Change 206 to 205
p. 223	Printed Molybdenum disulfide (MoS2), Should be Molybdenum disulfide (MoS_2),		2 is inferior

(continued)

(continued)

Location	Printed	Should be	Comment
p. 223	Printed Moment of inertia, see also rotary inertiaof area, 104 Should be Moment of inertia, see also rotary inertia, 104, 109		Delete "of area" and add 109
p. 224	Printed Plasma-enhanced chemical vapor deposition(PE-CVD), 39 Should be Plasma-enhanced chemical vapor deposition (PE-CVD), 39, 76		Insert a space after "deposition" and add 76
p. 224	Printed Polycrystalline line, 18, 20, 33, 34, 35, 36, 37, 74, 97 Should be Polycrystalline, 97 line, 18, 20, 33, 34, 35, 37, 74		Move 97 behind "Polycrystalline" and delete 36
p. 224	Rotary inertia, 105, 109, 114	Rotary inertia, 105, 114	Delete 109
p. 224	Printed Scanning electron microscope (SEM), 42, 44, 44, 47, profile, 162, 165 Should be Scanning electron microscope (SEM), 42, 44, 47, profile, 163, 165		Delete one 44 for scanning electron microscope Change 162 to 163 for scanning profile
p. 224	Silicon (Si) dioxide (SiO2),	Silicon (Si) dioxide (SiO_2),	2 is inferior
p. 224	Silicon (Si) nitride (Si3N4),	Silicon (Si) nitride (Si_3N_4),	3 and 4 are inferior
p. 224	Size-dependent, 144, 145, 169	Size-dependent, 144, 145	Delete 169
p. 225	Printed Stress migration (SM), 4, 6, 15, 19, 42, 43, 45, 46, 47, 48, 113 release, 11, 44, 46, 54 Should be Stress migration (SM), 4, 6, 15, 42, 43, 45, 46, 47, 48, 113 release, 11, 54		Delete 19 for stress migration Delete 44 and 46 for stress release

(continued)

(continued)

Location	Printed	Should be	Comment
p. 225	Printed Template assisted synthesis, 146 Should be Template assisted synthesis, 146 -based synthesis, 4, 5, 6		Add another expression
p. 225	Printed Tin dioxide (SnO2), Should be Tin dioxide (SnO_2),		2 is inferior

Back Cover

Location	Printed	Should be	Comment
line 18	expected are expected to be	expected to be	Delete "expected are"

Index

A

Acceleration test, 23, 25, 33, 38, 40, 42, 45, 49
Activation energy, 23, 42, 66, 69
Aluminum (Al), 3, 4, 8, 9, 16, 25, 28, 33, 35, 37, 39, 40, 43, 54, 60, 63, 65, 121, 154, 168, 210
  nitride (AlN), 205
Amorphous carbon (a-C), 95
Amplitude function, 118
Anisotropy, 8, 82, 212
Anode, 8, 12, 19, 35, 55, 57, 58, 60, 78
Anodic alumina, 5, 6, 145
Aspect ratio, 46, 57, 59, 159, 160
Atomic
  density, 17, 19, 20, 34, 37, 42, 67, 72, 73, 75, 78–80
  density gradient, 19, 20, 37, 42
  diffusion, 11, 15, 16, 18, 19, 35, 36, 43, 46, 54, 82, 212
  flux, 15, 17–21, 33, 36–38, 47, 48, 63, 66
  flux divergence, 11, 15, 20, 33, 36, 66
  force acoustic microscopy (AFAM), 93, 96, 98, 101, 107, 110
  force microscope (AFM), 6, 9, 12, 94, 98, 102, 113, 121, 143, 146, 161, 213

B

Back flow, 19, 69
Bamboo line, 18, 20, 56, 137
Beam vibration, 102, 104, 107
Belt, 65
Bending strength, 12, 93, 126
Bernoulli-Euler's formula, 104
Biaxial modulus, 107, 209
Black's equation, 16
Boltzmann transport equation, 144
Bond-order-bond-length-bond strength (bond-OLS) correlation, 129
Boron carbide (BC), 205
Brass (BS), 168
Buckling (BK) method, 95
Bulk material, 33, 129, 136, 176, 197

C

Cantilever, 94, 98, 105, 107, 110, 121, 129, 133, 152, 159, 162, 213
  bend (CB) method, 94
Capacitive sensor, 133
Carbon nanotube (CNT), 2, 3, 94
Characteristic
  constant, 15, 34, 35, 37, 39, 41, 42
  impedance, 162
Chromium (Cr), 95, 127, 210
Coaxial line, 163, 164, 167
Complete elliptic integral, 117, 125
Compressive stress, 7, 8, 19, 44, 46, 47, 48, 54, 55, 59, 87, 208
Concentrated-mass (CM) cantilever, 102, 107, 110
Conductometry, 169
Contact
  area, 10, 110, 180, 189
  resistance, 149, 151, 155
  resonance (CR), 94, 98, 109, 111
  stiffness, 101, 107, 110
Copper (Cu), 3, 16, 44, 45, 81, 120, 148, 210
Creep deformation, 213
Critical atomic density, 42, 73
Crystalline, 99, 112, 213
  structure, 143
  plane, 159, 160

M. Saka (ed.), *Metallic Micro and Nano Materials*, Engineering Materials,
DOI: 10.1007/978-3-642-15411-9,

C (*cont.*)
Cupric oxide (CuO), 95, 120, 127, 213
Current
density, 16, 17, 19, 21, 23, 26, 27, 30, 34, 38, 41, 55, 64, 72, 80, 180, 181, 183, 192
supply, 8, 12, 40, 76
Cutting, 10, 12, 184, 185, 187
Cyclic contact (CC), 101

D
Deposition, 3, 39, 63, 76, 94, 130, 205, 206, 210
Device under test (DUT), 149
Diameter dependence, 114, 126, 127
Diffuse scattering, 144
Diffusion coefficient, 17, 18
Divergence of atomic flux, see atomic flux divergence
DNA template, 146, 148, 169
Drift velocity, 35, 37–39, 41, 42, 66
Driving force, 8, 15, 17, 19, 44, 46, 48, 54, 55, 82, 211

E
Effective
mass, 100, 107, 110
Young's modulus, 110
Elastic
anisotropy, 212
modulus, 2, 12, 93, 101, 107, 110, 120, 130
Elastica, 93, 98, 113, 124
Electrical
conductivity, 144, 151, 157, 168, 177
property, 45
resistivity, 10, 17, 19, 31, 32, 144, 148, 196
Electrolyte, 147
Electromagnetic interaction, 200
Electromigration (EM), 4, 8, 15, 16, 18, 19, 23, 35, 40, 54, 55, 59, 63, 66, 67, 72
Electron, 17, 35, 43, 48
conductor, 144
distribution function, 144
–phonon interaction, 145
transport, 144
wind, 8, 15
–beam-evaporation technique (EBET), 210
–electron interaction, 145
Electronic mean free path, 144
Electroplating, 43, 145
Electro-thermal problem, 177, 180

Elliptic
function, 98, 114, 117
integral, 98, 116, 124
Empirical equation, 20
Enhanced bending, 206, 213
Equation of motion, 105
Equivalent plastic strain, 137
Etching mask, 159–161
Evaporation, 2, 3, 39, 63, 76, 146, 159, 161, 212
Extrinsic stress, 206

F
Fermi wavelength, 145
Finite element analysis (FEA), 24, 132, 180, 212, 213
Fixed end, 123, 131
Flexural rigidity, 104, 108, 114, 124
Focused ion beam (FIB), 37, 39, 40, 44, 76, 153, 162
–assisted chemical vapor deposition (FIB-CVD), 205
Force
gradient, 100, 123
modulation microscopy (FMM), 102
Four-point
AFM probe, 143, 151, 154
probe, 143, 147, 148, 151
Four-terminal sensing, 150
Fracture
mechanics, 182
stress, 93
test, 94, 122
Frank-van der Merwe (FvdM) growth, 207
Frequency equation, 106, 109

G
Gallium (Ga)
arsenide (GaAs), 159, 162, 165
nitride (GaN), 95, 127
Geometrical
effect, 152, 157
nonlinearity, 102
Gold (Au), 4, 5, 6, 95, 127, 181
Governing parameter
for electromigration damage, 19, 20, 33, 34, 35, 37, 72
for melting phenomenon, 182, 183
Grain
boundary, 9, 18, 20, 21, 25, 32, 36, 38, 41, 42, 66, 84, 85, 86, 137, 144, 208, 212

boundary scattering, 144, 145, 149
size, 8, 18, 20, 23, 25, 28, 37, 39, 46, 86, 97, 144, 145, 149, 207
structure, 20, 39, 40, 149

H
Hardening modulus, 130
Hardness, 95
Heat conduction, 24, 70, 72, 177, 187
Hillock, 16, 19, 33, 34, 42, 54
Hooke's law, 103
Huntington-Grone equation, 16, 17, 19, 33
Hydrostatic stress, 15, 46, 47, 82, 83, 84, 86, 89

I
Impurity, 4, 144, 206
In situ
measurement, 152, 157
observation, 188, 189
Incubation period, 40, 41, 42
Indium (In), 132
phosphide (InP), 205
Inhomogeneity, 151, 154
Integrated circuit (IC), 15, 16, 19, 35, 144, 158
Internal resistance, 149, 150
Intrinsic
impedance, 166
stress, 206, 211
Ion etching, 39, 154

J
Joule heating, 10, 12, 28, 55, 66, 67, 69, 72, 149, 177, 182, 189, 190

K
Kelvin
bridge, 150
sensing, 150

L
Lagrangian coordinate, 129
Large
bending (LB) test, 95, 113, 120
–amplitude frequency-modulation (LA-FM), 99
Lattice, 18, 44, 66, 128, 206, 208, 212
Lead (Pb), 3, 95, 127
Least-squares method (least-squaresfit), 38, 41, 112, 128
Lifetime, 16, 19, 20, 34, 62, 75, 78, 79, 80
Lift-off process, 160
Linear hardening elastic–plastic material, 130
Lithography, 146, 148, 152
Local current density, 180, 182
Low-dimensional material, 143
Lower-coordinated surface-atom, 128

M
Manipulation, 113, 194
Mean time to failure (MTF), 16
Mechanical
contact force, 155
diode (MD), 101
property, 45, 93, 126, 130, 136
resonance (MR), 94
Melting, 182, 183, 189
point, 60, 64, 176, 184, 187, 212
Micro
cantilever (CL), 98, 107, 213
ring, 197, 200
structure, 15, 18, 20, 36, 37, 38, 39, 40
Microwave
AFM, 12, 157, 164, 166
microscopy, 158
Mie scattering, 114
Misfit strain, 206, 208, 214
Misfit-strain-induced bending, 210
Mobility, 17, 207
Mode function, 105
Modification, 4, 9, 10, 11, 12, 19, 173, 210
Molecular combing, 148
Molybdenum disulfide (MoS2), 95
Moment
of inertia, see also rotary inertiaof area, 104
Multi-walled carbon nanotube (MWCNT), 94

N
Nanobelt (NB), 94, 143, 206
Nanocircuit, 151, 152, 157
Nanocoil (NC), 205, 210
Nanocontact, 180, 192
Nanodevice, 151, 157
Nanofilm, 143
Nanoindentation (NI), 94
Nanoparticle, 143
Nanorod, 98, 143
Nanoslit, 147
Nanotube (NT), 94, 206, 143
Nanowhisker, 143

N (*cont.*)
Nanowire (NW), 4, 5, 8, 10, 12, 45, 46, 57, 83, 84, 85, 93, 113, 120, 127, 143, 148, 154, 169, 174, 205, 210
Natural frequency (eigenfrequency), 100, 106
Near-field scanning optical microscope (NSOM), 157
Neutral surface, 103, 208, 214
Noncontact (NC), 100
  mode, 100
Non-Hookean elasticity, 128
Nonlinear
  effect, 128
  stress-strain relation, 127
Numerical simulation, 12, 42, 72, 73, 80

O
On-chip contamination, 144
Optical
  diffraction image, 122
  lithography, 147
  microscopy (OM), 41, 42, 95, 113, 120
Oxide, 3, 39, 44, 45, 46, 47, 54, 82, 83, 84, 87, 114, 120, 205
Oxidized membrane, 157

P
Passivation, 15, 19, 33, 42, 47, 55, 57, 63, 87, 88
Photolithography, 57, 152
Plasma-enhanced chemical vapor deposition(PE-CVD), 39
Platinum (Pt), 6, 10, 112, 132, 184, 187, 195, 201, 210
Point-mass model, 100, 109, 110
Polycarbonate membrane, 145
Polycrystalline
  line, 18, 20, 33, 34, 35, 36, 37, 74, 97
  metallic film, 144
Position-sensitive photodetector (PSD), 99
Potassium hydroxide (KOH), 112
Potential drop, 40, 41, 76, 153, 155, 196
Progressive scanning, 155
Protective layer, 11, 53, 54
Purity, 42, 65, 66

Q
Quantized effect, 145
Q-value, 162

R
Rayleigh scattering, 113
Reactive ion etching (RIE), 39
Recrystallization, 44, 89, 207
Reliability, 3, 9, 15, 16, 42, 43, 45, 94
Residual stress, 43, 206, 215
Resonant frequency, 94, 99, 107, 110, 158, 169
Reversible work, 129
Rotary inertia, 105, 109, 114

S
Sapphire, 113
Scanning
  electron microscope (SEM), 42, 44, 44, 47, 94, 108, 122, 145, 152, 159, 186, 188, 193, 205, 210
  evanescent microwave microscope(SEMM), 158
  local-acceleration microscopy (SLAM), 102
  microdeformation microscopy (SMM), 102
  probe microscope (SPM), 157
  profile, 162, 165
  tunneling microscope (STM), 1, 98, 146, 157
Seebeck effect, 201, 204
Self-completed mechanism, 189
Semi
  –empirical method, 147
  –infinite model, 157
Silicon (Si), 15, 19, 39, 46, 95, 99, 107, 112, 121, 127, 210
  carbide (SiC), 95, 127, 205
  dioxide (SiO2), 46, 58, 82, 87, 88, 89, 210
  nitride (Si3N4), 99, 152
Silver (Ag), 3, 4, 43, 86, 87, 88, 89, 95, 127, 132, 147, 148, 199, 205
Single-walled carbon nanotube (SWCNT), 95
Singularity, 28
Size-dependent, 144, 145, 169
Slit, 58, 59, 60, 63
Small-span bending test, 12, 93, 130
Solidification, 189
Spacer technique, 146
Specific temperature field, 181, 192
Specularity parameter, 144
Sphalerite structure, 159
Sphere, 64, 65, 66, 67, 70, 71
Spring constant, 99, 107, 108, 121, 153, 214
Sputtering, 56, 76, 146, 154, 210
Strain energy, 209
Stranski-Krastanov (SK) growth, 207

Stress, 19
  concentration, 81
  gradient, 37, 44, 46
  migration (SM), 4, 6, 15, 19, 42, 43, 45, 46, 47, 48, 113
  release, 11, 44, 46, 54
String deflection (SD) method, 94
Subharmonic (SH), 102
Surface, 44, 45, 47
  elasticity, 102, 128
  free energy, 129, 216
  roughness, 144, 192
  scattering, 144, 148
  stress, 126, 128, 207
Synthesis, 4, 5, 6, 9, 59, 205

T
Temperature, 177, 180, 204
  gradient, 24, 25, 27, 28, 29, 33, 54, 55, 56, 78
Template
  assisted synthesis, 146
Tensile load (TL), 94
Tetraethyl orthosilicate (TEOS), 39, 56, 57, 58, 74, 75, 76, 78
Theoretical cohesive strength, 127
Thermal
  expansion coefficient, 43, 46, 209, 212
  function, 183, 186
  stress, 19, 46, 81, 206
  vibration (TV), 95
Thermoelectric element, 203, 205
Thin film, 18, 33, 39, 45, 47, 58, 107, 206
Three-point bending (TPB), 94, 126
Threshold current density, 19, 20, 34, 37
Tin dioxide (SnO2), 63, 205
Titanium nitride (TiN), 35, 39, 59, 67, 76
Topography, 12, 143, 157, 162, 164
Transmission electron microscope (TEM), 4, 44, 46, 47, 94, 146, 184

U
Ultrasonic atomic force microscopy (UAFM), 101
Ultrasonic force microscopy (UFM), 101

V
Vacuum evaporation, 3, 39, 76
Vapor-liquid-solid (VLS), 3, 4
Vapor-solid (VS), 120
Void, 16, 33, 34, 35, 38, 40, 41, 42, 43, 72
Volmer-Weber (VW) growth, 207

W
Welding
  condition, 190, 193
  process, 188
Wet etching, 57, 67, 76, 159, 160, 161
Whisker, 3, 43, 44, 45, 46, 54, 86

X
X-ray diffraction (XRD), 149

Y
Yield stress, 95, 130, 136
Young's modulus, 94, 103, 111, 114, 121, 125, 127, 130, 136, 209, 212

Z
Zinc oxide (ZnO), 3, 63, 95, 127, 205
Zinc sulfide (ZnS), 6, 95

MIX
Papier aus verantwortungsvollen Quellen
Paper from responsible sources
FSC® C105338

If you have any concerns about our products,
you can contact us on
ProductSafety@springernature.com

In case Publisher is established outside the EU,
the EU authorized representative is:
Springer Nature Customer Service Center GmbH
Europaplatz 3, 69115 Heidelberg, Germany

Printed by Libri Plureos GmbH
in Hamburg, Germany